Tropical Forestry

Series Editor
Michael Köhl, Hamburg, Germany

More information about this series at
http://www.springer.com/series/5439

Walter Liese • Michael Köhl

Editors

Bamboo

The Plant and its Uses

 Springer

Editors
Walter Liese
Department of Wood Science
University of Hamburg
Hamburg
Germany

Michael Köhl
Department of Wood Science
University of Hamburg
Hamburg
Germany

Series Editor
Michael Köhl
Department of Wood Science
University of Hamburg
Hamburg, Germany

ISSN 1614-9785
Tropical Forestry
ISBN 978-3-319-14132-9   ISBN 978-3-319-14133-6   (eBook)
DOI 10.1007/978-3-319-14133-6

Library of Congress Control Number: 2015935200

Springer Cham Heidelberg New York Dordrecht London

Springer International Publishing AG Switzerland is part of Springer Science+Business Media (www.springer.com)

Preface

Bamboo is the fastest-growing and most versatile plant on Earth. For centuries, bamboo has played an indispensible part in the daily life of millions of people in tropical countries. In the last decades, it has gained increasing importance as a substitute for timber.

Bamboo is a unique group of tall grasses with woody jointed stems. Bamboo belongs to the subfamily Bambusoideae of the grass family Poaceae (or Graminaceae). There are about 75 genera with approximately 1,300 species and varieties covering 25 million hectares worldwide. In cooler and temperate regions of Central Asia, bamboo plants grow single stemmed apart from each other (leptomorph type) or in dense clumps (pachymorph type) in warm, tropical regions of Western Asia, Southeast Asia, and South America. The culm (stem) is mostly hollow and characterized by nodes with internodes in between. The nodes give the plant its strength. The culms arise from buds at the underground shoot–root system, the so-called rhizome. Shoots emerge with the rainy season and expand within a few months to their final length of 10–30 m and diameters from 5 to 30 cm.

During the growth of the culms, bamboo produces the highest amount of living biomass in the plant realm. Depending on type, location, and climate, the annual growth rate is about 5–12 metric tons of air-dried biomass per hectare.

After 3–4 years, the culms are selectively harvested. Bamboo is a self-regenerating, renewable raw material. Due to new shoots, which appear each year, its production continues after individual culms have been harvested. Depending on the species, the culms of one population flower after 40–80 years, mostly with a subsequent dying of the entire population across large regions. This simultaneous flowering can have substantial economic implications by depriving people of their basic natural resource.

There are over 1,000 described uses of bamboo. Bamboo provides food, renewable raw material, and regenerative energy. Culms have excellent technological properties and are used for construction, scaffolding, handicraft products, furniture, and as material for secondary products such as bamboo mats, boards, or flooring.

Over 1 billion people live in bamboo houses, from simple dwellings to four-story city houses and engineered structures.

Bamboo processing is often done at craft level associated with relatively low capital investment. The fibers are a valuable material for pulp and paper as a substitute for wood. Bamboo crops are also used to provide wind protection in farming and to stabilize riverbanks and hillsides by the interlocked rhizome system.

Due to the overexploitation of natural forests and the increasing demand for woody material, bamboo is gaining importance as a substitute resource material for timber. In many areas, the increase of plantation areas, the improvement of utilization, and the development of innovative uses is a widely accepted goal. However, problems with stand management, harvest, storage, and biological hazards and the need for preserving natural bamboo forests against overexploitation by expanding demands have to be considered. In times of declining forest populations, increased need for renewable resources and regenerative energies, and a change of an economic paradigm from market economy to green economy, bamboo has a promising future.

The current book is intended to be a primer on bamboo. The focus is on the relevant biological basis, production, and utilization of bamboo. Our intent is to give an introduction and overview about basic concepts and principles, which can be adapted for real-world situations.

Hamburg   Walter Liese
January, 2015   Michael Köhl

Acknowledgments

The current book is the result of 60 years of personal pursuit with the amazing plant bamboo. Many colleagues and friends accompanied me on this path for understanding and revealing the treasures and secrets of bamboos. There are innumerable companions who shared and supported my scientific, applied, and practical activities. Some of them have contributed to this book as authors of individual chapters; many of them have been cited in the text.

Michael Köhl, as coeditor of this book, handled most of the technical issues and reviewed manuscripts. I am grateful that he initiated this book and insisted on its completion.

Special thanks go to Yannick Kühl for providing the structure of the book and for establishing links to the authors and to Johannes Welling for his valuable support in the final phase of editing this book. GIZ generously made material from previous publications available. Annette Lindqvist from Springer Publishers, Heidelberg, was always a competent and reliable contact and professionally supported the editing of this book.

Walter Liese, Hamburg, 2015

Contents

Chapter 1
Bamboo Taxonomy and Habitat

L.G. Clark, X. Londoño, and E. Ruiz-Sanchez

Abstract Bamboos (subfamily Bambusoideae) comprise one of 12 subfamilies within the grass family (Poaceae) and represent the only major grass lineage to diversify in forests. Bamboos are distinguished by the presence of well-developed, asymmetrically strongly invaginated arm cells in the leaf mesophyll as seen in cross section and also generally exhibit relatively broad, pseudopetiolate leaf blades usually with fusoid cells flanking the vascular bundles. The nearly 1,500 described species of bamboos are classified into three tribes: Arundinarieae (temperate woody bamboos, 546 species), Bambuseae (tropical woody bamboos, 812 species), and Olyreae (herbaceous bamboos, 124 species). Relationships between the three tribes remain uncertain, but a much better understanding of evolutionary relationships within the tribes has been achieved based on analyses of DNA sequence data, which we summarize. We present synoptic descriptions for the three tribes and, for the Bambuseae and Olyreae, their currently accepted subtribes, as well as lists of included genera and comments. The history of bamboo classification goes back over 200 years; we provide an overview of the most important advances leading to the current phylogenetic classification of bamboos based on their inferred evolutionary relationships. Bamboos are native to all continents except Antarctica and Europe and have a latitudinal distribution from 47° S to 50° 30′ N and an altitudinal distribution from sea level to 4,300 m. Bamboos therefore grow in association with a wide variety of mostly mesic to wet forest types in both temperate and tropical regions, but some bamboos have adapted to more open grasslands or occur in more specialized habitats.

L.G. Clark (✉)
Department of Ecology, Evolution and Organismal Biology, Iowa State University, 251 Bessey Hall, Ames, IA 50011-1020, USA
e-mail: lgclark@iastate.edu

X. Londoño
Sociedad Colombiana del Bambú, A. A. 11574, Cali, Colombia
e-mail: ximelondo@gmail.com

E. Ruiz-Sanchez
Red de Biodiversidad y Sistemática, Instituto de Ecología, AC, Centro Regional del Bajío, Av. Lázaro Cárdenas 253, Pátzcuaro, Michoacán 61600, Mexico

© Springer International Publishing Switzerland 2015
W. Liese, M. Köhl (eds.), *Bamboo*, Tropical Forestry 10,
DOI 10.1007/978-3-319-14133-6_1

1

Keywords Arundinarieae • Bamboo classification • Bamboo diversity • Bamboo habitats • Bamboo phylogenetics • Bambuseae • Olyreae • Temperate woody bamboos • Tropical woody bamboos

1.1   Introduction

Woody bamboos, or the "tree grasses," are a cultural and ecological feature of many countries of Asia, America, and Africa, where bamboos can provide environmental, social, and economic benefits. Bamboo is a multipurpose plant—it can substitute for timber in many respects due to its lignified culms, and because of its fast growth, intricate rhizome system, and sustainability, it has become a plant with conservation value, able to mitigate phenomena that result from global climate change. Bamboo is also an essential resource for many other organisms, not just pandas. Bamboo, like rice, maize, wheat, and sugar cane, is another important grass inextricably linked to human livelihood, fulfilling needs for shelter, food, paper, and more; the range of its use is hardly rivaled in the plant kingdom—not for nothing is bamboo known as "the plant of a thousand uses." Bamboos are complex plants that can be difficult to identify or classify, but given their ecological and economic importance, correct identification is critical to their conservation and development and a robust phylogenetic classification system underpins identification. In this chapter, we present a history of bamboo classification, discuss bamboo habitats, and present an up-to-date classification of bamboos based on synthesis of the most recent systematic work in this fascinating and charismatic group of grasses.

1.2   Definition of Bamboo

Bamboos comprise the subfamily Bambusoideae, one of 12 subfamilies currently recognized within the grass family (Poaceae). Unlike the other grasses, bamboos are the only major lineage within the family to adapt to and diversify within the forest habitat (Judziewicz et al. 1999; Grass Phylogeny Working Group [GPWG] 2001; Bamboo Phylogeny Group [BPG] 2012). Molecular sequence data strongly support the bamboos as a distinct lineage, as does the presence of well-developed, asymmetrically strongly invaginated arm cells in the leaf mesophyll as seen in cross section (GPWG 2001; Kelchner et al. 2013). Bamboos also generally exhibit relatively broad, pseudopetiolate leaf blades, with fusoid cells flanking the vascular bundles [but fusoid cells are often lacking in sun plants, March and Clark (2011)]. Bamboos include ca. 1,482 described species classified in approximately 119 genera, which in turn are grouped into three tribes: Arundinarieae (known as the

temperate woody bamboos, even though some occur in the tropics at high elevations; ca. 546 species), Bambuseae (known as the tropical woody bamboos, even
though some occur outside of the tropics; 812 species), and Olyreae (herbaceous
bamboos, 124 species) (BPG 2012).

Within the Bambusoideae, the herbaceous bamboos (Olyreae) are easily distinguished by their lack of both well-differentiated culm leaves and outer ligules
(contraligules) combined with relatively weakly lignified culms, restricted vegetative branching, and unisexual spikelets. Additionally, all Olyreae except for
Buergersiochloa possess cross-shaped and crenate (olyroid) silica bodies and
virtually all Olyreae exhibit seasonal flowering (BPG 2012). In contrast, woody
bamboos (Arundinarieae and Bambuseae) commonly have complex rhizome systems, a tree-like habit with highly lignified, usually hollow culms, well-
differentiated culm leaves, well-developed aerial branching, and foliage leaf blades
with outer ligules. Culm development occurs in two phases: first, new, unbranched
shoots bearing culm leaves elongate to their full height; second, culm lignification
and branch development with production of foliage leaves take place. Woody
bamboos also have bisexual spikelets and typically exhibit gregarious flowering
cycles followed by death of the parent plants (monocarpy) (Dransfield and Widjaja
1995; Judziewicz et al. 1999; BPG 2012).

Woody bamboos have particular characteristics that make them unique
grasses and an important non-timber resource. In Table 1.1 we compare bamboos
with trees (wood, defined as 2° xylem) to highlight some of the unique features of
bamboo.

Table 1.1 Comparison between bamboo and trees (wood)

Bamboo	Trees (wood)
Underground parts consisting of rhizomes and roots	Underground parts consisting of roots
Culms (stems) usually hollow and segmented	Stems solid and not segmented
The hardest part of the culm is the periphery	The hardest part of the stem is in the center
There is no vascular cambium so the culm does not increase in diameter with age	A vascular cambium is present so the stem increases in diameter with age
The conducting tissues, phloem and xylem, are together inside each vascular bundle	The conducting tissues, phloem and xylem, are separated by the vascular cambium
Culms lack bark	Stems have bark (cork + 2° phloem)
No radial (lateral) communication in the culms except at the nodes	Radial (lateral) communication throughout the stem
Culms grow extremely fast (to as much as 36 m tall at 6 months), reaching full height in one growing season	Stems grow slowly in height and diameter over many seasons
Culms grow in an association from a network of rhizomes, such that each culm depends on the others and the harvest of a culm directly affects the rest of the community	Each stem usually grows as an independent individual, and the harvest of a stem does not directly affect the rest of the community

1.3   History of Bamboo Taxonomy

While the first uses of bamboo in arts and technology were documented by early Chinese scholars, early taxonomic studies of bamboos were dominated by the Western world [see Soderstrom (1985) for a detailed review]. Within the last century, however, much work by botanists in the regions where bamboo is most diverse (Asia, India, and Central and South America) has contributed greatly to a vastly improved understanding of bamboo diversity and evolution. DNA sequence data in combination with morphological and anatomical studies form the basis of the most recent comprehensive and phylogenetically based classification system for bamboos (BPG 2012). We have used Soderstrom (1985) and Bedell (1997) as primary sources and recommend them especially to those readers interested in the earliest phases of bamboo classification. We here summarize recent advances, but begin our overview with the more global perspective on bamboos that emerged starting in the mid-nineteenth century.

Munro (1868) published a world monograph on the representatives of the Bambuseae known at the time, which remains a useful reference to this day even though the divisions are clearly not natural. In this work, he described 170 species grouped into 21 genera divided into three divisions: (a) Triglossae or Arundinarieae (*Arthrostylidium*, *Arundinaria*, *Chusquea*, among others), (b) Bambuseae verae (e.g., *Bambusa*, *Gigantochloa*), and (c) Bacciferae (*Dinochloa*, *Melocanna*, and others with fleshy fruits). Munro included only woody bamboos in his treatment, a taxonomic concept of bamboos that persisted for nearly a century.

A few botanists in the early twentieth century (e.g., Arber 1927) examined the flowering structure of bamboos (mainly Asiatic ones) in more detail and suggested modifications in bamboo classification, but none conducted a comprehensive study. In Japan, Nakai (1925, 1933) described a number of new genera and species, while Takenouchi (1931a, b) examined morphology and development of bamboos, with a particular focus on vegetative structures. In one of his earlier papers, McClure (1934) analyzed the inflorescence structure of *Schizostachyum* and coined the term "pseudospikelet" to refer to the peculiar rebranching spikelet found in this and a number of other bamboo genera. The next influential work on bamboo classification from a global perspective was that of Holttum (1956), who critically examined inflorescence, spikelet, ovary, and fruit structure and proposed a classification scheme for bamboos based on perceived evolutionary trends. This, to our knowledge, represents the first attempt to produce a natural classification for bamboos.

The 1960s was a period of active bamboo research in many parts of the world. McClure (1966), in his exhaustive work on the bamboo plant, pointed out that all parts of the vegetative and the flowering structures should be used for bamboo classification. This was revolutionary in grass taxonomy, where floral characters often continue to be given undue weighting to this day. He offered a significant step forward in the taxonomic conquest of the bamboos of the Americas (McClure 1973) based on this philosophy of synthesizing all available knowledge. Meanwhile, the Argentinian agrostologist Parodi (1961) offered a broader concept of the

Bambusoideae, including the herbaceous grass tribes Olyreae, Phareae, and Streptochaeteae in addition to the woody bamboos (as the tribe Bambuseae).

The next phase in the history of bamboo systematics was led by T. R. Soderstrom, who made significant contributions to the systematics and evolution of grasses with particularly enlightening studies of bamboos. A hallmark of Soderstrom's approach was to study and collect bamboos in the field, and he strongly supported such efforts by others. In collaboration especially with C. E. Calderón, L. G. Clark, R. Ellis, E. J. Judziewicz, and X. Londoño, Soderstrom investigated bamboo diversity and evolution with a special focus on American and Sri Lankan woody bamboos (e.g., Calderón and Soderstrom 1973, 1980; Soderstrom and Ellis 1988; Soderstrom and Londoño 1988) as well as detailed studies of herbaceous grass groups including Olyreae (e.g., Soderstrom and Zuloaga 1989). Soderstrom (1981) placed much cytological and morphological data on bamboos in an evolutionary context. One of Soderstrom's most important contributions to bamboo systematics was his revised classification of bamboos based on leaf anatomical features analyzed in an evolutionary context (Soderstrom and Ellis 1987). This classification and that of Keng (1982–1984) and Clayton and Renvoize (1986) were the last global bamboo treatments published prior to the advent of molecular sequence data in plant systematics. Although they differ in many generic concepts and hypotheses of relationships, both classifications include the woody bamboos (as the Bambuseae) and several tribes of herbaceous grasses, building on the broader Bambusoideae of both Nees (1835) and Parodi (1961).

During the 1980s and 1990s, bamboo research in China began in earnest. Landmarks included the global generic revision of Keng (1982–1984) and the publication of an account of all the Chinese bamboos for the *Flora Republicae Popularis Sinicae* (Keng and Wang 1996), with contributions by a number of Chinese bamboo botanists. The history of bamboo classification in China was reviewed in detail by Zhang (1992), who especially highlighted the contributions of Keng Y.-L., considered to be the father of Chinese bamboo taxonomy, and his son Keng P.-C. (Geng B.J.), as well as Wen T.-H., Yi T.-P., and Hsueh C.-J. These masters trained the next generation of bamboo systematists in China, including Li D.-Z. and Xia N.-H., who have now trained a fourth generation. Two contrasting schools of thought on bamboo taxonomy in China clashed over generic recognition for 25 years (Stapleton, pers. comm.). The innovative classification system established by Keng and colleagues used vegetative characters extensively, while others continued to follow a more traditional system, which emphasized floral characters in a classic grass taxonomy approach. Phylogenetic information (see below) was incorporated to achieve a more modern treatment in the English-language version of the *Flora of China* (Li et al. 2006).

This period also saw the production of several compilations of knowledge about bamboos. Ohrnberger (1999) compiled the published names of bamboos of the world, reporting the occurrence of 110 genera and 1,110–1,140 species. A compendium of bamboos from India (Seethalakshmi and Kumar 1998), a compendium of Chinese bamboos (Zhu et al. 1994), and a compendium of American bamboos (Judziewicz et al. 1999) were published. Dransfield (1992, 1998), Dransfield and

Widjaja (1995), Stapleton (1994a, b, c), Widjaja (1987), and Wong (1993, 1995, 2005), among others, advanced knowledge of the diversity of Madagascan and Asiatic bamboos, including the description of many new genera. Yi et al. (2008) recently produced an updated and beautifully illustrated compendium of Chinese bamboos.

Even into the early 1990s, the Bambusoideae were defined as all perennial, forest-inhabiting grass groups with broad, pseudopetiolate leaf blades, usually with fusoid cells in the mesophyll (Clayton and Renvoize 1986; Soderstrom and Ellis 1987). The first comprehensive DNA sequence analysis of the grass family that included good representation of the various tribes of the subfamily clearly showed that this broader Bambusoideae, however, was not a natural group (Clark et al. 1995). These results were confirmed and extended by the GPWG (2001), and the concept of the Bambusoideae was restricted to the woody bamboos (as Bambuseae) and the herbaceous bamboos (as the Olyreae, including Buergersiochloeae and Parianeae). Herbaceous grass tribes formerly regarded as bamboos were transferred to the Anomochlooideae (Anomochloeae and Streptochaeteae), Ehrhartoideae (Ehrharteae, Phyllorachideae, Oryzeae), Pharoideae (Phareae), Pooideae (Brachyelytreae, Diarrheneae, Phaenospermateae), and Puelioideae (Guaduelleae and Puelieae), and Streptogyneae was placed without a fixed position in the Bambusoideae–Ehrhartoideae–Pooideae (BEP) clade (Clark and Judziewicz 1996; GPWG 2001). Additionally, the Anomochlooideae, Pharoideae, and Puelioideae were strongly supported as the three early-diverging lineages within the family, unequivocally indicating that grasses originated in the forest habitat.

The Bamboo Phylogeny Group, consisting of an international team of 21 bamboo taxonomists coordinated by L. G. Clark, was formed in 2005 primarily to generate a global evolutionary tree (phylogeny) for bamboos based on extensive chloroplast sequence data and to produce a revised tribal and generic classification based on the phylogeny. A number of papers addressed phylogenetic relationships within bamboo lineages (e.g., Fisher et al. 2009; Triplett and Clark 2010; Zhang et al. 2012), but both Sungkaew et al. (2009) and Kelchner et al. (2013) explicitly addressed broader relationships across the subfamily. All studies with sufficient sampling resolved three strongly supported lineages which are now recognized as tribes (BPG 2012): temperate woody bamboos (Arundinarieae), tropical woody bamboos (Bambuseae), and the herbaceous bamboos (Olyreae). A review of phylogenetic work in the bamboos and the revised tribal, subtribal, and generic classification, which we follow here with a few updates, can be found in BPG (2012).

1.4 Bamboo Habitat

Bamboos, both woody and herbaceous, are well known as forest grasses, even though some species have radiated into open, grassy, or shrubby habitats at high elevations in montane systems of the tropics (Soderstrom and Calderón 1979;

Soderstrom and Ellis 1988; Judziewicz et al. 1999; Judziewicz and Clark 2007). Native to all continents except Antarctica and Europe, bamboos have a latitudinal distribution from 47° S to 50°30′ N and an altitudinal distribution from sea level to 4,300 m (Soderstrom and Calderón 1979; Judziewicz et al. 1999; Ohrnberger 1999). Bamboos therefore occupy a broad range of habitat types, especially forests, from temperate to tropical climatic zones and bamboos are often dominant or highly visible elements of the vegetation. We here summarize these habitats, and note that although there has been some important recent work on bamboo ecology (BPG 2012 and references cited therein), much more needs to be done.

With some exceptions, the Arundinarieae occupy temperate deciduous forests or mixed coniferous and deciduous forests or coniferous forests in the temperate to subtropical zones of the Northern hemisphere in Eastern Asia and Eastern North America (Stapleton 1994a, b, c; Li and Xue 1997; Taylor and Qin 1997; Triplett et al. 2006; Dai et al. 2011). Temperate bamboos are common in the understory and often form the dominant element on wetter sites (Taylor and Qin 1997; Noguchi and Yoshida 2005; Tsuyama et al. 2011). In Chinese montane forests, species of *Bashania*, *Chimonobambusa*, *Fargesia*, *Indosasa*, and *Yushania* are characteristic (Li and Xue 1997; Taylor and Qin 1997), while in the more seasonally dry areas of the central Himalayas clump-forming bamboos are more prevalent, especially *Thamnocalamus* and *Drepanostachyum*, with spreading bamboos of *Yushania*, *Sarocalamus*, and *Chimonobambusa* restricted to the wetter ends of the mountain range (Stapleton 1994a, b, c and pers. comm.). In wetter forests of E China, Korea, and Japan, rampant species of *Sasa* and *Sasamorpha* are especially aggressive and dominant in the understory (Noguchi and Yoshida 2005; Tsuyama et al. 2011). Some temperate bamboos in Asia, such as *Acidosasa*, *Drepanostachyum*, *Indosasa*, and *Sinobambusa*, extend into dry or evergreen subtropical forests as well (Stapleton 1994a, b, c; Li and Xue 1997). In the Eastern USA, *Arundinaria* occurs in the Southeastern Coastal Plain in woodlands and forests, and often along water courses. Switch cane (*A. tecta*) is notable for often occurring in swamps, and like other Arundinarieae known to grow in wetter sites (see below), air canals are a prominent feature of its rhizomes (Triplett et al. 2006). Although the extensive canebrakes in the Southeastern USA have virtually disappeared (Judziewicz et al. 1999), temperate bamboo-dominated habitats in China are classified as bamboo forests and may form a significant portion of the vegetation in some regions (Yang and Xue 1990; Dai et al. 2011).

Many Bambuseae, especially genera of larger stature such as *Bambusa*, *Dendrocalamus*, *Eremocaulon*, *Guadua*, *Gigantochloa*, and *Schizostachyum*, grow in lowland moist tropical forests or lower montane forests up to ca. 1,500 m in elevation in both the Old and New Worlds (Soderstrom and Calderón 1979; Seethalakshmi and Kumar 1998; Judziewicz et al. 1999). It is common to see these bamboos in valleys or along rivers or streams, especially in secondary forest, often to the exclusion of other vegetation. However, a number of lowland tropical bamboos, including species of *Alvimia*, *Chusquea*, *Dinochloa*, *Hickelia*, *Neomicrocalamus*, and *Racemobambos*, have smaller culms that twine around or scramble over trees and shrubs or form beautiful curtains of hanging foliage

(Soderstrom and Londoño 1988; Dransfield 1992, 1994); species of *Ochlandra* may form dense, reed-like thickets along stream banks (Seethalakshmi and Kumar 1998; Gopakumar and Motwani 2013). Some lowland species or genera (e.g., *Dendrocalamus strictus* in India, *Guadua paniculata* in Latin America, *Otatea* in Mexico and Colombia, *Perrierbambus* in Madagascar) are well adapted to drier forest types, and some populations of *Otatea acuminata* inhabit xerophytic scrub, often on calcareous substrates, or they may occupy early successional sites created by forest clearing (Soderstrom and Calderón 1979; Gadgil and Prasad 1984; Rao and Ramakrishnan 1988; Seethalakshmi and Kumar 1998; Ruiz-Sanchez et al. 2011b). Some lowland bamboos, such as *Actinocladum* and *Filgueirasia* in Brazil (Soderstrom and Calderón 1979; Judziewicz et al. 1999), and *Vietnamosasa* in Indochina (Stapleton 1998), are drought tolerant and fire adapted for survival in their grassland habitats. Natural tropical bamboo forests are known from some regions (Li and Xue 1997; Judziewicz et al. 1999); probably the most extensive of these are the *Guadua*-dominated forests of the Amazon Basin (Judziewicz et al. 1999).

A significant portion of tropical bamboo species diversity, however, is associated with moist subtropical montane forests above 1,500 m in elevation, especially in the Neotropics and Asia (Li and Xue 1997; Judziewicz et al. 1999; Uma Shaanker et al. 2004). Although a few montane forest species of *Chusquea* with erect to arching culms attain diameters of up to 7 cm, most montane tropical bamboos have culms not more than 2–3 cm in diameter and are often smaller, in keeping with a general decrease in size with increasing elevation. These bamboos are usually scandent or scrambling, in moist ravines arching over streams or hanging from sometimes steep slopes, but they may also occupy ridges or form part of the understory (Dransfield 1992; Wong 1993; Judziewicz et al. 1999). In the Neotropics, *Aulonemia*, *Chusquea*, and *Rhipidocladum* are characteristic of Andean montane forests, whereas *Chusquea* and *Merostachys* are the most common bamboos of the Atlantic forests of Brazil (Judziewicz et al. 1999). Some species of *Chusquea* extend northward to Mexico in cloud forest and pine–oak–fir forests, but others extend southward into *Nothofagus* or *Araucaria* forests in Chile and Argentina (Judziewicz et al. 1999). Some tropical bamboos form characteristic belts of vegetation within montane forests—*Nastus borbonicus* on Réunion Island is a good example (T. Grieb, pers. comm.) while others, including many species of *Chusquea*, commonly invade gaps formed by treefalls or landslides (Judziewicz et al. 1999). Species of genera such as *Holttumochloa* and *Racemobambos* are characteristic of montane forests in South-east Asia (Dransfield 1992; Wong 1993). And as noted for both Arundinarieae and lowland Bambuseae, some tropical bamboos can form bamboo forests in montane systems (Yang and Xue 1990).

Members of both Arundinarieae and Bambuseae occur above treeline in high elevation grasslands or shrublands, where they are characteristic or often dominant plants (Soderstrom and Calderón 1979; Soderstrom and Ellis 1988; Judziewicz et al. 1999). High elevation bamboos are usually erect and have a shrubby habit, sometimes lacking aerial branching (e.g., species of *Chusquea*) and giving the appearance of non-bambusoid grasses. In the temperate mountains of China, species of *Fargesia* (Arundinarieae) are the most common above treeline, but a few species of *Yushania* (Arundinarieae) also occur at high elevations in relatively open habitats (Li et al. 2006). Some Arundinarieae occur in tropical mountain grasslands and shrublands, notably *Bergbambos* from South Africa and species of *Kuruna* from Sri Lanka and India (Soderstrom and Ellis 1982, 1988). Both *K. densifolia* from Sri Lanka and *B. tessellata* from South Africa grow in wetter habitats and have air canals in their roots. Interestingly, all of the high elevation, open-habitat Arundinarieae possess pachymorph rhizomes. In the tropical Americas, the high montane open habitats known as *páramos*, *subpáramos*, and *campos de altitude* are populated mainly by species of *Chusquea*, which may form extensive and some-times impenetrable stands (Judziewicz et al. 1999). Species of *Aulonemia* may also form mono-dominant stands at high elevations, as can *Cambajuva* in southern Brazil (Judziewicz et al. 1999; Viana et al. 2013). The very odd *Glaziophyton*, resembling a giant rush (*Juncus*), is endemic to rocky mountaintops near the city of Rio de Janeiro (Fernandez et al. 2012). Among Arundinarieae, *Fargesia yulongshanensis* reportedly reaches 4,200 m in elevation, and there may be other Chinese or Himalayan species with comparable elevational ranges (Li et al. 2006). Among Bambuseae, the species with the highest documented elevational ranges are (note that only the high end of the range is cited): *Chusquea acuminatissima* (to 4,000 m, Clark & Londoño pers. obs.), *Chusquea aristata* (to 4,200 m, TROPICOS), *Chusquea guirigayensis* (to 4,000 m, Clark & F. Ely, pers. obs.), *Chusquea tessellata* (to ca. 4,200 m, Judziewicz et al. 1999), and *Chusquea villosa* (4,250–4,400 m, TROPICOS).

Olyreae usually occupy the understory of humid, lowland tropical forests at elevations from sea level to ca. 1,000 m, with *Pariana* often occurring in the periodically flooded *várzea* in Amazonian Brazil. Some herbaceous bamboos, especially strongly rhizomatous ones such as *Pariana*, may even be dominant in the herbaceous layer (Judziewicz et al. 1999). Some species of *Cryptochloa*, *Lithachne*, *Pariana*, *Raddiella* (*R. esenbeckii*), and *Olyra* (*O. latifolia*) occur in lower montane forests at up to 1,500 m in elevation, although *Olyra standleyi* may extend up to 2,200 m in elevation (Judziewicz et al. 1999; Judziewicz and Clark 2007). Other Olyreae are found in more specialized habitats in savannas or wet cliff faces associated with waterfalls or in semi-deciduous seasonal forests, whereas *Ekmanochloa* is a serpentine endemic (BPG 2012 and references cited therein; Ferreira et al. 2013). A few species, especially of *Lithachne* or *Olyra*, may become weedy (Judziewicz et al. 1999; Judziewicz and Clark 2007). Olyreae exhibit their greatest species diversity from 7 to 10° N and 12 to 18° S, with minimal diversity near the equator. The monotypic *Reitzia* is the only member of the tribe with a strictly extratropical distribution in the southernmost extension of the Atlantic forests (Judziewicz et al. 1999). The highest level of endemism for Olyreae is in the Atlantic forests of Brazil; many species of Olyreae and woody bamboos are endangered due to the continuing loss of these and other types of forests (Soderstrom et al. 1988; Ferreira et al. 2013).

1.5   Phylogenetic Relationships Within the Bamboos

As noted previously, that the Bambusoideae all share a common ancestor (i.e., the subfamily is monophyletic) is well established based on molecular sequence data, primarily from the plastid genome (GPWG 2001; GPWG II 2012). The presence of strongly asymmetrically invaginated arm cells in the chlorenchyma, as seen in cross section, appears to be uniquely derived in this lineage (GPWG 2001). All recent studies with sufficient sampling also strongly support Bambusoideae as the sister lineage to the Pooideae (bluegrass or wheat subfamily) in the BEP clade (Fig. 1.1; GPWG II 2012; Wu and Ge 2012), but no unique structural feature that supports or diagnoses this relationship has been identified.

Within Bambusoideae, three major lineages are resolved in all studies to date with sufficient sampling [Fig. 1.1; Kelchner et al. (2013); see also the more detailed review in BPG (2012)]: Arundinarieae (the temperate woody bamboos); Bambuseae (the tropical woody bamboos); and Olyreae (herbaceous bamboos). However, the relationships between these three lineages are not known with certainty. Analyses of sequence data from the chloroplast genome (represented by up to five markers) consistently indicate paraphyly of the woody bamboos with strong statistical support for branches; that is, there are two distinct lineages of woody bamboos and they are not each other's closest relative (Bouchenak-Khelladi

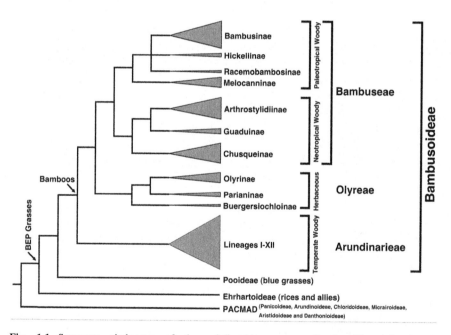

Fig. 1.1 Summary phylogeny of the relationships among Bambusoideae (bamboos), Ehrhartoideae (rices and allies) and Pooideae (wheat and allies), and among the tribes and subtribes of the Bambusoideae.

et al. 2008; Sungkaew et al. 2009; Kelchner et al. 2013). But tests of alternate relationships show that the possibility of a single lineage of woody bamboos cannot be rejected based on the chloroplast sequence data (Kelchner et al. 2013). Sequence data from the nuclear genome is only now becoming available for a reasonable sampling across the bamboos. Preliminary findings reveal that a single origin of woody bamboos may be supported, but that their evolutionary history is more complex than previously suspected and involves ancient hybridization events and allopolyploidy (Triplett et al. 2014).

Whether Olyreae is the closest lineage to the tropical woody bamboos (Bouchenak-Khelladi et al. 2008; Sungkaew et al. 2009; Kelchner et al. 2013) or a distinct, basically diploid lineage (Triplett et al. 2014) within the Bambusoideae, the herbaceous bamboos are strongly supported as monophyletic in all analyses of molecular sequence data. Notably, the herbaceous bamboos also show rates of sequence evolution at many loci much higher than those of woody bamboos and more similar to other grasses (Gaut et al. 1997). *Buergersiochloa*, a monotypic endemic of New Guinea, is consistently supported as sister to the remaining Olyreae, which are all native to the New World, at least based on morphology and chloroplast sequence data (Fig. 1.1; Kellogg and Watson 1993; Zhang and Clark 2000; Kelchner et al. 2013; Oliveira et al. 2014). *Pariana*, *Eremitis*, and *Parianella* form a lineage sister to the remaining olyroid genera (Ferreira 2013; Oliveira et al. 2014), but a comprehensive phylogenetic analysis is still lacking for the tribe, so evolutionary relationships within the Olyreae remain obscure. However, close relationships between *Raddia* and *Sucrea* on the one hand and *Raddiella* and *Parodiolyra* on the other hand are evident, and it is likely that *Olyra* is not monophyletic (Judziewicz et al. 1999; Zhang and Clark 2000; Oliveira et al. 2014).

The temperate woody bamboos (Arundinarieae) were resolved as a distinct phylogenetic group from the earliest molecular investigations onwards [see BPG (2012) and references cited therein] and form a robustly supported lineage in all recent molecular analyses (e.g., Bouchenak-Khelladi et al. 2008; Peng et al. 2008; Sungkaew et al. 2009; Triplett and Clark 2010; Zeng et al. 2010; Zhang et al. 2012). Eleven lineages have been identified within the Arundinarieae, and a twelfth has now been characterized (Attigala et al. 2014), but inferred relationships within and among them are poorly supported at best (Peng et al. 2008; Triplett and Clark 2010; Zeng et al. 2010; Zhang et al. 2012; Yang et al. 2013). The often decades-long generation times in temperate woody bamboos may explain the low rate of sequence evolution in this group (Gaut et al. 1997), which in turn may be partially responsible for the lack of resolution among the recognized phylogenetic lineages. A long and presumably largely isolated evolutionary history followed by recent, rapid radiation has also been suggested as an explanation for the lack of molecular resolution (Hodkinson et al. 2010). More complete and rigorous molecular analyses may reveal relationships in more detail, and improved knowledge of the fossil record (e.g., Wang et al. 2013) should help to better understand divergence times. Both ancient and recent (and ongoing) reticulation are important in the evolutionary history of the Arundinarieae, further complicating efforts to reconstruct the phylogeny of this group (Triplett et al. 2010; Yang et al. 2013; Triplett et al. 2014).

The tropical woody bamboos (Bambuseae) form two (Paleotropical woody and Neotropical woody) lineages (Fig. 1.1; Sungkaew et al. 2009; Kelchner et al. 2013). The Paleotropical woody bamboos (PWB) consistently receive strong support, and hexaploidy appears to be the general condition (Soderstrom 1981; Li et al. 2001), but support for the monophyly of the Neotropical woody bamboos (NWB), which as far as is known are all tetraploid, is moderate at best (Kelchner et al. 2013) and no defining character for the NWB has been identified. The rate of sequence evolution in the Bambuseae is mostly comparable to that of the Arundinarieae, although sequence evolution in the Chusqueinae appears to be somewhat accelerated (Kelchner et al. 2013). Seven subtribes (three in the NWB, four in the PWB) based on morphological and anatomical differences have been recognized within the Bambuseae in the recent literature (BPG 2012 and references cited therein) and these are, to a large extent, supported by molecular sequence data.

Within the NWB, Arthrostylidiinae and Guaduinae are consistently supported as sister to each other and each is well supported as monophyletic (Ruiz-Sanchez et al. 2008, 2011a; Fisher et al. 2009; Sungkaew et al. 2009; Ruiz-Sanchez 2011; Tyrrell et al. 2012; Kelchner et al. 2013). The presence of refractive papillae on the leaf epidermises may be a diagnostic feature supporting the sister relationship of these two subtribes (Ruiz-Sanchez et al. 2008), but further investigation is needed. Within the Arthrostylidiinae, the *Glaziophyton* clade, defined by erect, tessellate leaf blades, is sister to the remainder of the subtribe, which exhibit reflexed leaf blades (Tyrrell et al. 2012). Within this group, three other subclades are also identified and well supported, but the monophyly of larger genera such as *Arthrostylidium*, *Aulonemia*, and *Rhipidocladum* is not supported (Tyrrell et al. 2012). Relationships within the Guaduinae are less well understood, but evidence to date supports a sister relationship between *Guadua* and *Eremocaulon* and *Otatea* and *Olmeca*, respectively (Ruiz-Sanchez et al. 2011a, b). Within the Chusqueinae, the two clades of the species formerly recognized as *Neurolepis* form the earliest diverging branches, with *Chusquea* subg. *Rettbergia* as the next diverging lineage sister to the highly diverse Euchusquea clade. *Chusquea* subg. *Rettbergia*, plus the Euchusquea clade, comprises *Chusquea* in the strict (and traditional) sense, but the concept of the genus is now expanded to include the two *Neurolepis* clades, which will be recognized as subgenera (Fisher et al. 2009; Fisher et al. 2014).

The PWB are more diverse than the NWB in terms of both number of genera and number of species (Table 1.2), but despite their diversity and enormous ecological and economic importance (Dransfield and Widjaja 1995), an understanding of broad relationships within the PWB lags behind that of the NWB. Sungkaew et al. (2009) resolve the Melocanninae as robustly monophyletic and sister to the remainder of the PWB, a finding confirmed by Kelchner et al. (2013) with sampling from all four subtribes. Molecular phylogenetic studies of the PWB to date have focused on the Melocanninae and Bambusinae, but resolution of relationships within each subtribe is still tentative, with a few exceptions (Yang et al. 2007, 2008, 2010; Goh et al. 2010, 2013). Within Melocanninae, *Melocanna* and *Pseudostachyum* are supported as distinct genera, but the relationship between

Table 1.2 Diversity of Bambusoideae by tribe and subtribe

Taxon	Number of genera	Number of species
Arundinarieae	**31**	**546**
Bambuseae	**66**	**812**
Neotropical	21	405
Arthrostylidiinae	15	183
Chusqueinae	1	172
Guaduinae	5	50
Paleotropical	45	407
Bambusinae	27	268
Hickeliinae	8	33
Melocanninae	9	88
Racemobambosinae	1	17
Olyreae	**22**	**124**
Buergersiochloinae	1	1
Parianinae	3	38
Olyrinae	18	85
Total for subfamily	**119**	**1,482**

Cephalostachyum and *Schizostachyum* requires further study (Yang et al. 2007, 2008). The core of the Bambusinae consists of *Bambusa*, *Dendrocalamus*, and *Gigantochloa* and a few other small genera [the BDG complex of Goh et al. (2013)], and although there is good support for the monophyly of *Bambusa* in some analyses, species of *Dendrocalamus* and *Gigantochloa* are completely interdigitated (Yang et al. 2008, 2010; Goh et al. 2013). Hybridization and introgression among species of *Dendrocalamus* and *Gigantochloa* are documented and clearly contribute to the evolutionary and taxonomic complexity of the core Bambusinae (Wong and Low 2011; Goh et al. 2013). *Dinochloa* and several other clambering Bambusinae form a lineage distinct from the BDG complex (Yang et al. 2008; Goh et al. 2010, 2013). Racemobambosinae (*Racemobambos*) and Hickeliinae are supported as monophyletic, but their relationships to other PWB remain unclear (Goh et al. 2013; Kelchner et al. 2013).

Despite recent efforts to understand phylogenetic relationships among bamboos at the tribal and subtribal levels, to date there is no phylogenetic study that clearly shows well-resolved internal relationships between genera within subtribes. Future work in bamboos will require adding more taxa from the different recognized subtribes, especially targeting genera not previously included in molecular analyses, and sampling more plastid markers to generate increased internal resolution. Sequencing of low-copy nuclear loci is also needed, but this must be undertaken with the complex reticulate history of the woody bamboos in mind. Detailed morphological analyses of many bamboos are still needed to understand their phylogenetic relationships, but also to facilitate identification and classification, all of which will ultimately guide conservation and development decisions.

·ibal and Subtribal Classification of the Bamboos

_ ᴜⲥᴛailed description of the Bambusoideae can be found in BPG (2012). We here present synoptic descriptions for the currently recognized tribes and subtribes of bamboos, followed by additional comments about each group. Features character-istic of all or most members of a particular group are underlined. Although we follow the treatment presented in BPG (2012), the description of new species and genera continues, so we have updated numbers and lists accordingly; bamboo diversity is summarized in Table 1.2. We also list the included genera within each tribe or subtribe, with the number of species for each in parentheses, after the tribal or subtribal descriptions. For more detailed descriptions, see BPG (2012). A key to the bamboo genera of the world will be included in the forthcoming volume on Poaceae by E. A. Kellogg, which will be published as part of the series *The Families and Genera of Vascular Plants* edited by K. Kubtizki.

1.6.1   Herbaceous Bamboos: Tribe Olyreae

Description: Plants with rhizomes, these sometimes only weakly developed, or only pachymorph rhizomes present. Culms herbaceous to weakly lignified, with limited aerial branching. Culm leaves usually absent, sometimes present in taxa with larger culms (*Olyra*). Foliage leaves with the outer ligule absent; sheaths sometimes bearing fimbriae (*Eremitis*, *Pariana*) and/or blister like swellings at or near the summit (*Pariana*), more often these or auricular appendages absent; blades with epidermal silica cells usually with cross-shaped silica bodies in the costal zone and crenate (olyroid) silica bodies in the intercostal zone (these absent in *Buergersiochloa*). Flowering usually annual or seasonal for extended periods, very rarely gregarious and monocarpic. Synflorescences usually lacking well-developed bracts, apparently determinate. Spikelets unisexual, dimorphic, and 1-flowered with no rachilla extension, the plants monoecious. Female spikelets with 2 glumes, the floret usually leathery. Male spikelets usually smaller than the females, glumes usually absent or rarely 2 and well developed, the floret membra-nous. Caryopsis basic.

The Olyreae are the herbaceous bamboos. This group of 22 genera and 124 described species is native to tropical America, with two exceptions: *Buergersiochloa*, a rare monotypic bamboo endemic to New Guinea and Papua New Guinea, and *Olyra latifolia*, a widespread American species presumably introduced into Africa and Sri Lanka (Judziewicz and Clark 2007; BPG 2012). Members of Olyreae typically occur in rain forests or less commonly in lower montane forests up to 1,500 m in elevation. The four centers of diversity for Olyreae are (1) Bahia, in eastern Brazil; (2) northern Brazil (Amapá) and the Guianas; (3) the Chocó region of Colombia and Panama; and (4) Cuba (Soderstrom and Calderón 1979; Soderstrom et al. 1988). Herbaceous bamboos often develop strikingly beautiful synflorescence colorations, including bright displays of often numerous stamens, suggesting pollination by insects (Soderstrom and Calderón 1971).

Molecular data combined with traditional morphological and anatomical evidence have shown the herbaceous bamboos to be well supported as a lineage within the Bambusoideae. However, there is no single unique feature that diagnoses the Olyreae, although the presence of functionally unisexual spikelets and the lack of outer ligules distinguish Olyreae from the woody bamboos (Judziewicz and Clark 2007; BPG 2012), in addition to the limited aerial branching and less lignified stems of the Olyreae. Preliminary molecular data support three lineages (recognized as subtribes Buergersiochloinae, Parianinae, and Olyrinae, below) (Kelchner et al. 2013; Oliveira et al. 2014).

Amerindian tribes in Central and South America have reportedly used certain herbaceous bamboos as antifungal agents, an ointment for head lice, a snakebite remedy, for alleviation of general body aches, and to combat fevers, headaches, and coughs (Londoño 1990; Judziewicz et al. 1999). The herbaceous bamboos also have great potential value as ornamental plants.

1.6.1.1 Subtribe Buergersiochloinae

Description: Foliage leaf sheaths bearing fimbriae at the apex; blades lacking cross-shaped and crenate (olyroid) silica bodies in both epidermises. Synflorescences paniculate. Female lemmas awned. Stamens 2–3. Endemic to New Guinea/Papua New Guinea.

Included genus: *Buergersiochloa* (1).

1.6.1.2 Subtribe Parianinae

Description: Foliage leaf sheaths bearing fimbriae at the apex; blades with cross-shaped and crenate (olyroid) silica bodies in the epidermises. Synflorescences spicate. Female lemmas unawned. Stamens 2, 3, or 6 (to 36–40). Costa Rica and Trinidad, northern South America to Amazonian Bolivia and Atlantic Brazil (Bahia).

Included genera: *Eremitis* (3), *Pariana* (33), *Parianella* (2).

1.6.1.3 Subtribe Olyrinae

Description: Foliage leaf sheaths lacking fimbriae at the apex; blade with cross-shaped and crenate (olyroid) silica bodies in the epidermises. Synflorescences paniculate or racemose. Female lemmas usually unawned (awned only in *Agnesia*, *Ekmanochloa*). Stamens 2–3. Mexico and the West Indies, Central America, northern South America to Argentina and southern Brazil.

Included genera: *Agnesia* (1), *Arberella* (7), *Cryptochloa* (8), *Diandrolyra* (3), *Ekmanochloa* (2), *Froesiochloa* (1), *Lithachne* (4), *Maclurolyra* (1), *Mniochloa* (1), *Olyra* (24), *Parodiolyra* (5), *Piresia* (5), *Piresiella* (1), *Raddia* (9), *Raddiella* (8), *Rehia* (1), *Reitzia* (1), and *Sucrea* (3).

1.6.2   Temperate Woody Bamboos: Tribe Arundinarieae

Description: Rhizomes well developed, some taxa with pachymorph rhizomes only. Culms woody, usually hollow; branch development beginning at the apex and continuing toward the base (basipetal); aerial vegetative branching complex, usually derived from a single bud per node (multiple, subequal buds per node in *Chimonobambusa*). Culm leaves usually well developed. Foliage leaves with an outer ligule; sheaths often bearing fimbriae and/or auricular appendages at the summit. Flowering usually cyclical, gregarious, and monocarpic. Synflorescences with well-developed bracts or not, determinate (spikelets) or indeterminate (pseudospikelets). Spikelets (or spikelets proper of the pseudospikelets) bisexual with 1 to many bisexual florets; glumes (0–1) 2–4; lemmas and paleas similar in texture to the glumes. Caryopsis basic, uncommonly baccate (e.g., *Ferrocalamus*). Base chromosome number $x = 12$; $2n = 48$.

Included genera: *Acidosasa* (11), *Ampelocalamus* (13), *Arundinaria* (3 + ca. 6 of uncertain placement), *Bashania* (2), *Bergbambos* (1), *Chimonobambusa* (37), *Chimonocalamus* (11), *Drepanostachyum* (10), *Fargesia* (90), *Ferrocalamus* (2), *Gaoligongshania* (1), *Gelidocalamus* (9), *Himalayacalamus* (8), *Indocalamus* (23), *Indosasa* (15), *Kuruna* (6), *Oldeania* (1), *Oligostachyum* (15), X*Phyllosasa* (=*Hibanobambusa*) (1), *Phyllostachys* (51), *Pleioblastus* (40), *Pseudosasa* (19), *Sarocalamus* (3), *Sasa* (40), *Sasaella* (13), *Sasamorpha* (5), *Semiarundinaria* (10), *Shibataea* (7), *Sinobambusa* (10), *Thamnocalamus* (3), and *Yushania* (80).

The Arundinarieae are the temperate woody bamboos, a diverse clade of 30 genera and ca. 546 species, distributed primarily in forests of the northern temperate zone, but also in some high elevation tropical regions of both northern and southern hemispheres [see Fig. 2 in Kelchner et al. (2013)] (Triplett and Clark 2010). The center of diversity is in East Asia (ca. 430 spp.), with areas of endemism in Southwestern China (ca. 180 spp.), Japan (ca. 80 spp.), Southeast Asia (ca. 60 spp.), Madagascar (ca. six spp.), Africa (two spp.), and Sri Lanka (five spp.) (Ohrnberger 1999; Triplett and Clark 2010; BPG 2012). The diversity of Arundinarieae in East Asia and the three species of *Arundinaria* native to North America represent a classic, if asymmetrical, disjunction pattern between East Asia and eastern North America, potentially indicating a past migration across the Bering Land Bridge (Stapleton et al. 2004; Triplett and Clark 2010).

The recognition of Arundinarieae as a distinct lineage within the Bambusoideae is well supported by molecular phylogenetic studies (Bouchenak-Khelladi et al. 2008; Sungkaew et al. 2009; Kelchner et al. 2013). Although a formal morphological analysis has not been done, basipetal branch development and a chromosome number of $2n = 48$ have been identified as putative defining characters and thus support recognition of this lineage at the tribal level (BPG 2012).

Species of what is now recognized as the Arundinarieae were traditionally classified in up to three subtribes, the Arundinariinae, Shibataeinae, and Thamnocalaminae, based on the presence or absence of pseudospikelets and rhizome structure (Zhang 1992; BPG 2012). Recent studies have provided strong evidence that none of these three subtribes is a natural group, so this subtribal classification has been abandoned. Eleven numbered lineages have been resolved, some at the generic level, some possibly subtribal, but several cutting across phenetically based genera or groups of genera (Triplett and Clark 2010; Zeng et al. 2010; BPG 2012). Relationships between and within these lineages have not been clearly revealed by molecular studies, although genomics tools are now being used in an attempt to obtain resolution. Intergeneric hybridization certainly plays a role in generating some of the taxonomic confusion, but other evolutionary processes (e.g., incomplete lineage sorting) also are factors (Triplett et al. 2010; Yang et al. 2013). The 12 lineages currently recognized are (Triplett and Clark 2010; Zeng et al. 2010; Yang et al. 2013; Attigala et al. 2014): (I) Bergbamboes, (II) African Alpine bamboos, (III) *Chimonocalamus*, (IV) *Shibataea* clade, (V) *Phyllostachys* clade, (VI) *Arundinaria* clade, (VII) *Thamnocalamus*, (VIII) *Indocalamus wilsonii*, (IX) *Gaoligongshania*, (X) *Indocalamus sinicus*, (XI) *Ampelocalamus calcareus*, and (XII) *Kuruna*. Five of these clades (I, VIII, IX, X, and XI) consist of a single species each. The *Arundinaria* and *Phyllostachys* clades, as defined based on data primarily from the chloroplast genome and discussed below, are by far the most diverse, including about 85 % of total species in the Arundinarieae. We note that ongoing studies, especially those including data from the nuclear genome (e.g., Yang et al. 2013), will undoubtedly reveal additional complexity and suggest additional phylogenetic lineages.

Arundinaria clade (VI)

With at least ten genera and more than 130 species, this is the second most speciose lineage in Arundinarieae. The morphology-based taxonomy of *Arundinaria* and its relatives has been especially problematic, but the phylogenetic study of Triplett and Clark (2010) revealed that some of the taxonomy of this group was inconsistent with its evolutionary history. The *Arundinaria* clade is united by rhizome type (leptomorph) but exhibits significant morphological diversification (e.g., spikelets, pseudospikelets, various numbers of branches, and three or six stamens). The concept of *Arundinaria* itself is limited to the three species native to North America, and work continues to place the other species still classified in *Arundinaria* in the broad sense (Stapleton 2013). Other genera minimally included in the *Arundinaria* clade are *Acidosasa*, *Oligostachyum*, X*Phyllosasa* (=*Hibanobambusa*), *Pleioblastus* (in part), *Pseudosasa*, *Sasa* (in part), *Sasaella*, *Sasamorpha*, and *Semiarundinaria*. At least four of these (X*Phyllosasa*, *Pseudosasa*, *Sasamorpha*, and *Seminarundinaria*) are wholly or partly derived through intergeneric hybridization (Triplett and Clark 2010).

***Phyllostachys* clade (V)**

This is the largest clade in Arundinarieae with ca. 16 genera and more than 330 species. The clade containing *Phyllostachys* and allies comprises about 50 % of the temperate genera and more than 70 % of the temperate bamboo species. The clade unites members from all three of the earlier morphology-based subtribes, combining plants with true spikelets or pseudospikelets, bracteate or ebracteate synflorescences, and pachymorph or leptomorph rhizomes. Clade V includes at least four genera from the traditional Shibataeinae (*Brachystachyum, Chimonobambusa, Phyllostachys,* and *Sinobambusa,* with ebracteate, indeterminate synflorescences), six from the traditional Thamnocalaminae (*Ampelocalamus, Drepanostachyum, Fargesia, Himalayacalamus, Thamnocalamus,* and *Yushania,* with ebracteate to bracteate determinate synflorescences and pachymorph rhizomes), and five from the traditional Arundinariinae (*Bashania, Gelidocalamus, Indocalamus, Pleioblastus* in part and *Sarocalamus,* with semelauctant inflorescences) (Triplett and Clark 2010). Chinese *Sasa* also fall within this clade. The *Phyllostachys* clade is remarkable for contrasting high morphological diversity with low chloroplast DNA variation. Sequences in this group are nearly identical, differing by only a few point mutations or indels, most of which are found in only one taxon (Triplett and Clark 2010; Zeng et al. 2010).

1.6.3   Tropical Woody Bamboos: Tribe Bambuseae

Description: Rhizomes well developed, usually pachymorph but some taxa amphimorph. Culms woody, usually hollow (solid in most *Chusquea* and a few species of other genera); branch development from the base to the apex (acropetal) or bidirectional; aerial vegetative branching complex (but absent in a few taxa), usually derived from a single bud per node (multiple, subequal buds per node in *Apoclada, Filgueirasia, Holttumochloa*; multiple, dimorphic buds in most of *Chusquea*). Culm leaves usually well developed, sometimes poorly differentiated from foliage leaves or absent. Foliage leaves with an outer ligule; sheaths often bearing fimbriae and/or auricular appendages at the summit; blades usually pseudopetiolate, deciduous. Flowering usually cyclical, gregarious, and monocarpic. Synflorescences with well-developed bracts or not, determinate (spikelets) or indeterminate (pseudospikelets). Spikelets (or spikelets proper of the pseudospikelets) bisexual with 1 to many bisexual florets; glumes (0–) 1 to 4 (–6), sometimes very reduced; lemmas and similar in texture to the glumes. Caryopsis usually basic, sometimes baccate (e.g., *Alvimia, Dinochloa, Melocanna, Ochlandra, Olmeca,* at least one species of *Guadua*) or nucoid (e.g., *Actinocladum, Merostachys, Pseudostachyum*). Base chromosome numbers $x = 10$, (11), and 12; $2n = (20)$ 40, (44), 46, 48, 70, 72.

The tribe Bambuseae comprises the Paleotropical and Neotropical woody bamboos, widespread in both the Old World and New World. It includes seven subtribes, 66 genera, and 812 species (BPG 2012). The recognition of Bambuseae as a

distinct lineage within the Bambusoideae is well supported by molecular phylogenetic studies (Bouchenak-Khelladi et al. 2008; Sungkaew et al. 2009; Kelchner
et al. 2013). Although a formal morphological analysis has not been done, acropetal
or bidirectional branch development has been suggested as a possible defining
character for Bambuseae, including both the Paleotropical and Neotropical lineages
(BPG 2012). Recent analyses suggest that sympodial, pachymorph rhizomes and
determinate spikelets are likely ancestral within the tribe (Clark et al. 2007;
Kelchner et al. 2013), but it is clear that patterns of morphological evolution within
the Bambuseae are complex and much work remains to be done to characterize this
tribe. The two major groups within the Bambuseae are the Neotropical and
Paleotropical woody bamboos. We here discuss the Neotropical and Paleotropical
groups and their respective subtribes separately.

1.6.3.1 Neotropical Woody Bamboos

Neotropical woody bamboos are a moderately well-supported subclade within the
Bambuseae, with three well-supported subtribes: Arthrostylidiinae, Chusqueinae,
and Guaduinae. The NWB comprise 21 genera and at least 405 species (BPG 2012),
and new genera and new species continue to be discovered and described. The
NWB have a geographical distribution from Mexico along Central America to
South America and also in the Caribbean Islands, with an altitudinal range from
sea level to 4,300 m (Judziewicz et al. 1999; BPG 2012).

1.6.3.1.1 Subtribe Arthrostylidiinae

Description: Rhizomes necks short to somewhat elongated; internodes of the aerial
culms usually hollow, all subequal or sometimes very short internodes alternating
in various combinations with elongated internodes. Aerial branching derived from a
single bud per node; thorns absent. Culm leaves with sheaths usually bearing
fimbriae or fimbriate auricles; oral setae absent. Foliage leaf sheaths usually bearing
fimbriae or fimbriate auricles at the summit, oral setae absent; blades with a simple,
abaxially projecting midrib; intercostal sclerenchyma usually present; adaxial epidermis lacking stomates and papillae or these infrequent and poorly developed;
abaxial epidermis usually with a green stripe along the narrow-side margin, with
stomates common and papillae usually well developed on at least some long cells;
stomatal apparatus with papillae absent from the subsidiary cells but usually
overarched by papillae from adjacent long cells. Synflorescences usually without
bracts, indeterminate (pseudospikelets) or determinate (spikelets), paniculate or
racemose. Spikelets (or spikelets proper of the pseudospikelets) consisting of 2–3
glumes, 1 to many female-fertile florets, and a rachilla extension bearing a rudimentary floret; palea keels wingless. Stamens (2) 3 (6). Ovary glabrous, with a short
style; stigmas 2 (3). Caryopsis basic, uncommonly baccate (*Alvimia*) or nucoid
(*Actinocladum*, *Merostachys*).

Included genera: *Actinocladum* (1), *Alvimia* (3), *Arthrostylidium* (32), *Athroostachys* (1), *Atractantha* (6), *Aulonemia* (47), *Cambajuva* (1), *Colanthelia* (7), *Didymogonyx* (2), *Elytrostachys* (2), *Filgueirasia* (2), *Glaziophyton* (1), *Merostachys* (48), *Myriocladus* (12), and *Rhipidocladum* (18).

The Arthrostylidiinae can be distinguished from other woody bamboo subtribes using branch leaf micromorphology and anatomy (Soderstrom and Ellis 1987). The leaf blades possess a unique combination of intercostal sclerenchyma fibers in the mesophyll of the blades and simple vasculature in the midrib, and the leaf blades are basically hypostomatic with papillae usually developed on the abaxial epidermis (Tyrrell et al. 2012). With 15 genera and 183 species, Arthrostylidiinae comprises 70 % of the genera and 45 % of the total diversity in NWB (BPG 2012). Arthrostylidiinae is arguably the most morphologically diverse subtribe of the NWB.

1.6.3.1.2   Subtribe Chusqueinae

Description: Rhizomes with short necks, sometimes leptomorph rhizomes present; internodes of the aerial culms usually solid, all subequal. Aerial branching when present derived from a multiple, dimorphic bud complement, absent in two clades (=*Neurolepis*) but a single bud per node usually present in these; thorns absent. Culm leaf sheaths usually lacking fimbriae or fimbriate auricles; oral setae absent. Foliage leaf sheaths usually bearing cilia at the summit, rarely well-developed fimbriae present, oral setae absent, auricles absent; blades with a complex, abaxially projecting midrib; intercostal sclerenchyma absent; adaxial epidermis lacking stomates and papillae or these infrequent and poorly developed; abaxial epidermis usually lacking a green stripe along the narrow-side margin, with stomates common and papillae usually well developed on at least some long cells; stomatal apparatus bearing two papillae per subsidiary cell and also often overarched by papillae from adjacent long cells. Synflorescences usually without bracts, determinate (spikelets), paniculate, or rarely racemose. Spikelets consisting of 4 glumes and 1 female-fertile floret, rachilla extension absent; palea keels lacking wings. Stamens (2) 3. Ovary glabrous, with a short style; stigmas 2. Caryopsis basic.

Included genus: *Chusquea* Kunth (172).

Chusqueinae, which includes the single, well-supported yet very diverse genus *Chusquea*, can be distinguished from other woody bamboo subtribes by the presence of two papillae on each subsidiary cell of the foliar stomatal apparatus and spikelets consisting of four glumes, one fertile floret, and no rachilla extension (Fisher et al. 2009). Species of Chusqueinae are characteristic of montane forests throughout Mexico, Central and South America, and the Caribbean, but a number of species inhabit high altitude grasslands and a few species occur in lowland tropical forest or in temperate forests at higher latitudes (both north and south) (Fisher et al. 2009). Species of *Chusquea* range from sea level to 4,300 m in elevation, giving this genus the broadest altitudinal range of any bamboo. *Chusquea* species tend to form a visible and sometimes dominant component of the vegetation

(Judziewicz et al. 1999). Although including only one genus, with 172 described species the Chusqueinae has 42 % of the total species diversity in NWB (BPG 2012).

1.6.3.1.3   Subtribe Guaduinae

Description: Rhizomes with necks short to elongated; internodes of the aerial culms hollow to solid, all subequal. Aerial branching usually derived from a single bud per node (1–4 subequal buds per node in *Apoclada*); thorns absent or present (*Guadua*). Culm leaves with sheaths often bearing fimbriae or fimbriate auricles at the sheath summit; oral setae usually present (absent in *Apoclada* and *Guadua*). Foliage leaf sheaths often with fimbriae or fimbriate auricles at the summit; oral setae present; blades with a complex, abaxially projecting midrib; intercostal sclerenchyma absent; <u>adaxial epidermis usually with abundant stomates and well-developed papillae</u>, rarely these lacking or infrequent and poorly developed; abaxial epidermis usually lacking a green stripe along the narrow-side margin, with stomates present and abundant (absent in *Apoclada*) and papillae absent to well developed; stomatal apparatus with papillae absent from the subsidiary cells but usually overarched by papillae from adjacent long cells. Synflorescences with bracts or not, indeterminate (pseudospikelets) or determinate (spikelets), paniculate. Spikelets (or spikelets proper of the pseudospikelets) consisting of (0–) 1 to 4 (–7) glumes, 1 to many female-fertile florets, and a rachilla extension bearing a rudimentary floret; palea keels wingless to prominently winged. Stamens 3 or 6. Ovary glabrous or hairy, with a short style; stigmas 2 or 3. Caryopsis basic, uncommonly baccate (some species of *Olmeca* and *Guadua sarcocarpa*).

Included genera: *Apoclada* (1), *Eremocaulon* (4), *Guadua* (32), *Olmeca* (5), *Otatea* (8).

Guaduinae can be distinguished from other woody bamboo subtribes by the presence of abundant stomates on both adaxial and abaxial foliage leaf blade surfaces, often combined with the presence of papillae on the adaxial surface whether or not papillae are present on the abaxial surface (Judziewicz et al. 1999; Ruiz-Sanchez et al. 2008). Species of Guaduinae inhabit low to mid-elevation wet or dry tropical forests and form extensive mono-dominant *Guadua* forests in part of the Amazon basin (Judziewicz et al. 1999). With 5 genera and 50 species, this subtribe represents slightly less than 25 % of the generic diversity and only 12 % of the total species diversity in NWB (BPG 2012). Despite the relatively low diversity of this subtribe, it occupies an estimated area of 11 million ha. from Mexico to Argentina and it has great economic importance due to the utility of *G. angustifolia* and a few other species of the genus (Judziewicz et al. 1999).

1.6.3.2 Paleotropical Woody Bamboos

The PWB or Old World bamboos include 45 genera and 407 species and are grouped into four subtribes: Bambusinae, Hickeliinae, Melocanninae, and Racemobambosinae (Table 1.2). They are distributed throughout South-East Asia, northern Australia, India, Sri Lanka, Africa, and Madagascar (Soderstrom and Ellis 1987; Dransfield and Widjaja 1995; Ohrnberger 1999; BPG 2012).

The areas with greatest diversity of Old World bamboos are (1) the region including southern China, northern Burma (Myanmar), Thailand, and Vietnam, and (2) Madagascar, where almost all known native species and genera are endemic (Dransfield and Widjaja 1995). The largest numbers of species occur in the largest countries, China and India. *Bambusa* is the most widespread genus of bamboo in tropical and subtropical Asia, and several species of it and the related genera *Dendrocalamus* and *Gigantochloa* have been introduced to Central and South America where they can play an important role in local economies.

1.6.3.2.1 Subtribe Bambusinae

Description: Rhizomes with necks short to slightly elongated; internodes of the aerial culms hollow or solid, all subequal; nodes of the aerial culms with or without a patella. Aerial branching derived from a single bud per node (multiple buds in *Holttumochloa*); thorns usually absent, sometimes present (*Bambusa*). Culm leaves with sheaths bearing fimbriae or fimbriate auricles at the summit or neither; oral setae present or absent. Foliage leaf sheaths often with fimbriae or fimbriate auricles at the summit; oral setae present or absent; blades with a complex or simple, abaxially projecting midrib; intercostal sclerenchyma absent; adaxial epidermis with or without stomates, with or without papillae; abaxial epidermis usually lacking a green stripe along the narrow-side margin, usually with abundant stomates and well-developed papillae; stomatal apparatus with papillae absent from the subsidiary cells but usually overarched by papillae from adjacent long cells. Synflorescences bracteate or not, indeterminate (pseudospikelets) or less commonly determinate (spikelets), paniculate. Spikelets or spikelets proper of the pseudospikelets consisting of (0–) 1 to several glumes, 1–10 or more female-fertile florets, and sometimes a rachilla extension bearing 1–3 rudimentary florets; palea keels wingless to prominently winged. Stamens 6, filaments free or fused. Ovary glabrous or hairy, usually with a short style; stigma 1, 2, or 3. Caryopsis basic or baccate (*Cyrtochloa, Dinochloa, Melocalamus, Sphaerobambos*).

Included genera: *Bambusa* (100), *Bonia* (5), *Cyrtochloa* (5), *Dendrocalamus* (41), *Dinochloa* (31), *Fimbribambusa* (2), *Gigantochloa* (30), *Greslania* (4), *Holttumochloa* (3), *Kinabaluchloa* (2), *Maclurochloa* (2), *Melocalamus* (5), *Mullerochloa* (1), *Neololeba* (5), *Neomicrocalamus* (5), *Oreobambos* (1), *Oxytenanthera* (1), *Parabambusa* (1), *Pinga* (1), *Pseudobambusa* (1),

Pseudoxytenanthera (12), *Soejatmia* (1), *Sphaerobambos* (3), *Temochloa* (1), *Temburongia* (1), *Thyrsostachys* (2), and *Vietnamosasa* (3).

No single feature has been identified to define this subtribe, but a core Bambusinae [also known as the BDG complex, Goh et al. (2013)] defined primarily by molecular evidence clearly includes *Bambusa, Dendrocalamus, Gigantochloa, Maclurochloa, Melocalamus, Oreobambos, Oxytenanthera, Phuphanochloa* (if recognized as distinct from *Bambusa*), *Soejatmia, Thyrsostachys,* and *Vietnamosasa* (Yang et al. 2008; Sungkaew et al. 2009; Goh et al. 2013), although these have not all been included in a comprehensive molecular analysis and some genera remain unsampled. A distinct clade of primarily climbing bamboos includes at least *Dinochloa, Mullerochloa, Neololeba,* and *Sphaerobambos*, and may merit recognition at the subtribal level once additional studies are completed (Goh et al. 2010, 2013; Chokthaweepanich 2014). This subtribe is notable for its paramount economic importance in tropical Asia, due to the cultivation and use of many species of *Bambusa, Dendrocalamus*, and *Gigantochloa* (e.g., Dransfield and Widjaja 1995; Lucas 2013). Many species of all three genera are also cultivated widely in tropical and subtropical regions around the globe (Dransfield and Widjaja 1995; Ohrnberger 1999).

1.6.3.2.2 Subtribe Hickeliinae

Description: Rhizomes with necks short to elongated; internodes of the aerial culms usually hollow or rarely solid, all subequal. Aerial branching derived from a single bud per node (multiple buds in *Nastus productus*), central branch dominant; thorns absent. Culm leaves with sheaths bearing fimbriae or fimbriate auricles or neither; oral setae absent. Foliage leaf sheaths with fimbriae or fimbriate auricles present or absent; oral setae absent; blades with a complex, sometimes adaxially projecting midrib; intercostal sclerenchyma and fiber-like epidermal cells sometimes present; adaxial epidermis lacking stomates and papillae or these infrequent and poorly developed; abaxial epidermis usually lacking a green stripe along the narrow-side margin, with stomates common and papillae usually well developed on at least some long cells; stomatal apparatus with papillae absent from the subsidiary cells but usually overarched by papillae from adjacent long cells. Synflorescences determinate (spikelets), bracteate or ebracteate, paniculate, racemose, or capitate. Spikelets consisting of <u>4–6 glumes and 1 female-fertile floret</u>; rachilla extension present or absent, if present well developed or much reduced bearing a rudimentary or reduced floret; palea usually 2-keeled (without keels when rachilla extension absent), keels wingless. Stamens 6, filaments usually free. Ovary glabrous or hairy, with long or short style; stigmas 3. Caryopsis basic.

Included genera: *Cathariostachys* (2), *Decaryochloa* (1), *Hickelia* (4), *Hitchcockella* (1), *Nastus* (20), *Perrierbambus* (2), *Sirochloa* (1), and *Valiha* (2).

The Madagascan and Réunion Island Hickeliinae are strongly supported as a distinct lineage within the PWB, although relatively few species have been sampled in molecular analyses to date (Clark et al. 2007; Kelchner et al. 2013). Adaxially projecting midribs have been proposed as a diagnostic character for this subtribe (Soderstrom and Ellis 1987), but further analyses do not support this (Stapleton 1994c; Chokthaweepanich 2014). All of the diversity in this subtribe, except for one species of *Hickelia* and several Asiatic species of *Nastus*, are endemic to Madagascar and Réunion Island (Ohrnberger 1999; Clark et al. 2007), indicating a remarkable level of generic diversity in Madagascar. Dransfield (1994, 1998) has provided detailed morphological descriptions for many of these taxa and continues to study this unique bamboo radiation.

1.6.3.2.3   Subtribe Melocanninae

Description: Rhizomes with necks short or elongated; internodes of the aerial culms moderately long or very long, hollow, with thin walls; nodes of the aerial culms lacking a patella. Aerial branching derived from a single bud per node; thorns absent. Culm leaves with sheaths bearing fimbriae or fimbriate auricles at the summit or neither; oral setae usually absent. Foliage leaf sheaths bearing fimbriae or small fimbriate auricles or neither; oral setae present or absent; blades with a complex, abaxially projecting midrib; intercostal sclerenchyma absent; adaxial epidermis lacking stomates or these infrequent and poorly developed, papillae often present; abaxial epidermis with (usually) or without a green stripe along the narrow-side margin, with stomates common and papillae usually well developed on at least some long cells; stomatal apparatus with papillae absent from the subsidiary cells but usually overarched by papillae from adjacent long cells. Synflorescences indeterminate (pseudospikelets), spicate, or capitate. Spikelets proper consisting of (0) 2 (or 4) glumes, one female-fertile floret (3 in *Schizostachyum grande*), with or without rachilla extension, if present bearing a rudimentary floret; palea keels wingless or winged. Stamens 6 (15–120 in *Ochlandra*), filaments free or fused. Ovary glabrous, with a long, slender, hollow style; stigmas (2–) 3. Caryopsis basic or baccate (*Melocanna, Ochlandra, Stapletonia*) or nucoid (*Pseudostachyum*).

Included genera: *Cephalostachyum* (14), *Davidsea* (1), *Melocanna* (2), *Neohouzeaua* (7), *Ochlandra* (9), *Pseudostachyum* (1), *Schizostachyum* (51), *Stapletonia* (1), and *Teinostachyum* (2).

With nine genera and 88 species, Melocanninae is the second most diverse subtribe of PWB. *Melocanna* is perhaps the best known representative of this subtribe, primarily due to its large, fleshy, pear-shaped fruits that are implicated in rat population explosions when the bamboo flowers gregariously approximately every 48 years (Singleton et al. 2010). Holttum (1956) hypothesized that *Cephalostachyum, Pseudostachyum, Schizostachyum,* and *Teinostachyum* should be combined into one genus, but molecular data support maintaining at least *Pseudostachyum* as a distinct genus (Yang et al. 2007, 2008; Sungkaew et al. 2009) and Yang et al. (2008) present evidence in support of maintaining *Cephalostachyum* and *Schizostachyum* as separate genera. Given the well-supported position of this subtribe as sister to the remaining PWB, a more detailed morphological and molecular analysis of this subtribe, including all of the recognized genera, is needed.

1.6.3.2.4   Subtribe Racemobambosinae

Description: Rhizomes with necks short or elongated; internodes of the aerial culms hollow, all subequal; nodes of the aerial culms without a patella. Aerial branching derived from a single bud per node; thorns absent. Culm leaves with sheaths usually bearing small fimbriate auricles at the summit or rarely efimbriate and exauriculate; oral setae absent. Foliage leaf sheaths usually bearing small fimbriate auricles at the summit or rarely efimbriate and eauriculate; oral setae absent; blades with an abaxially projecting midrib; blade anatomy and micromorphology unknown. Synflorescences bracteate, determinate (spikelets), racemose. Spikelets consisting of 2–3 glumes, 3–8 female-fertile florets and a rachilla extension bearing one rudimentary floret; palea keels wingless. Stamens 6, filaments free. Ovary usually hairy toward the apex, usually with a short style; stigmas 3. Caryopsis basic.

Included genus: *Racemobambos* (17).

Two genera initially included within this subtribe, *Neomicrocalamus* and *Vietnamosasa*, are clearly allied with the Bambusinae based on chloroplast sequence data (Yang et al. 2008; Goh et al. 2013) and have been transferred to that subtribe (BPG 2012). *Racemobambos* is resolved as a distinct lineage among the PWB, but its affinities are uncertain and no unique structural character has yet been found to diagnose this genus (or subtribe) (Goh et al. 2013).

Acknowledgments We thank Dr. Emmet J. Judziewicz and Dr. Chris Stapleton for providing constructive comments that greatly improved this chapter.

References

Arber A (1927) Studies in the Gramineae. II. Abnormalities in *Cephalostachyum virgatum* Kurz, and their bearing on the interpretation of the bamboo flower. Ann Bot 41:47–74

Attigala LR, Kaththriarachchi H, Clark LG (2014) A new genus and a major temperate bamboo lineage of the Arundinarieae (Poaceae: Bambusoideae) from Sri Lanka based on a multi-locus plastid phylogeny. Phytotaxa 174(4):187–205

Bamboo Phylogeny Group [BPG] (2012) An updated tribal and subtribal classification of the bamboos (Poaceae: Bambusoideae). In: Gielis J, Potters G (eds) Proceedings of the 9th world bamboo congress, Antwerp, Belgium, 10–12 Apr 2012, pp 3–27

Bedell PE (1997) Taxonomy of bamboos. APC Publications, New Delhi, p 150

Bouchenak-Khelladi Y, Salamin N, Savolainen V, Forest F, van der Bank M, Chase MW, Hodkinson TR (2008) Large multi-gene phylogenetic trees of the grasses (Poaceae): progress towards complete tribal and generic level sampling. Mol Phylogenet Evol 47:488–505

Calderón CE, Soderstrom TR (1973) Morphological and anatomical considerations of the grass subfamily Bambusoideae based on the new genus *Maclurolyra*. Smithson Contrib Bot 11:1–54

Calderón CE, Soderstrom TR (1980) The genera of Bambusoideae (Poaceae) of the American continent: keys and comments. Smithson Contrib Bot 44:1–27

Chokthaweepanich H (2014) Phylogenetics and evolution of the paleotropical woody bamboos (Poaceae: Bambusoideae: Bambuseae). Unpubl. Ph.D. Dissertation, Iowa State University, p 22

Clark LG, Judziewicz EJ (1996) The grass subfamilies Anomochlooideae and Pharoideae. Taxon 45:641–645

Clark LG, Zhang W-P, Wendel JF (1995) A phylogeny of the grass family (Poaceae) based on ndhF sequence data. Syst Bot 20:436–460

Clark LG, Dransfield S, Triplett JK, Sánchez-Ken JG (2007) Phylogenetic relationships among the one-flowered, determinate genera of Bambuseae (Poaceae: Bambusoideae). Aliso 23:315–332

Clayton WD, Renvoize SA (1986) Genera Graminum, Grasses of the world, vol XIII, Kew bulletin additional series. Her Majesty's Stationery Office, London, p 389

Dai L-M, Wang Y, Su D-K, Zhou L, Yu D-P, Lewis BJ, Qi L (2011) Major forest types and the evolution of sustainable forestry in China. Environ Manag 48:1066–1078

Dransfield S (1992) The bamboos of Sabah. Sabah forest records no. 14, Forestry Department, Sabah, Malaysia, p 94

Dransfield S (1994) The genus Hickelia (Gramineae-Bambusoideae). Kew Bull 49:429–443

Dransfield S (1998) Valiha and Cathariostachys, two new bamboo genera (Gramineae-Bambusoideae) from Madagascar. Kew Bull 53:375–397

Dransfield S, Widjaja EA (eds) (1995) Plant resources of South-East Asia No. 7: bamboos. Backhuys, Leiden, p 189

Fernandez EP, Avila Moraes M, Martinelli G (2012) New records and geographic distribution of Glaziophyton mirabile (Poaceae: Bambusoideae). Check List 8:1296–1298

Ferreira FM (2013) Filogenia da subtribo Parianinae e sistemática de Eremitis Döll (Poaceae: Bambusoideae: Olyreae). D. Phil. thesis, Universidade Estadual de Feira de Santana, Brasil, p 210

Ferreira FM, Dórea MC, Leite KRB, Oliveira RP (2013) Eremitis afimbriata and E. magnifica (Poaceae, Bambusoideae, Olyreae): two remarkable new species from Brazil and a first record of blue iridescence in bamboo leaves. Phytotaxa 84:31–45

Fisher AE, Clark LG, Kelchner SA (2014) Molecular phylogeny estimation of the Chusquea bamboos (Poaceae: Bambusoideae: Bambuseae) and description of two new bamboo subgenera. Syst Bot 39(3):829–844

Fisher A, Triplett JK, Ho C-S, Schiller A, Oltrogge K, Schroder E, Kelchner SA, Clark LG (2009) Paraphyly in the Chusqueinae (Poaceae: Bambusoideae: Bambuseae). Syst Bot 34:673–683

Gadgil M, Prasad SN (1984) Ecological determinants of life history evolution of two Indian bamboo species. Biotropica 16:161–172

Gaut BS, Clark LG, Wendel JF, Muse SV (1997) Comparisons of the molecular evolutionary process at rbcL and ndhF in the grass family (Poaceae). Mol Biol Evol 14:769–777

Goh W-L, Chandran S, Lin R-S, Xia N-H, Wong KM (2010) Phylogenetic relationships among Southeast Asian climbing bamboos (Poaceae: Bambusoideae) and the Bambusa complex. Biochem Syst Ecol 38:764–773

Goh W-L, Chandran S, Franklin DC, Isagi Y, Koshy KC, Sungkaew S, Yang H-Q, Xia N-H, Wong KM (2013) Multi-gene region phylogenetic analyses suggest reticulate evolution and a clade of Australian origin among paleotropical woody bamboos (Poaceae: Bambusoideae: Bambuseae). Plant Syst Evol 299:239–257

Gopakumar B, Motwani B (2013) Adaptive strategies of reed bamboos, Ochlandra spp., to the Western Ghat habitats of India. Bamboo Sci Cult 26:33–40

Grass Phylogeny Working Group [GPWG] (2001) Phylogeny and subfamilial classification of the grasses (Poaceae). Ann Missouri Bot Gard 88:373–457

Grass Phylogeny Working Group [GPWG] II (2012) New grass phylogeny resolves deep evolutionary relationships and discovers C_4 origins. New Phytol 193:304–312

Hodkinson TR, Chonghaile GN, Sungkaew S, Chase MW, Salamin N, Stapleton CMA (2010) Phylogenetic analyses of plastid and nuclear DNA sequences indicate a rapid late Miocene radiation of the temperate bamboo tribe Arundinarieae (Poaceae, Bambusoideae). Plant Ecol Divers 3:109–120

Holttum RE (1956) The classification of bamboos. Phytomorphology 6:73–90

Judziewicz EJ, Clark LG (2007) Classification and biogeography of New World grasses: Anomochlooideae, Pharoideae, Ehrhartoideae and Bambusoideae. Aliso 23:303–314

Judziewicz EJ, Clark LG, Londoño X, Stern MJ (1999) American bamboos. Smithsonian Institution, Washington, DC, p 392

Kelchner SA, Bamboo Phylogeny Group (2013) Higher level phylogenetic relationships within the bamboos (Poaceae: Bambusoideae) based on five plastid markers. Mol Phylogenet Evol 67:404–413

Kellogg EA, Watson L (1993) Phylogenetic studies of a large data set. I. Bambusoideae, Andropogonodae, and Pooideae (Gramineae). Bot Rev 59:273–343

Keng P-C (1982–1984) A revision on the genera of bamboos from the world. J Bamboo Res 1 (1):1–19; 1982; l.c. 1(2):31–36, 1982; l.c. 2(1):11–27, 1983; l.c. 2(1):11–27, 1983; l.c. 3 (1):22–42, 1984; l.c. 3(2):1–22, 1984

Keng P-C, Wang Z-P (1996) Flora Reipublicae Popularis Sinicae, delectis Florae Reipublicae Popularis Sinicae agendae Academiae Sinicae edita; vol. 9, pt. 1 (Gramineae 1: Bambusoideae). Science Press, Beijing, p 761 + 215 figs

Li D-Z, Xue J-R (1997) The biodiversity and conservation of bamboos in Yunnan, China. In: Chapman GP (ed) The bamboos. Academic, London, pp 83–94

Li X-L, Lin R-S, Fung H-L, Qi Z-X, Song W-Q, Chen R-Y (2001) Chromosome numbers of some bamboos native in or introduced to China. Acta Phytotax Sin 39:433–442

Li D-Z, Wang Z-P, Zhu Z-D, Xia N-H, Jia L-Z, Guo Z-H, Yang G-Y, Stapleton C (2006) Bambuseae. In: Wu Z-Y, Raven PH, Hong D-Y (eds) Flora of China: Poaceae, vol 22. Science Press, Missouri Botanical Garden Press, Beijing, St. Louis, pp 7–180

Londoño X (1990) Estudio botánico, ecológico, silvicultural, y económico-industrial de las Bambusoideae de Colombia. Cespedesia 16(17):51–78

Lucas S (2013) Bamboo. Reaktion Press, London, p 182

March R, Clark LG (2011) Sun-shade variation in bamboo (Poaceae: Bambusoideae) leaves. Telopea 13:93–104

McClure FA (1934) The inflorescence in *Schizostachyum* Nees. J Wash Acad Sci 24:541–548

McClure FA (1966) The bamboos: a fresh perspective. Harvard University, Cambridge, MA, p 347

McClure FA (1973) Genera of bamboos native to the New World (Gramineae: Bambusoideae). In: Soderstrom TR (ed) Smithson Contrib Bot 9:1–148

Munro W (1868) A monograph of the Bambusaceae, including descriptions of all the species. Trans Linn Soc Lond 26:1–157

Nakai T (1925) Two new genera of Bambusaceae with special remarks on the related genera growing in eastern Asia. J Arnold Arbor 6:145–153

Nakai T (1933) Bambusaceae in Japan proper. J Jpn Bot 9:77–95

Nees von Esenbeck CGD (1835) Bambuseae Brasilienses: recensuit et alias in India Orientalis provenientes adjecit. Linnaea 9:461–494

Noguchi M, Yoshida T (2005) Factors influencing the distribution of two co-occurring dwarf bamboo species (*Sasa kurilensis* and *Sasa senanensis*) in a conifer-mixed broadleaved stand in northern Hokkaido. Ecol Res 20:25–30

Ohrnberger D (1999) The bamboos of the world: annotated nomenclature and literature of the species and the higher and lower taxa. Elsevier, Amsterdam, p 585

Oliveira RP, Clark LG, Schnadelbach AS, Monteiro SHN, Longhi-Wagner HM, van den Berg C (2014) A molecular phylogeny of *Raddia* (Poaceae, Olyreae) and its allies based on noncoding plastid and nuclear spacers. Mol Phylogenet Evol 78(105):117

Parodi LR (1961) La taxonomia de las Gramineae Argentinas a la luz de las investigaciones más recientes. Recent Adv Bot 1:125–130

Peng S, Yang H-Q, Li D-Z (2008) Highly heterogeneous generic delimitation within the temperate bamboo clade (Poaceae: Bambusoideae): evidence from GBSSI and ITS sequences. Taxon 57:799–810

Rao KS, Ramakrishnan PS (1988) Architectural plasticity of two bamboo species, *Neohouzeaua dulloa* Camus and *Dendrocalamus hamiltonii* Nees in successional environment in North-East India. Proc Indian Acad Sci (Plant Sci) 98:121–133

Ruiz-Sanchez E (2011) Biogeography and divergence time estimates of woody bamboos: insights in the evolution of Neotropical bamboos. Bol Soc Bot Méx 88:67–75

Ruiz-Sanchez E, Sosa V, Mejía-Saules MT (2008) Phylogenetics of *Otatea* inferred from morphology and chloroplast DNA sequence data and recircumscription of Guaduinae (Poaceae: Bambusoideae). Syst Bot 33:277–283

Ruiz-Sanchez E, Sosa V, Mejía-Saulés MT (2011a) Molecular phylogenetics of the Mesoamerican bamboo *Olmeca* (Poaceae: Bambuseae): implications for taxonomy. Taxon 60:89–98

Ruiz-Sanchez E, Sosa V, Mejía-Saulés MT, Londoño X, Clark LG (2011b) A taxonomic revision of *Otatea* (Poaceae: Bambusoideae: Bambuseae) including four new species. Syst Bot 36:314–336

Seethalakshmi KK, Kumar MSM (1998) Bamboos of India: a compendium. Kerala Forest Research Institute, Peechi, India and International Network for Bamboo and Rattan, New Delhi, p 342

Singleton GR, Belmain SR, Brown PR, Hardy B (2010) Rodent outbreaks: ecology and impacts. International Rice Research Institute, Manila, p 289

Soderstrom TR (1981) Some evolutionary trends in the Bambusoideae (Poaceae). Ann Missouri Bot Gard 68:15–47

Soderstrom TR (1985) Bamboo yesterday, today and tomorrow. J Am Bamb Soc 6:4–16

Soderstrom TR, Calderón CE (1971) Insect pollination in tropical rain forest grasses. Biotropica 3:1–6

Soderstrom TR, Calderón CE (1979) V. Ecology and phytosociology of bamboo vegetation. In: Numata M (ed), Ecology of grasslands and bamboolands in the world. VEB Gustav Fischer Verlag, Jena, pp 223–236

Soderstrom TR, Ellis RP (1982) Taxonomic status of the endemic South African bamboo, *Thamnocalamus tessellatus*. Bothalia 14:53–67

Soderstrom TR, Ellis RP (1987) The position of bamboo genera and allies in a system of grass classification. In: Soderstrom TR, Hilu KW, Campbell CS, Barkworth ME (eds) Grass systematics and evolution. Smithsonian Institution, Washington, DC, pp 225–238

Soderstrom TR, Ellis RP (1988) The woody bamboos (Poaceae: Bambuseae) of Sri Lanka: a morphological-anatomical study. Smithson Contrib Bot 72:1–75

Soderstrom TR, Londoño X (1988) A morphological study of *Alvimia* (Poaceae: Bambuseae), a new Brazilian bamboo genus with fleshy fruits. Am J Bot 75:819–839

Soderstrom TR, Zuloaga F (1989) A revision of the genus *Olyra* and the new segregate genus *Parodiolyra* (Poaceae: Bambusoideae: Olyreae). Smithson Contrib Bot 69:1–79

Soderstrom TR, Judziewicz EJ, Clark LG (1988) Distribution patterns of neotropical bamboos. In: Vanzolini PE, Heyer RE (eds) Proceedings of a workshop on neotropical distribution patterns, Academia Brasileira de Ciencias, Rio de Janeiro, 12–16 Jan 1987, pp 121–157

Stapleton CMA (1994a) The bamboos of Nepal and Bhutan Part I: *Bambusa, Dendrocalamus, Melocanna, Cephalostachyum, Teinostachyum,* and *Pseudostachyum* (Gramineae: Poaceae, Bambusoideae). Edinb J Bot 51:1–32

Stapleton CMA (1994b) The bamboos of Nepal and Bhutan Part II: *Arundinaria, Thamnocalamus, Borinda* and *Yushania* (Gramineae: Poaceae, Bambusoideae). Edinb J Bot 51:275–295

Stapleton CMA (1994c) The bamboos of Nepal and Bhutan Part III: *Drepanostachyum, Himalayacalamus, Ampelocalamus, Neomicrocalamus* and *Chimonobambusa*. Edinb J Bot 51:301–330

Stapleton CMA (1998) Form and function in the bamboo rhizome. J Am Bamb Soc 12(1):21–29

Stapleton CMA (2013) *Bergbambos* and *Oldeania*, new genera of African bamboo (Poaceae, Bambusoideae). PhytoKeys 25:87–103

Stapleton CMA, Ní Chonghaile G, Hodkinson TR (2004) *Sarocalamus*, a new Sino-Himalayan bamboo genus (Poaceae–Bambusoideae). Novon 14:345–349

Sungkaew S, Stapleton CMA, Salamin N, Hodkinson TR (2009) Non-monophyly of the woody bamboos (Bambuseae; Poaceae): a multi-gene region phylogenetic analysis of Bambusoideae s.s. J Plant Res 122:95–108

Takenouchi Y (1931a) Systematisch-vergleichende Morphologie und Anatomie der Vegetationsorgane der japanischen Bambus-Arten. Taihoku Imp Univ (Formosa) Fac Sci Mem 3:1–60

Takenouchi Y (1931b) Morphologische und entwicklungsmechanische Untersuchungen bei japanischen Bambus-Arten. Mem Coll Sci Kyoto Imp Univ Ser B 6:109–160

Taylor AH, Qin Z-S (1997) The dynamics of temperate bamboo forests and panda conservation in China. In: Chapman GP (ed) The bamboos. Academic, London, pp 189–203

Triplett JK, Clark LG (2010) Phylogeny of the temperate woody bamboos (Poaceae: Bambusoideae) with an emphasis on *Arundinaria* and allies. Syst Bot 35:102–120

Triplett JK, Clark LG, Fisher AE, Wen J (2014) Independent allopolyploidization events preceded speciation in the temperate and tropical woody bamboos. New Phytol 204:66–73. doi:10.1111/nph.12988

Triplett JK, Oltrogge KA, Clark LG (2010) Phylogenetic relationships and natural hybridization among the North American woody bamboos (Poaceae: Bambusoideae: Arundinarieae). Am J Bot 97:471–492

Triplett JK, Weakley A, Clark LG (2006) Hill cane (*Arundinaria appalachiana*), a new species of woody bamboo (Poaceae: Bambusoideae) from the Southern Appalachian Mountains. SIDA 22:79–95

Tsuyama I, Nakao K, Matsui T, Higa M, Horikawa M, Kominami Y, Tanaka N (2011) Climatic controls of a keystone understory species, *Sasamorpha borealis*, and an impact assessment of climate change in Japan. Ann For Sci 68:689–699

Tyrrell CD, Santos-Gonçalves AP, Londoño X, Clark LG (2012) Molecular phylogeny of the arthrostylidioid bamboos (Poaceae: Bambusoideae: Bambuseae: Arthrostylidiinae) and new genus *Didymogonyx*. Mol Phylogenet Evol 65:136–148

Uma Shaanker R, Ganeshaiah KN, Srinivasan K, Ramanatha Rao V, Hong LT (2004) Bamboos and rattans of the Western Ghats. Ashoka Trust for Research in Ecology and the Environment (ATREE), Bangalore, p 203

Viana P, Filgueiras T, Clark LG (2013) A new genus of woody bamboo (Poaceae: Bambusoideae: Bambuseae) endemic to Brazil. Syst Bot 38:97–103

Wang L, Jacques FMB, Su T, Xing Y-W, Zhang S-T, Zhou Z-K (2013) The earliest fossil bamboos of China (middle Miocene, Yunnan) and their biogeographical importance. Rev Palaeobot Palynol. doi:10.1016/j.revpalbo.2013.06.004

Widjaja EA (1987) A revision of Malesian *Gigantochloa* (Poaceae-Bambusoideae). Reinwardtia 10:291–389

Wong KM (1993) Four new genera of bamboos (Gramineae: Bambusoideae) from Malesia. Kew Bull 48:517–532

Wong KM (1995) The morphology, anatomy, biology and classification of Peninsular Malaysian bamboos. Univ Malaya Bot Monogr 1:1–189

Wong KM (2005) *Mullerochloa*, a new genus of bamboo (Poaceae: Bambusoideae) from Northeast Australia and notes on the circumscription of *Bambusa*. Blumea 50:425–441

Wong KM, Low YW (2011) Hybrid zone characteristics of the intergeneric hybrid bamboo × *Gigantocalamus malpenensis* (Poaceae: Bambusoideae) in Peninsular Malaysia. Gard Bull Singap 63(1&2):375–384

Wu Z-Q, Ge S (2012) The phylogeny of the BEP clade in grasses revisited: evidence from the whole-genome sequences of chloroplasts. Mol Phylogenet Evol 62:573–578

Yang Y-M, Xue J-R (1990) A preliminary study on the natural bamboo forests in the Dawei Mountain of southeastern Yunnan. J Southwest For Coll 10:22–29

Yang H-Q, Peng S, Li D-Z (2007) Generic delimitations of *Schizostachyum* and its allies (Gramineae: Bambusoideae) inferred from GBSSI and *trnL-F* sequence phylogenies. Taxon 56:45–54

Yang H-Q, Yang J-B, Peng Z-H, Gao J, Yang Y-M, Peng S, Li D-Z (2008) A molecular phylogenetic and fruit evolutionary analysis of the major groups of the paleotropical woody bamboos (Gramineae: Bambusoideae) based on nuclear ITS, GBSSI gene and plastid *trnL-F* DNA sequences. Mol Phylogenet Evol 48:809–824

Yang J-B, Yang H-Q, Li D-Z, Wong K-M, Yang Y-M (2010) Phylogeny of *Bambusa* and its allies (Poaceae: Bambusoideae) inferred from nuclear GBSSI gene and plastid *psb*A-*trn*H, *rpl*32-*trn*L and *rps*16 intron DNA sequences. Taxon 59:1102–1110

Yang H-M, Zhang Y-X, Yang J-B, Li D-Z (2013) The monophyly of *Chimonocalamus* and conflicting gene trees in Arundinarieae (Poaceae: Bambusoideae) inferred from four plastid and two nuclear markers. Mol Phylogenet Evol 68:340–356

Yi T-P, Shi J-Y, Ma L-S, Wang H-T, Yang L (2008) Iconographia Bambusoidearum Sinicarum. Science Press, Beijing, p 766

Zeng C-Z, Zhang Y-X, Triplett JK, Yang J-B, Li D-Z (2010) Large multi-locus plastid phylogeny of the tribe Arundinarieae (Poaceae: Bambusoideae) reveals ten major lineages and low rate of molecular divergence. Mol Phylogenet Evol 56:821–839

Zhang W-P (1992) The classification of Bambusoideae (Poaceae) in China. J Am Bamb Soc 9:25–42

Zhang W-P, Clark LG (2000) Phylogeny and classification of the Bambusoideae (Poaceae). In: Jacobs SWL, Everett J (eds) Grass systematics and evolution. CSIRO, Melbourne, pp 35–42

Zhang Y-X, Chun C-X, Li D-Z (2012) Complex evolution in Arundinarieae (Poaceae: Bambusoideae): incongruence between plastid and nuclear GBSSI gene phylogenies. Mol Phylogenet Evol 63:777–797

Zhu S-L, Ma N-X, Fu M-Y (1994) A compendium of Chinese bamboo. China Forestry Publication House, Beijing, p 241

Chapter 2
Priority Species of Bamboo

Andrew Benton

Abstract There are over 1,250 species in approximately 75 genera of woody bamboos in the world. Bamboos are native to Africa, the Americas, Asia and Oceania and have been introduced into Europe. This chapter summarizes a revised list of 'Priority Species of Bamboo and Rattan', compiled by the International Network for Bamboo and Rattan (INBAR).

Keywords Arundinaria ssp. • Bambusa ssp. • Cephalostachyum ssp. • Chusquea ssp. • Dendrocalamus ssp. • Gigantochloa ssp. • Guadua ssp. • Melocanna ssp. • Ochlandra ssp. • Oxytenanthera ssp. • Phyllostachys ssp. • Schizostachyum ssp. • Thyrsostachys ssp.

There are over 1,250 species of woody bamboos in the world, in approximately 75 genera. They are native to Africa, the Americas, Asia and Oceania and have been introduced into Europe. Bamboos are naturally found as understorey plants in forests and grow in habitats from the humid tropics, through a range of humid sub-tropical forest types, to temperate regions including northern parts of China, Japan, Korea and the foothills of the Himalayas. They vary in stature from 50 cm (*Sasa borealis* in Japan) to 40 m or more (*Dendrocalamus giganteus* in tropical Asia). Most do not comprise the dominant vegetation unless they are cultivated, such as the huge areas of *Phyllostachys pubescens* in China, but the *Melocanna baccifera* forests of Northeast India, Bangladesh and Myanmar are a major exception, as are the *Guadua* forests of western Amazonia, which cover 120,000 ha (Dransfield and Widjaja 1995; Judziewicz et al. 1999).

With 75 genera of bamboos growing in a wide range of different habitats, it is not surprising that there is much variation in the characteristics they exhibit. Growth habits vary from clump forming to grove forming, culm wall thickness varies from solid in *Chusquea culeo* to just a couple of centimetres in *Bambusa textilis*, internode length between a few centimetres and one metre, and culms may be very straight or zigzag—the latter an important consideration if machine

A. Benton (✉)
International Network for Bamboo and Rattan (INBAR), 8, Futong Dong Da Jie, Wangjing, Chaoyang District, P. O. Box 100102-86, Beijing 100102, P. R. China
e-mail: andrew@inbar.int, http://www.inbar.int

© Springer International Publishing Switzerland 2015
W. Liese, M. Köhl (eds.), *Bamboo*, Tropical Forestry 10,
DOI 10.1007/978-3-319-14133-6_2

Table 2.1 Bamboo characteristics suitable for various end uses (adapted from Hoogendoorn et al. 2013)

Plant part used	Examples of uses	Characteristics required
Whole plant	Ornamentals	Appearance, growth habit
Whole plant	Environmental services—erosion control and watershed protection	High growth and establishment rate
Whole plant	Environmental services—carbon sequestration	Rapid establishment, high processing value to ensure sustainable management of forest
Round-pole uses	Houses, agricultural implements, etc. (usually lower value uses)	Sturdy, strong (and long) poles
Round-pole uses	Construction	Sturdy and strong, uniformity along the culms
Split pole—slabs	Processing into laminates/lumber (high-value uses)	Straight, thick-walled culms
Split poles—splits	Weaving and stick products (high value–low volume or low value–high volume uses)	Straight culms with long internodes
Shoots	Food	Thick culms with low hydrocyanin content, good flavour
Whole-pole pulping	Paper and rayon	High biomass productivity
Extracts	Tar oil, medicines, 'beer'	No clear specific requirements

processing. Selection of species for particular uses has been done by communities and producers for centuries, and these days 16 of the 20 most commercially important species are thought to have been domesticated (Rao and Ramanatha Rao 1998). Bamboos have traditionally been used for a wide range of uses from construction to weaving, farm implements to fodder for farm animals, paper making to musical instruments, stick products to tar oil and alcohol. Innovation has been a driving force behind expanding the uses of bamboos, particularly in China, where glued and laminated bamboo products and production systems developed for specific species (usually *Phyllostachys pubescens*) have been developed and contribute hugely to the nation's bamboo sector (Table 2.1).

Because bamboos are so different, the technologies used for processing them also need to differ—bamboos and the technologies for processing them often fall into species-product groupings (e.g. *Phyllostachys pubescens* for chopsticks and laminates, *Guadua* for construction and *Dendrocalamus asper* and *D. latiflorus* for bamboo shoots), and this has to be taken into account when developing new bamboo businesses (Hoogendoorn et al. 2013).

2.1 Priority Species of Bamboo

In 1998, the International Network for bamboo and Rattan released a revised list of 'Priority Species of Bamboo and Rattan', that included 20 taxa (species and genera) of particular economic importance and a further 18 taxa of importance (Rao and Ramanatha Rao 1998). INBAR has not updated the list since that time, but since then, many more species of bamboos have been trialled and tested under different growing conditions in different countries, and much international germplasm exchange has taken place. In part, this has been enabled by INBAR's species-to-site matching software, which enables a user in one location to see which bamboo species would be able to grow at their own site with the click of a button, based on climatic conditions and includes information on import/export regulations for such species. Although outside the scope of this short report, it would seem timely to update the priority species list and the software to reflect the current realities.

Selection and, in some cases, breeding of bamboos have been possible. Two examples where selection followed by mass multiplication have reaped commercial benefits are highlighted by Hoogendoorn et al. (2013) and shown below:

1. Thick-walled Moso (*Phyllostachys pubescens* cv Pachyloen).

    Discovered in southeastern China's Jiangxi province in 1995, the culm wall is almost twice the thickness of the normal phenotype of the species and yields are correspondingly higher. It has been shown to be a genetic trait and is now grown on thousands of hectares in Jiangxi province (Guo et al. 2003).

2. Beema bamboo

    Beema bamboo is a selection of *Bambusa balcooa* from Bihar in India discovered over 10 years ago. Its culm walls are very thick and culms weigh three times that of a normal *B. balcooa,* with a reported annual productivity of over 100 tonnes per hectare, three or four times higher than a normal *B. balcooa.* It is produced on a huge scale—200,000 seedlings per annum by micropropagation and currently cultivated commercially in India.

2.2 Priority Species of Bamboo: List

Brief introductions to each of the priority taxa are given in the list, as per INBAR's enumerations (Rao and Ramanatha Rao 1998), with updates and amendments where available.

Note that all species are 'sympodial', thereby forming discrete clumps, unless otherwise stated.

Note also that flowering of bamboos often results in the death of the flowered part, particularly if seed is set, but that the flowering habits of most bamboos are poorly understood. In the list, lack of information on gregarious flowering (and therefore subsequent death) indicates only a lack of information rather than a confirmation that such flowering does not happen. Sporadic flowering of a clump or a population of clumps may also result in death. It is thus important for growers to record, where possible, the date the seed that gave rise to their bamboo plants was produced, in order to enable prediction of possible flowering dates decades to come.

Bambusa arundinacea (B. bambos)

B. bambos is a tough, vigorous and widespread multipurpose bamboo of South Asia but with limited value for high-value products due to its thick culm internodes that do not enable easy splitting and its thorny nature.

Brief description—thorny bamboo with culms erect up to 30 m tall, to 18 cm in diameter, thick walled up to 15 cm, nodes up to 40 cm long and slightly swollen. Lower branches bear recurved spines, forming dense, often impenetrable, thicket. *B. bambos* clumps are variable with some races almost thorn-less and others very straight culmed.

Native to India, Bangladesh, Myanmar, Thailand and China. Grows at up to about 1,200 m altitude and can tolerate −2 °C.

Introduced to Nepal, Indonesia, Vietnam, Philippines and elsewhere.

Propagation—Seed and offsets.

Lifecycle—clumps can often be found in flower—either whole clumps or a portion of the culms—and produce copious quantities of seeds which exhibit high levels of diversity in the next generation.

Uses—*B. bambos* is planted for land rehabilitation and riverbank stabilisation and is used locally for handicrafts and low-value construction. Culms are often used for pulp and supplies 20 % of India's bamboo pulp demand.

Bambusa balcooa

B. balcooa is used for construction as well as agricultural implements, round-pole furniture and pulp but is not used for high-value products.

Brief description—culms are up to 24 m tall and 15 cm in diameter; new culms are grey-green to light white bloom, nodes 30–45 cm long and thick walled up to 2.5 cm.

Native to N.E. India and Bangladesh.

Introduced to Indonesia and Australia.

Propagation—offsets and seed.

Uses—construction and farm use.

Bambusa blumeana

Large, thorny bamboo with limited commercial value but often used locally.

Brief description—densely tufted, culms 15–25 m tall, up to 20 cm diameter, internodes 25–60 cm long, green and glabrous with prominent nodes. Young shoots with yellowish-green sheaths and blades.

Distribution—believed to be native to Sumatra, Java, Lesser Sunda Islands and Borneo. Introduced to Papua New Guinea, Peninsular Malaysia, Thailand, Vietnam, Philippines and southern China.

Life cycle—gregarious flowering perhaps once in 20–30 years. Sporadic flowering is also known.

Propagation—culm cuttings, rhizome cuttings, layering and marcotting. Seeds/caryopses are not available.

Uses and value—good for rehabilitating degraded lands and as borders to agricultural areas. Used locally for low-quality furniture, chopsticks, handicrafts and, occasionally, shoots are used after processing.

Bambusa polymorpha

A medium to large bamboo.

Brief description—densely tufted bamboo with culms 15–25 m tall and up to 15 cm in diameter, relatively thick walls of 1 cm, occasionally solid.

Distribution—native to Myanmar, extending to Bangladesh, India and Thailand. Introduced to germplasm collections elsewhere but not known to be cultivated outside its distribution area.

Life cycle—flowering after 50–60 years.

Propagation—cuttings, rhizome cuttings, branch cuttings, layering and marcotting.

Uses and value—locally used for building and structural uses and for baskets and low-quality furniture. The shoots are edible.

Bambusa textilis

A delicate thin-walled bamboo ideal for weaving.

Brief description—medium-sized bamboo with straight culms and long internodes. Culms are up to 15 m tall, 3–5 cm in diameter, internodes up to 60 or more cm long and culms are thin walled.

Cultivars and varieties—there are three botanical varieties that are widely grown in China: cv Albo striata, var glabra and var gracilis.

Distribution—southern China.

Introduced to other provinces of China.

Life cycle—not known to flower gregariously.

Propagation—offsets and seed are the two most common methods of propagation.

Uses and value—*Bambusa textilis* culms split easily and very finely, providing high-quality bamboo for woven items. *B. textilis* and some of its varieties are also often used for landscaping. Shoots are edible but small.

Bambusa tulda

Brief description—culms are up to 30 m tall, usually 20 m, 5–10 cm diameter, internodes 40–70 cm long, culm walls to up 1 cm thick.

Distribution—India, Bangladesh, Myanmar and Thailand.

Introduced to Nepal, Indonesia, Vietnam and Philippines.

Life cycle—reports of flowering after 25–40 years.

Propagation and cultivation—culm cuttings, marcotting, rhizomes and macroproliferation.

Uses and value—multiple uses. Culms are used structurally, for furniture, pulp and handicrafts.

Bambusa vulgaris

Brief description—vigorous medium-large bamboo with relatively open clumps. Culms are up to 20 m tall, internodes 25–35 cm long, 5–10 cm diameter, walls up to 1.5 cm thick. Culms are not straight, internodes often zigzagging.

Cultivars and varieties—B. vulgaris var vulgaris (Yellow culms—very widely grown as an ornamental plant) and B. vulgaris cv Wamin (Bhudda's belly bamboo, very widely grown as an ornamental plant).

Distribution—global tropics. *B. vulgaris* is known as the only pan-tropical bamboo.

Life cycle—gregarious flowering not seen, sporadic flowering occurs only very rarely.

Propagation and cultivation—culm cuttings, rhizomes, branch cuttings, layering and marcotting. *B. vulgaris* is one of the easiest bamboos to propagate and is extremely vegetatively vigorous.

Uses and value—widely used for local purposes including construction, furniture and handicrafts, but its non-straight culms limit its uses. Good for pulping. In Brazil, it is grown for pulp in large plantations and harvested mechanically on a 3-year rotation.

Cephalostachyum pergracile

Brief description—medium-sized bamboo with straight culms that keep the culm sheaths. Up to 30 m tall, thin walled, internodes up to 45 cm.

Distribution—N.E. India, Myanmar, Northern Thailand, Yunnan province and S.W. China. Also cultivated in botanic gardens.

Introduced to southern China, Java.

Life cycle—*C. pergracile* is not known to flower gregariously.

Propagation—offsets and seed are the two most common methods of propagation. Seedlings can be collected from the forest. Offsets are taken from 1 to 2-year old culms, keeping 1–1.5 m of the culm and planted directly in situ.

Uses and value—light construction, basketry. Outer layer can be split very finely and used for handicrafts. Attractive, ornamental with glaucous culms and brownish sheaths.

Dendrocalamus asper

Brief description—Large bamboo culms up to 20–30 m tall, internodes 20–45 cm long with a diameter of 8–20 cm and thick walls up to 2 cm.

Distribution—N.E. India, Nepal, Bangladesh, Myanmar, northern Thailand, Laos and Vietnam.

Introduced to southern China, Malaysia, Indonesia and Philippines.

Cultivars—six are known including cv betung wulung, cv Tahi green and cv Phai Tong Dam.

Life cycle—gregarious flowering not known.

Propagation—culm and branch cuttings and offsets.

Uses and value—*D. asper* is a multipurpose bamboo with a wide range of uses. Structural use bamboo with large and strong culms, and a very useful bamboo for construction in rural areas with durable culms. Shoots are edible and are good, tasty and sweet. Plantation for bb shoots established of this bamboo in Thailand. Used for good quality furniture, musical instruments, chopsticks, household utensils and handicrafts.

Dendrocalamus giganteus

Brief description—large, sometimes huge bamboo, culms 25–60 m green to dark bluish green, internodes 40–50 cm long, 10–20 cm in diameter and thick walls of 2.5 cm.

Distribution—southern Myanmar and northern Thailand.

Introduced to India, Sri Lanka, Bangladesh, Nepal, Thailand, Vietnam, China, Indonesia, Malay Peninsular, Philippines and Kenya.

Life cycle—not known to flower gregariously.

Propagation—culm cuttings, rhizome planting, branch cuttings, layering, marcotting and macroproliferation.

Uses and value—structural bamboo, strong, for building and for bamboo boards. Also for pulp and for household items. Used for furniture too, and shoots are of good quality.

Dendrocalamus latiflorus

Brief description—medium-sized bamboo 14–25 m tall, internodes 20–70 cm long, 8–20 cm diameter and walls 0.5–3 cm.

Cultivars—cv Meimung in China.

Distribution—Myanmar, South China and Taiwan. Like high rainfall.

Introduced to Philippines, Indonesia, Thailand, India, Vietnam and Japan.

Life cycle—D. latiflorus is not known to flower gregariously.

Propagation and cultivation—culm cuttings, layering and marcotting.

Uses and value—structural uses of medium quality; very good for shoots, also for high-quality furniture, crafts, baskets, pulp and thatching; leaves are used to wrap rice for cooking.

Dendrocalamus strictus

Brief description—medium-sized bamboo culms are 8–20 m tall, internodes 30–45 cm long, 2.5–8 cm diameter, thick walls but slightly zigzag.

Distribution—India, Nepal, Bangladesh, Myanmar and Thailand.

Introduced to many countries in SE Asia but is of limited value outside these regions.

Life cycle—report indicates 20–40 years flowering cycle.

Propagation—culm cuttings, rhizome planting, layering, marcotting and macroproliferation of seedlings.

Uses and value—structural uses, medium to light quality, edible shoots but of poor quality. Pulp use, thick walled and ok for boards, agricultural implements and household utensils.

Gigantochloa apus

Brief description—large bamboo 8–30 m, culms 4–13 cm diameter, strongly tufted, internodes 35–45 cm long, wall thickness 1.5 cm and flexible culms.

Distribution—Myanmar, Thailand, Indonesia and Malaysia. Known to survive in drier areas.

Introduced to N.E. India.

Life cycle—not known to flower gregariously.

Propagation—culm cuttings, offsets.

Uses and value—structural uses of medium quality, and for furniture of medium and good quality, also for handicrafts, musical instruments, utensils and baskets. Shoots are edible but are of poor quality, very bitter in taste.

Gigantochloa levis

Brief description—large bamboo, culms up to 30 m tall, 5–16 cm in diameter, walls 1–1.2 cm thick and internodes up to 45 cm long.

Distribution—origin unknown, cultivated in Philippines, Eastern Indonesia, Northern and Western Kalimantan, east Malaysia, China and Vietnam. Common in the homesteads and gardens of the Philippines.

Life cycle—gregarious flowering not known.

Propagation—culm cuttings and offsets.

Uses and value—structural, shoots are edible and of good quality, also for utensils, furniture, craft paper, fencing and other subsistence uses.

Gigantochloa pseudoarundinacea

Brief description—culms 7–30 m tall, internodes 35–45 cm, 5–13 cm diameter, medium to think wall of 2 cm, strong.

Distribution—native to Java, cultivated in Java and Sumatra.

Introduced to China, Malaysia, India and Vietnam.

Life cycle—gregarious flowering not known.

Propagation and cultivation—culm cuttings and branch cuttings.

Uses and value—structural, water pipes, handicrafts, good quality furniture, household articles, chopsticks and toothpicks; good quality edible shoots.

Guadua angustifolia

Brief description—large bamboo culms up to 30 m tall, dark green with white bands at the nodes, diameter up to 20 cm.

Distribution—extending from Mexico to Argentina.

Introduced to India, Bangladesh, China and many other countries.

Life cycle—gregarious flowering not known.

Propagation and cultivation—offsets, culm cuttings and seed.

Uses and value—*G. angustifolia* is the most widely used bamboo in Latin America. Culms are large and strong and of superior quality. Excellent for construction, furniture, pulping and laminates.

Melocanna baccifera (Grove Forming)

Brief description—Culms to 10–20 m tall, very open, thin walls of 0.5–1.2 cm, internodes 20–50 cm long, 5–7 cm in diameter, culm tips are pendulous.

Distribution—*M. baccifera* covers huge swathes of northeast India and adjoining parts of Myanmar and Bangladesh.

Introduced to Indonesia and China.

Life cycle—48 years.

Propagation and cultivation—fruits and culm cuttings.

Uses and value—roofing, thatching, matting, pulp, paper and rayon; shoots are locally eaten and used for preparing liquor.

Phyllostachys pubescens (Grove Forming)

Brief description—Medium to large monopodial bamboo with culms 10–20 m tall, approximately 18–20 cm diameter and internodes up to 45 cm long, young culms bearing a noticeable waxy white covering.

Distribution—China.

Introduced to Japan (eighteenth century), Korea, Vietnam, the USA and Europe amongst others.

Life cycle—*P. pubescens* is not known to flower gregariously, though sporadic flowering does occur regularly in pockets.

Propagation and cultivation—offsets and seed.

Uses and value—*P. pubescens* is the most economically valuable bamboo species in the world and contributes the vast majority of the value of the 19bn yuan annual bamboo economy in China.

Ochlandra spp.

Brief description—there are about ten species of *Ochlandra*, all between 5 and 10 m tall with culms up to 5 cm diameter.

Distribution—native to the Western Ghats of southern India and southwestern Sri Lanka.

Introduced to East Africa.

Life cycle—*O. travancorica* is reported as flowering once in every 7 years.

Propagation and cultivation—culm cuttings.

Uses and value—often for pulp and paper, but local uses include non-structural construction uses such as walling and handicrafts.

Thyrsostachys siamensis

Brief description—graceful bamboo, densely clumped, culms 8–16 m tall, internodes 15–30 cm long, culm sheaths persistent, white ring below the nodes.

Distribution—Myanmar and Indochina. Naturally in pure or mixed forests in monsoonal areas.

Introduced to many Asian countries where it is cultivated—adaptable to more humid areas on good soils.

Life cycle—both gregarious (48 years) and sporadic.

Propagation and cultivation—seeds, macroproliferation and offsets.

Uses and value—Pulp, shoots, general quality handicrafts, furniture, light construction and for fences and windbreaks, as well as ornamentals.

2.3 Suggested Other Taxa of Value

Arundinaria alpina—large areas in Ethiopia and forming forests in neighbouring Uganda, Kenya, Rwanda and Burundi, as well as further afield; the 'highland bamboo', *A. alpina*, is the foremost bamboo in Africa, where it is used for subsistence uses such as agricultural implements and basic construction.

Arundinaria amabalis—*A. amabalis* is also known by its English name of Tonkin cane and has been used for over a century to produce top-quality fly fishing rods in North America and Europe. Cultivation is limited to southwest Guangdong province in China.

Bambusa atra—*B. atra* is up to 10 m tall and native to eastern Indonesia.

Bambusa heterostachya—Culms are up to 16 m tall, internodes 30–80 cm in length and walls thick. It is known to be from Malaysia and Indonesia, where it has value as poles for picking oil palm fruit bunches.

Bambusa nutans—Culms are up to 15 m tall, internodes up to 45 cm long. *B. nutans* has value as a source of pulp from the lower Himalayas south to Bangladesh south to Thailand. It is also used for furniture and construction.

Bambusa chungii—native to China, *B. chungii* is an open-clumped sympodial bamboo that when matures, forms groves with young waxy white culms that are very attractive. It has considerable potential as an ornamental in the tropics and subtropics.

Bambusa oldhamii is a medium-sized bamboo up to about 12 m tall. It is native to China, where it is grown as a windbreak and is used as a medium-quality structural bamboo and sometimes for furniture making.

Bambusa pervariabilis is a medium-sized bamboo that is native to southern China where it is used for furniture, thatching and weaving, as well as for agricultural implements.

Chusquea spp.—the 200 or so species of *Chusquea* are native to Latin America and many are unusual in having solid culms which gives them potential for a huge range of uses.

Dendrocalamus brandisii is a large bamboo up to 35 m tall, with culms up to 20 cm in diameter and internodes up to 40 cm. It grows in southern and northeastern India, Myanmar and in adjoining parts of China. It is often used for construction, baskets, furniture and handicrafts. Its shoots are edible.

Dendrocalamus hamiltonii grows to 20 m tall, with culms up to 18 cm diameter and internodes up to 50 cm. It is native to Northeast India, Myanmar, Thailand, Laos, Vietnam and Yunnan province in China, where it has much value for construction, baskets, handicrafts, household utensils, fuel, fodder, rafts and edible shoots.

Dendrocalamus hookeri grows to 20 m with culms up to 15 cm in diameter. It is distributed in Northeast India, Myanmar and Nepal and is used for construction and simple woven articles.

Dendrocalamus membranaceous grows to 25 m tall with thin internodes up to 10 cm diameter. It is native to northeast India, Laos, Thailand, Vietnam and China and has value for construction, furniture, pulp, handicrafts and as edible shoots.

Gigantochloa albociliata grows to 10 m tall, in northeast India, Myanmar, Thailand and China, where it is used for its edible shoots, furniture and construction.

Gigantochloa atroviolacea grows up to 15 m tall and is used in its native Java for musical instruments, handicrafts, furniture and edible shoots.

Gigantochloa balui grows to 12 m tall with internodes up to 40 cm long. It grows in Thailand, Indochina, Sarawak and parts of Indonesia.

Gigantochloa hassarkliana grows to 10 m tall with culms just 6 cm in diameter in its native western Indonesia, where it has value for furniture and for erosion control due to its fast growth and establishment.

Guadua amplexifolia and *Guadua chacoensis*—both these bamboos have culms up to 20 m or more tall and are widely used for construction and other uses in Latin America.

Oxytenanthera abyssinica—the 'lowland bamboo of Africa', grows in drier parts of the continent, where its thick-walled, sometimes solid culms are used for construction, weaving and household utensils.

Phyllostachys spp.—there are over 100 species of *Phyllostachys* native to China and Japan, and some have become some of the most widely grown ornamentals in the world. *P. aurea* has been transplanted across the globe, and it is often used for making furniture in places as far afield as Madagascar and Latin America. The shoots of many *Phyllostachys* species are edible, supplying the small shoots industry in China, with shoots of *P. pubescens* shoots used for canned and fresh large shoots.

Schizostachyum spp.—Schizostachyums are native to Southeast Asia. A number of species of *Schizostachyum* have local value in Southeast Asia with potential for enhancement. Species including *S. diffusum, S. dulooa, S. glaucifolium, S. lumampao* are small- to medium-sized bamboos with culms up to 15 m tall that are used for construction, weaving, pulping and for bamboo laminates.

References

Dransfield S, Widjaja EA (eds) (1995) Plant resources of Southeast Asia no 7 bamboos. Backhuys, Leiden, 198 pp

Guo XM, Chen GS, Liu DK et al (2003) Impacts of balanced fertilization on the growth and yields of bamboo shoots. Acta Griculturae Jiangxi 25(1):48–53

Hoogendoorn J, Lou YP, Fu JH, Li YX, Benton AJ (2013) Genetic diversity of bamboos around the world. Acta Horticult (ISHS) 1003:97–106

Judziewicz EJ, Clark LG, Londono X, Stern MJ (1999) American bamboos. Smithsonian Institution Press, Washington, DC

Rao AN, Ramanatha Rao V (eds) (1998) Priority species of bamboo and rattan. IPGRI-APO, Serdang

Chapter 3
Morphology and Growth

Ratan Lal Banik

Abstract The morphological characteristics of different organs like clump habit and culm nature, branches, leaves and rhizomes including sheathing organs in various groups of bamboos are presented. The emergence of culms starts from spring and continues up to autumn, and there also exists natural mortality of emerging culms which vary with the nature of clump and culm wall thickness. The number of new culms that develop from a clump varies by species, soil and climatic conditions, harvesting method, age and size of clump, overhead cover, etc. In most of the species, both the height and diameter at breast height of full-grown culm produced during 5–7 years of clump age are maximum if not felled, and after that period, all the increments including clump girth are very little and remain more or less static. The clump girth in *Melocanna baccifera* shows continuous rapid expansion even after 10 years of age. Morphological characteristics such as the presence or absence of culm sheath, culm texture and colour, node nature, branching habits, etc., are found important to diagnose the age of a culm in the field. The growth form and development of branch, leaf and rhizome both in seedling and adult stages are discussed. All these knowledge of growth periodicity are important in scientific management of the bamboo clump.

Keywords Bamboo • Clump habit • Culm nature • Culm sheath • Culm emergence • Juvenile mortality • Branching • Culm growth • Culm age determination • Rhizome type • Rhizome growth periodicity • Below- above- ground ratio

R.L. Banik (✉)
NMBA (National Mission on Bamboo Applications), New Delhi, India

INBAR, New Delhi, India

BFRI, Chittagong, Bangladesh
e-mail: bamboorlbanik@hotmail.com; rlbanik.bamboo@gmail.com

© Springer International Publishing Switzerland 2015
W. Liese, M. Köhl (eds.), *Bamboo*, Tropical Forestry 10,
DOI 10.1007/978-3-319-14133-6_3

3.1   Introduction

Bamboos are the member of grass family with the tree-like habit, well-developed rhizome system, woody and hollow culms, branching pattern, petiolate leaves and specialised sheathing organs. Their tree-like growth habit is itself probably a necessary specialisation, as suggested by Arber (1934), since it is entirely different from the growth habit of any other trees, either monocotyledonous or dicotyledonous groups. Unlike trees there is no central trunk or main axis in the basic frame of a bamboo plant.

3.2   Morphological Form

A bamboo plant may be of a bushy (clump) or non-bushy (non-clump) type. The plant has many jointed tall, cylindrical aerial stems (culms) with distinct nodes and internodes, the internodes being hollow and a complex underground elaborate rhizome system with a prominent sheathing organ at each node. The culms are the main parts of a bamboo plant utilised for various purposes by man since prehistoric times.

Bamboos are, usually, classified into tall bamboos, dwarf bamboos and climbing bamboos. Most bamboos are tall 'tree form' generally found in deciduous to semi-deciduous forests and cultivated in homesteads and farms, and those which form dense clumps on the edges of the forest have rigidly erect culms, the upper part often curving outwards or the slender tip drooping. Those with shrubby habits grow specifically in the forests of higher altitude up to snowline, and a few are climbers. All species of the genus *Arundinaria* and *Sasa* bamboos are shrubby in nature. The herbaceous bambusoid grasses usually have softer culms that are not long-lived and nodes with less complex branching.

There are some bamboos which cannot grow efficiently in isolation, because their culms are not strong enough to bear the weight of leaves and branches. Such bamboos only grow in the forest, examples are *Dendrocalamus pendulus* and *Schizostachyum grande*, both of which are very abundant in the foothills of the Main Range of Malaya (Holttum 1958). Young culm of these species, being slender and unbranched, grows vertically upwards, bending later with the weight of lateral branches, which rest on any nearby trees. The climbing bamboos (*Dinochloa scandens*, *Melocalamus compactiflorus*) are even more dependent on the support of trees for their upward development.

3.2.1   The Rhizome

The rhizome system in a bamboo is well developed and constitutes the structural foundation of the plant. It is subterranean and highly branched. The individual axes of the rhizome system are referred as rhizome segments. Each individual branch or axis of the rhizome system is known as a rhizome. The rhizome consists of two parts, the rhizome neck and the rhizome proper. The rhizome neck may be short or sometimes elongated. Basically, there are two types of rhizomes. They are determinate and indeterminate or *sympodial* and *monopodial*. The determinate rhizome is called pachymorph rhizome, and indeterminate rhizome is known as leptomorph rhizome.

Pachymorph or Sympodial Rhizome The *pachymorph* (Gr. *Pachys* = thick; *Morphē* = form) rhizome has 6–7 large lateral buds on either side of the thick rhizome proper, and the buds grow up to new bamboos with a short rhizome neck. The rhizome internodes are broader than long, solid and asymmetrical and nodes are not elevated. The underground rhizome consists typically of two parts: the *rhizome proper* and the *rhizome neck*. The neck is basal to the rhizome proper, generally shorter in length and obconical in shape, and connects the new rhizome to the mother rhizome (Plate 3.1). Rhizomes are usually more or less curved shaped and rarely straight, with maximum thickness typically somewhat greater than that of the culm. Bamboos of this type produce rhizome which grows horizontally from the base of an existing culm and then turns up to form the new shoots. Due to this type of branching form (sympodial) and growing nature, standing bamboos aggregate closely, composing a clump. Bamboos with *short-necked pachymorph* rhizomes grow in discrete, compactly *caespitose* clumps (growing in tufts) species of *Bambusa, Dendrocalamus, Cephalostachyum, Gigantochloa, Ochlandra, Oxytenanthera, Schizostachyum, Thyrsostachys*, etc. This type of clump is called *sympodial* clump. Pachymorph rhizomes with a *slightly elongated neck* produce a less compact clump (*Bambusa cacharensis, B. polymorpha, B. vulgaris*, etc.). However, *Melocanna baccifera* has an open and *diffuse* type of *pachymorph* rhizome system with *very long rhizome necks* (1.0–2.0 m) forming a widespread ramified rhizome system below the ground (Plate 3.1a–c). By the development of a *long-necked rhizome*, the distribution of standing culms shows a real diffuse form and defined by Watanabe (1986) as the *woody-pachymorph-diffuse* type (Fig. 3.1). As a result, emerged culms have been standing solitarily, not clustered, making the formation of *loose, open* and *diffuse* clump in *M. baccifera*. No buds are present on any node of the elongated neck and bear some roots only in a few spots, but large buds exist on the nodes of thick rhizome properly connected under the culm base.

Two kinds of buds are observed on the pachymorph rhizomes, the scaly pointed buds and flat buds. The former develops into rhizomes and the latter into culms. The scaly buds are formed during the summer months, while the culm buds develop during the winter months.

Leptomorph or Monopodial Rhizome The leptomorph (Gr. *Leptos* = thin; *Morphē* = form) rhizome is long and slender and has the following associated characteristics: a cylindrical or subcylindrical form, with a diameter usually less than that of the culm originating from it, internodes longer than broad, relatively uniform in length, rarely solid, typically hollow with interrupted at each node by a diaphragm; nodes in some genera usually elevated or inflated, in others not; lateral buds in the dormant state boat-shaped (McClure 1966). This type of rhizome is found in running bamboos (e.g. *Phyllostachys* sp.) and exhibits basically a monopodial branching pattern. The culms develop from the lateral buds. Some of the lateral buds give rise to rhizomes.

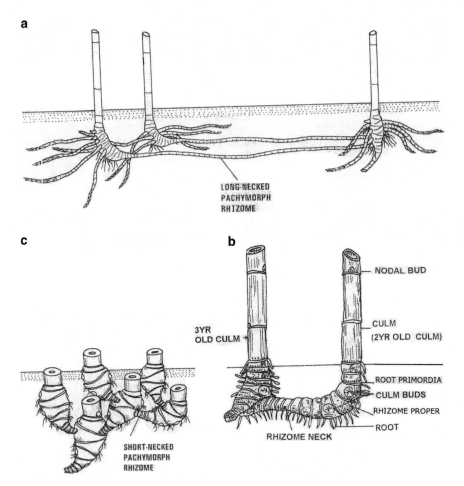

Plate 3.1 Growth form of rhizome system in bamboos, (**a**) *Melocanna baccifera*: the *woody-pachymorph-diffuse* type with long rhizome necks; (**c**) typical short-necked rhizome of *Bambusa vulgaris* resulted in (**b**) compact *caespitose* clump

3.2.2   *Roots*

In appearance, the roots are typically symmetrical in size and shape. They form at the base of the culm from the rhizome nodes and generally go no deeper than 70 cm below the surface. A single ring of roots arises at the nodal region of rhizome. In both the well-developed adult clumps of *Bambusa tulda* and *Melocanna baccifera*,

Fig. 3.1 A *long-necked rhizome* system known as the *woody-pachymorph-diffuse* type in *M. baccifera* resulted in *loose, open* and *diffuse* clump where emerged culms are standing solitarily and not clustered

Table 3.1 Distribution of root of a well-developed clump of *B. tulda* and *M. baccifera* at different soil depth

Soil depth from surface to bottom (cm)	Average amount of root present per 10 cubic cm below the ground			
	B. tulda[a]		*M. baccifera*[b]	
	Biomass (g)	Percentage	Biomass (g)	Percentage
0–33	Not reported	83.00	19.94	70.38
33–66	NR	12.0	6.74	23.80
66–100	NR	4.0	1.25	4.41
100–150	NR	1.0	0.40	1.41
	NR	100	28.33	100

[a]White and Childers (1945)
[b]Banik (2010a)

the rhizomes are distributed horizontally under the ground and laid up to the depth of 25.0–65.0 cm, anchoring and supporting the culms in clump. Sometimes the dorsal side of the whole rhizome system is visible within 1.0–5.0 cm of the soil surface. About 70.00–80.00 % of roots and other plant materials are present in the upper 33 cm of soil, 15.00–24.00 % in 33–66 cm, 4.0–6.0 % in 66–100 cm and the remaining 1.0 % between 100 and 150 cm deep from the surface as reported in *B. tulda* and *M. baccifera* (Table 3.1). Thus, bamboo produces maximum amount of roots within 33 cm of the ground, and the root diameters usually vary from 0.04 to 0.48 cm (Banik 2010). In a pure *M. baccifera* forest, several thousands of clumps have been growing side by side forming continuous rhizome networks with compact root mass just below the surface of the soil and provide an important role in soil-water conservation on the hills.

3.2.3 Clump and Culm Character

A bamboo plant may be either *clump forming* or *non-clump forming*, producing only shoots which emerge from the underground rhizomes and elongate into jointed cylindrical stem known as culm (Latin: *culmus* = stalk, stem). In pachymorph bamboos, the culms are derived from the terminal buds, while in the case of leptomorph bamboo, these developed from the lateral buds of the rhizomes.

In habit, culms vary from strictly erect, erect with pendulous tips or ascending, nearly straight to strongly zigzag, straggling or semi-scandent and clambering. There are culms of species as thin as a pencil, others almost 20 cm thick in diameter, yellow bamboo, pale green bamboo, dark purple brown bamboo, a bamboo with stripes on the wall, and another with internodes that ballooned like a grotesquely swollen bottle called Buddha's belly bamboo. One of the most popular garden bamboos, Black Bamboo (*Phyllostachys nigra*), is unique in the fact that the culms exhibit a nearly jet black colour.

Small culms generally taper gradually from base to tip. In larger ones, the lower half or so is roughly cylindrical or imperceptibly tapered, and the rest of the culm more sharply so. In the largest ones from vigorous plants of many species (e.g. *Dendrocalamus giganteus*, *Phyllostachys* spp.), there may be an appreciable increase in the diameter for some distance, beginning at or near the base, and then a gradual decrease in diameter, and, finally, in the upper 1/3 or 1/4 of its length, the culm is more sharply tapered.

Culm Internode The clump height depends on the length of the culms, and it varies from 2 to 37 m depending on the species character (McClure 1966). The clump height along with the length of the internodes of bamboo culms varies from species to species and has a stable genetic basis. Banik (1991) has graphically represented the march of increase and decrease in internode length of culms of ten different bamboo species by plotting the length of each internode against its serial number, beginning from base to the tip (Fig. 3.2a–c). In all the ten species, the curves are usually steeper on the 'left' side and more gently sloped on the 'right' side, indicating the larger length increments in successive internodes from lower to the middle parts, and then internode length decreases gradually in the upper part of the culms. Among the four *Bambusa* species (Fig. 3.2a), *B. polymorpha* and *B. tulda* have more elongated internodes at the mid-culm zones, while both *B. balcooa* and *B. vulgaris* have a series of mid-culm internodes of more or less equal length. Among the three species of *Dendrocalamus* genus (Fig. 3.2b), *Dendrocalamus longispathus* possesses distinctly elongated internodes at the mid-culm region. Both *D. giganteus* and *D. strictus*, tall and large-sized bamboo

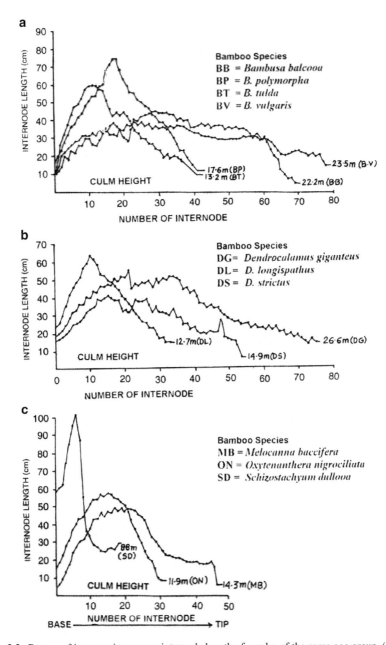

Fig. 3.2 Pattern of increase in average internode length of a culm of the same age group, (**a**) four species of *Bambusa*, (**b**) three species of *Dendrocalamus*, (**c**) three other thin-walled species of bamboos (values at the end of each curve indicate the length of the culm)

species, have a series of more or less equal length internodes in the mid-culm zones similar to other tall and large-sized *Bambusa* species (*B. balcooa* and *B. vulgaris*). *Schizostachyum dullooa*, a thin-walled bamboo species growing naturally in the forests of Northeast India and Chittagong Hill Tract (CHT), has very long internodes increasing successively from base to the mid-culm zone followed by a sharp decrease in length towards the culm tip (Fig. 3.2c). *Bambusa polymorpha* has comparatively long internodes than other common *Bambusa* species and grows naturally as an associate and understorey of the *Xylia* (pyinkado) forests of Myanmar. The presence of elongated internodes in the culms of *Schizostachyum dullooa* and *D. longispathus* thus indicates that these species are more or less shade tolerant, and, in fact, they grow along the moist banks of streams associating with tree shade inside the forest. These species are usually not found on the open hilltops. *Schizostachyum dullooa* grows in gullies, near the bank of the hilly streams and more relatively shade tolerant. On the contrary, *Melocanna baccifera* has a series of more or less equal length of internodes in the mid-culm zone, and *Gigantochloa andamanica* (syn. *Oxytenanthera nigrociliata*) represents an intermediate pattern of internodal length (Fig. 3.2c). Both the species are light demanders and prefer to grow on the hilltops and open slopes. Thus, it appears bamboos having elongated internodes can thrive well as understorey in the forest. It is evident that the assessment of internode length and their distribution pattern in the culm of a species may help to ascertain the required habitat condition of some bamboo species (Banik 2000).

A characteristic of many bamboos is the appearance of white exudates on the surface of the culm internodes. This varies from a barely perceptible bloom (seen on some kind of plums and grapes) to a conspicuously abundant, fluffy, flourlike deposit that more or less completely conceals the green surface of the internodes. The time of the appearance of the exudates in relation to the development of the culm and its texture, abundance, distribution and persistence are variables that may be, in conjunction of other characteristics, useful in the identification of some species of bamboos in the field. In *Dendrocalamus giganteus*, the internode is covered with white waxy scurf when young. The exudate from a young culm of *Lingnania* (*Bambusa*) *chungii* was analysed, and one component, comprising about 25 % of the powder, was identified as a triterpenoid ketone, identical with or closely resembling friedelin, a constituent of the wax of corks (McClure 1966). In *Bambusa polymorpha*, the young culms are white scurfy; after 1 year they turn grey to greyish green.

For intact use of culm as decoration, conservation of natural colour, like green or yellow, is important because the price value would increase with the stable colour.

The culm nodes also show more or less strongly marked variation in appearance as between different species. The culm sheath arises from the nodal region. It encloses a branch bud. The lowest portion of the node has the nodal line or a sheath scar formed as a result of fall of culm sheath. In certain species, the supra nodal ridge is prominent. An important feature of culm nodes is the sheath scar. The sheath scar is a transverse circumaxial offset in the surface of culm, marking the circumaxial locus of insertion of the culm sheath (McClure 1966). Bamboo species belonging to the close clump type frequently have a ring of fine roots at the basal culm nodes.

Branch Bud on Culm Node The bud is usually inserted just above the nodal line. Branch buds emerge on alternate sides of the culm just above the sheath scar at successive nodes. Each bud stands in a position median to the base of the sheath that subtends it. With the exception of those of the genus *Chusquea* (a species naturally found in South America), most of the bamboos have the primary buds solitary at culm nodes; as in *Bambusa vulgaris* in *Chusquea*, the principal bud at each culm node (above ground) is flanked by two to many smaller ones.

Culm Diameter and Wall Thickness In South Asia, some bamboo species like *Bambusa balcooa*, *B. bambos*, *B. cacharensis*, *B. polymorpha*, *B. nutans*, *B. tulda*, *B. vulgaris*, *D. giganteus*, *D. hamiltonii*, *D. strictus*, etc., possess very-wide-diameter culms with thin to thick walls, while others such as *Bambusa glaucescens*, *M. baccifera*, *Ochlandra travancorica*, *O. stridula*, *Schizostachyum dullooa*, etc., have small-diameter culms and thin to thin walls. However, culm diameter may further increase or decrease depending on the site conditions and the age of the plant. The diameter of the successively emerged culms increases with the age of the clumps. The culms are hollow and thickness of the wall varies with species. Thus, bamboos can be grouped into thin- and thick-walled species. However, in one species, the cavities of the culms are almost, if not quite, absent. This is especially the case with a certain number of culms in each clump in the 'male bamboos' (*D. strictus*) when it is found growing in a dry locality with poor soil, as, for instance, in the Siwalik Hills near Hardwar in India.

The culm diameter and wall thickness ratio of ten bamboo species have been studied, and it was reported (Banik 2000) that *D. giganteus*, *D. strictus* and *B. balcooa* have maximum (3.45, 3.34 and 2.82 cm) wall thickness in the basal portion of the culms, while *S. dullooa* possesses culms with thinnest wall (1.2–6.8 mm) (Table 3.2).

However, when the ratio between wall thickness (Wt) and culm diameter (Dia) is considered, the concept of the so-called grouping of 'thick'- and 'thin'-walled bamboos becomes different. The ratio values of *S. dullooa* and *G. andamanica* were higher (up to mid-culm zone) than those of 'thick'-walled species like *B. vulgaris* and *D. giganteus* (Table 3.2). Among all the species, *D. strictus* has the highest ratio values at base, mid and top positions of a culm. However, in all the ten species, the value of such ratio is found to be within 0.52. Thus, it appears that the ratio between wall thickness and culm diameter might be an important consideration in classifying bamboos as 'thick'- and 'thin'-walled species (Banik 1991, 2000). It is interesting to note that the value of ratio is found to be always lowest (0.06–0.13) in the mid-zones than those of base and top culm portions in all the bamboo species (Table 3.2). This indicates that in relation to the diameter, mid-culm zones have the thinnest walls than those of basal and top portion of the culms. The presence of elongated internodes with lowest ratio values of wall thickness and

Table 3.2 Ratio of wall thickness (Wt)/diameter (Dia) measurement at three different culm positions in some common bamboo species of Asia

Species	Base Dia (cm)	Wt (cm)	Ratio	Mid Dia (cm)	Wt (cm)	Ratio	Top Dia (cm)	Wt (cm)	Ratio
B. bal	9.15	2.82	0.308	7.82	0.92	0.117	1.24	0.36	0.290
B. pol	9.25	1.19	0.128	7.78	0.74	0.095	1.35	0.22	0.163
B. tul	5.75	1.17	0.203	4.52	0.41	0.091	1.02	0.21	0.205
B. vul	8.32	1.72	0.206	7.65	0.62	0.082	0.70	0.22	0.314
D. gig	18.72	3.45	0.187	11.00	0.70	0.063	0.42	0.10	0.238
D. lon	5.30	1.79	0.337	4.10	0.36	0.088	0.88	0.19	0.215
D. str	6.52	3.34	0.512	5.00	0.54	0.130	0.63	0.22	0.350
M. bac	4.05	2.28	0.316	3.25	0.29	0.089	0.60	0.16	0.267
G. and	4.01	0.96	0.239	3.04	0.31	0.101	0.69	0.10	0.145
S. dul	2.87	0.68	0.237	2.65	0.24	0.090	0.86	0.12	0.139

Source Banik (2000)

Species: B. bal = *Bambusa balcooa*, B. pol = *B. polymorpha*, B. tul = *B. tulda*, B. vul = *B. vulgaris*, D. gig = *Dendrocalamus giganteus*, D. lon = *D. longispathus*, D. str = *D. strictus*, M. bac = *Melocanna baccifera*, G. and = *Gigantochloa andamanica*, S. dul = *Schizostachyum dullooa*

diameter at mid-culm zones might be an architectural adaptation in bamboos to provide more elasticity against severe stormy winds.

3.2.4   Sheathing Organs

Every node of every segmented vegetative axis of a bamboo plant bears a sheathing organ, which embraces the developing internode(s) distal to its insertion. The term 'sheath' is commonly used in a comprehensive sense, to include both the sheath proper and its appendages. Culm sheaths are modified leaves and have the same parts as leaves, but the parts are differently proportioned. In the bamboos, the principal appendages of the *sheath* proper are the *blade*, the *ligule* and the *auricles* (Fig. 3.3). Auricles and oral setae may or may not be present. The auricles when developed are typically lobe-like or rim-like structures occurring on each side of the base of the culm sheath blade. Ligule is present on the inner side of the sheath, where the blade joins. The *sheath* part is usually large and rigid and protects the growing culm and bud on the node. The culm sheath blade is highly variable in size, position and morphology. It can be erect, spreading or reflexed. In some species, the blade is triangular or dome shaped as in the genus *Bambusa*. In some cases, it is inflated, lanceolate or narrowly linear. It may be persistent or deciduous. The blades of sheaths at the base of the culm are quite small; on the sheaths near the top of the culm, they are leaf-like and occasionally green (as in *B. tulda*, *B. cacharensis*). There is thus not one unique shape and size of culm sheath on a bamboo. The culm

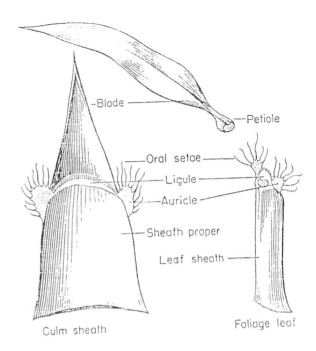

Fig. 3.3 Diagram showing the different parts in a typical culm sheath and foliage leaf of bamboo

sheath at about 150–200 cm above the ground or at the fifth node of the culm is considered as the one which has all the characteristics typical of the species. Morphology of the culm sheaths is species specific and, therefore, is used in the identification of bamboos at the level of a species (Chatterjee and Raizada 1963). In most species of *Arundinaria, Phyllostachys, Thyrsostachys, Oxytenanthera, Dendrocalamus, Melocanna* and *Schizostachyum*, the imperfect blade is comparatively narrow, frequently recurved and long, while in *Bambusa, Gigantochloa, Dinochloa* and *Cephalostachyum*, it is broad, triangular and much decurrent (meaning extending down and adnate to the stem). Culm sheaths may be persistent or soon fall off from the culm internodes, and this depends on the species nature. In *M. baccifera*, culm sheaths start dislodging basipetally with simultaneous bud break on the culm nodes. Sheaths persist on the lower two-thirds of the culm, and sheath blades are loosely fitted. In the next year, culm sheaths further dislodge, and bud break and branching continue up to two-thirds portion of the culm.

There are great variations in colour, shape and size of sheath characteristics of emerging shoots of different bamboo species (Banik 1997a, 2000; Table 3.3). The shoots of *D. longispathus, S. dullooa* and *T. oliveri* are somewhat cylindrical, while in others more or less conical in shape (Table 3.3). More commonly, very young shoots are harvested and used as gourmet vegetable by the people of different countries of Asia.

The basal diameters of shoots of *M. baccifera* and *S. dullooa* were comparatively smaller (4–5 cm) than those of the remaining species. The shoots of *D. giganteus, D. hamiltonii, B. polymorpha, B. bambos* and *B. vulgaris* have higher diameter

Table 3.3 Colour and sheath characteristics of shoots of some bamboo species of the Indian subcontinent

Species	Colour and shape	Avg. basal diameter (cm)
Bambusa bambos	Metallic purplish green; growing apex blunt, sheath usually coriaceous, glabrous to pubescent with dark brown hairs, ligule continuous with sheath top, auricle inconspicuous	8.6
B. longispiculata	Greyish green with dark brown hairs; sheath blade triangular-acuminate, leathery and mucronate, auricles long, ciliate	6.2
B. polymorpha	Golden purple or yellowish green, sheath auricles biserrate, lower blade brown, other blades green and cup shaped	9.8
B. tulda	Green surface usually with yellow stripes, calcareous band on one side of culm node; sheath asymmetric or oblique; auricles, at least one of them situated laterally	5.4
B. vulgaris	Dark brown to greenish, apex green, auricle distinct, sheath blade somewhat triangular, acute	10.3
Dendrocalamus giganteus	Purplish in colour and very big like a column, culm sheath very large, blade broadly triangular, ligule serrate	17.6
D. hamiltonii	Black tomentum, sheath blade stiff and pointed	14.5
D. longispathus	Green to brownish, elongated; sheath papyraceous with stiff appressed brown hairs	5.6
Melocanna baccifera	Yellowish green to yellowish brown, sheath margin and top pinkish, ligule horseshoe shaped, blades flagellate and glabrous	4.7
Schizostachyum dullooa	Greenish brown; sheath thick, leathery and striate with scattered white appressed hairs prominent above, blade narrow, linear lanceolate with dense brown hairs recurved, not auricled	4.2
Thyrsostachys oliveri	Dull green to yellowish; sheath fibrous, clothed on the back with thick white stiff pubescence, rounded at the top, blades linear, basal nodes of the shoots usually beared with aerial roots	6.3

ranged slightly above from 10 to 18 cm. *Bambusa tulda*, *B. longispiculata*, *D. longispathus* and *T. oliveri* have medium-sized shoots with basal diameter ranging from 6 to 8 cm.

The branch sheaths are the series of sheaths clothing each branch of the culm. The dimensions are progressively reduced, and the form of sheath is progressively generalised, in the successive orders of branches.

Leaf sheath is one of the leaf-bearing sheaths inserted at the distal nodes of each aerial vegetative axis of a bamboo, whether culm, branch or twig. In *M. baccifera*, leaf sheath is thick with ash colour, prominent, very long erect wavy bristles, 8–25 mm long auricles, short ligule and narrow petioles.

Rhizome sheaths of the rhizome proper vary least from node to node within a given axis and usually persistent or tardily abscissile. A series of sheath from the successive nodes of a pachymorph rhizome shows progressive change only in size that corresponds to changes in the diameter of the axis. In a leptomorph rhizome, the sheaths are, however, uniform in all respect.

Prophyll is a sheath, which occurs at the base of the node of each vegetative branch, representing the first leaf of that branch. It is a one- or two-keeled structure, which encloses and protects the branch primordia. The back of the prophyll closely adheres to the axis from which the branch emerges. The margins of the prophyll cover the branch primordia. The morphology of the prophyll varies in different species and, therefore, is considered taxonomically very important.

3.2.5 Branches

The bamboo has more complicated branching system than ordinary grasses. Under every sheath on the main culm is a bud, enclosed in a flat sheath of its own, the sheath backs on the culm itself, and has two inflexed edges which embrace the inner part of the buds. When a culm reaches its full height, the lateral buds on the nodes begin to grow and form branches. Branch buds emerge on alternative sides of the culm just above the sheath scar at successive nodes. Initiation of branching on the culm nodes may be in the same year of culm emergence (*syllepsis*) as in *Bambusa bambos* and *Thyrsostachys oliveri* or may be in the following year of culm emergence (*prolepsis*) as observed in *M. baccifera*. In most of the species of genera *Bambusa*, *Dendrocalamus* and *Gigantochloa*, the primary branch emerges and remains strongly dominant. Subsequent orders of axes develop from the small adventitious buds at the same culm nodes. This stout branch resembles the culm itself in varying degrees, and the resemblance is especially noticeable in many sympodial bamboos that have pachymorph rhizomes (Fig. 3.4a). In *Dendrocalamus hamiltonii*, culms are usually unbranched below and much branched above. Straw-creamy-coloured big branch buds on the culm nodes. Small, thin curving branchlets are on the younger culms. Strong, stout and long branches develop from the big buds on the nodes of older culms. Such branches are sometimes of similar diameter size of culms (Fig. 3.4b). The buds at the lower culm nodes usually remain dormant. But in some *Bambusa* and *Dendrocalamus*, the branches develop from the basal culm nodes as well. In *Bambusa bambos*, *B. bambos* var. *spinosa* and *B. blumeana*, these branches form a thicket around the base of the culm and are spiny. In *Melocanna* and *Schizostachyum*, a dense cluster of thin, widely radiating, numerous, subequal branches arise at each culm node which develops from a single primary branch axes at node. Such array of branches is called branch complements (Fig. 3.4c-i, c-ii). In *Phyllostachys*, the typical primary mid-culm branch is binary, the two axes being more or less strongly unequal.

Fig. 3.4 Branch characteristics of some bamboo species. (**a**) A stout branch resembles the culm itself in varying degrees, and the resemblance is especially noticeable in many sympodial bamboos that have pachymorph rhizomes; (**b**) *Dendrocalamus hamiltonii*: Strong, stout and long branches are sometimes of similar diameter size of culms; (**c-i**) *Melocanna* baccifera and (**c-ii**) *Schizostachyum dullooa* a dense cluster of thin, widely radiating, numerous, subequal branches arise at each culm node, called *branch complements*

3.2.6   Leaves

In bamboos the leaf organ diverged into two types: cauline leaf and foliage leaf (i.e. green leaf having photosynthesis). A foliage leaf (a leaf proper) is an append-age to a leaf sheath proper. Leaf blades differ from culm sheath blades and branch sheath blades being stalked or petiolate. Bamboo leaves are usually linear, lance-olate or oblong-lanceolate in shape; they have usually a short petiole into which the base, which is frequently unequal cut, extends; the point is usually long acuminate, often scabrous; and the side is glabrous or softly hairy. Leaf blades are generally thinner than culm sheath blades and often show more marked dorsiventrality. The leaves of all bamboos are very similar in general appearance, for although some species have usually large leaves and others quite small, the size depends much on the part of the plant from which they are taken. In *D. hamiltonii*, the leaves of young shoots and the end leaves of strong branches are usually very large, while those of medium branches are moderate in size and those of thin shoots from lower nodes are quite small. During a 12-month cycle (annual calendar), clumps of *B. balcooa*, *Dendrocalamus asper* and *D. hamiltonii* produce big- and comparatively small-sized leaves in warm humid (March–August) and in the beginning of dry winter season (October–November), respectively. Bamboos that grow in moist habitat generally have larger leaves than those species growing in drier sites.

Venation of the leaves is parallel, and veins of three orders, namely, the midrib, secondary veins and tertiary veins, are present. In certain genera, the adjacent veins are connected by transverse distinct veinlets.

Leaf dimorphism is often observed with some bamboo species depending with age and also due to annual seasonal variations. Leaves at the seedling stage of *M. baccifera* (Fig. 3.5) and *G. andamanica* are bigger in size than those produced in adult clumps, whereas leaves are smaller in seedlings of *B. tulda* compared to well-grown clumps.

3.3 Culm Growth

While shrubs and trees tend to grow slower, the competitive bamboos have rapid rates of dry matter production, continuous stem extension and leaf production during the growing period and rapid phenotypic adjustments in leaf area and shoot morphology in response to shade.

3.3.1 Culm Emergence

Knowledge of culm emergence periodicity is important in managing the bamboo clump. Banik (1993a) studied the culm emergence periodicity of 13 bamboo species and observed that the culm emergence mostly takes place at the beginning of rainy season, either from May or June, and may continue up to 6 or 7 months ending either in October or November (Table 3.4). The number of shoots emerging at the beginning (April–May) of growing season was low. After that it sharply increased during June to August and gradually decreased in the following months. It is suggested that felling of culms from the clumps should not be done during the culm emergence period. In improving the yield of a bamboo clump, fertilisers

Fig. 3.5 Leaves at the
seedling stage of
M. baccifera are big in size
than those produced in the
branch of adult clumps

should be applied about a month before sprouting periods so that the effects appear at the time of sprouting and growth (Uchimura 1980).

When the young culm bud on an underground rhizome or juvenile shoot first begins to develop, a conical growth is seen protruding from the ground, covered and protected by numerous overlapping rigid sheaths, often of a bright colour and furnished with blades. This bud grows slowly at first, forming the rudiments of all the parts of a fully developed culm, as it were in contracted condition. This slowly developing young shoot consists of a short massive little-differentiated stem packed with food materials. When its development is complete, the shoot may rest for a time for 7–10 days (Banik 1991, 2000), but in Malaya probably such resting often does not occur (Holttum 1958). Depending on the rainy weather, the cone gradually lengthens, the sheaths separate, the nodes appear and finally a full-grown culm is developed. Usually, one by one, the sheaths drop off, the buds on the culm nodes put out branches and these produce their leaves. Each species of bamboo, in general, has definite periodicity for culm emergence. The shoot emergence and

Table 3.4 The average number of culm emerged in different months from the clumps of 13 bamboo species in relation to monthly climatic condition at Chittagong

Species	Jan	Feb	Mar	Apr	May	Jun	Jul	Aug	Sep	Oct	Nov	Dec	Emrg Perd (mth)
B. bal	0	0	0	0	0.2	0.9	0.5	0.3	0.1	0.3	0.2	0	7
B. glu	0	0	0	0	0.9	0.7	2.0	0.9	1.4	0.5	0.4	0	7
B. lon	0	0	0	0	0	0.9	1.1	0.7	1.2	0.2	0.1	0	6
B. nut	0	0	0	0	0	0.7	1.4	0.5	0.5	0.6	0	0	5
B. pol	0	0	0	0	0	0.7	0.9	0.9	0.6	0.7	0	0	5
B. tul	0	0	0	0	0.07	0.5	0.7	0.6	1.2	1.1	0.6	0	7
B. vul	0	0	0	0	0	0.7	0.9	1.0	0.3	0.3	0.07	0	6
D. gig	0	0	0	0	2.6	3.8	4.2	0.6	0	0	0	0	4
D. str	0	0	0	0	0	0.1	0.8	1.0	0.4	0.7	0.3	0	6
G. ad	0	0	0	0	0.2	0.4	3.7	6.1	4.4	1.4	1.0	0	7
M. bec	0	0	0	0	0.1	0.7	1.7	3.7	2.2	0.9	0.2	0	7
S. dul	0	0	0	0	0.2	0.6	1.8	2.8	1.2	0.09	0	0	6

Climatic conditions

Soil temp (°C) in different depth

	Jan	Feb	Mar	Apr	May	Jun	Jul	Aug	Sep	Oct	Nov	Dec
50 cm	20.8	22.4	26.5	29.2	29.7	29.3	29.0	29.0	29.1	29.6	25.9	21.7
100 cm	22.1	22.8	25.3	28.3	29.1	28.9	28.7	28.6	28.5	28.5	26.5	23.5
200 cm	24.2	23.8	24.9	26.9	28.2	28.3	28.5	28.6	28.6	28.7	27.7	25.3
Average	22.3	23.0	25.5	28.1	29.0	28.8	28.7	28.7	28.7	28.9	26.7	23.4
Air temp (°C)	18.5	20.9	25.3	26.6	26.9	26.6	26.0	26.5	25.3	25.5	20.9	16.2
Relative Hum (%)	56.5	55.5	55.3	62.2	65.1	67.4	70.8	67.2	66.0	66.0	63.1	55.9
Total rainfall (mm)	17.9	12.7	33.6	117.5	375.6	566.3	783.0	426.1	321.6	247.1	95.1	13.6

Species: B. bal = *Bambusa balcooa*, B. glu = *B. glaucescens*, B. lon = *B. longispiculata*, B. nut = *B. nutans*, B. pol. = *B. polymorpha*, B. tul = *B. tulda*, B. vul = *B. vulgaris*, D. gig = *Dendrocalamus giganteus*, D. lon = *D. longispathus*, D. str = *D. strictus*, M. bec = *Melocanna baccifera*, G. ad = *Gigantochloa andamanica*, S. dul = *Schizostachyum dullooa*, Emrg Perd (mth) = Emergence period (month). *Relative Hum % =* Relative Humidity %
Source: Banik (1993a, 2000)

vigour are dependent on the activation of culm buds on the rhizomes and the amount of stored food in them (Ueda 1960). The climatic factors like soil temperature, relative humidity and total rainfall have been found to be important determinants of culm emergence in bamboos. In winter and dry season (from December to April), most of the species do not produce any culms as the climate is comparatively dry and both soil and air temperatures are also low. The enduring low temperatures in winter lower the function of photosynthesis and absorption of water by roots and make the ordinary growth of bamboo impossible. About 28 °C soil (50–200 cm depth), 26 °C air temperature, above 60 % mean relative humidity and monthly mean total rainfall within 100–800 mm seem to be the minimum requirement for culm emergence in South Asian countries (Banik 2000).

One of the most important climatic factors influencing the production of new culm is the rainfall and its seasonal distribution. If the monsoon comes in due time and the rainfall is sufficient in amount and is well distributed, the production of new culms is good. But if the monsoon comes at the abnormal time, or if there is a break after the first heavy showers, the production of new culms is unfavourably affected. Above 63 % mean relative humidity and monthly mean total rainfall within 100 mm (November) to 783 mm (July) seem to be the minimum requirement for culm emergence in South Asian countries (Banik 2000). When rainfall is early, *Bambusa vulgaris* starts producing culms in the month of April, whereas the species begins sprouting in June when rainfall is delayed.

In northern India (Uttarakhand, UP, Himachal, etc.), when the southwest monsoon begins, both *Bambusa bambos* and *D. strictus* send up their new culms in June or July. But in south India, as may be seen on the eastern slopes of the Nilgiris (e.g. in the Coonoor valley), the new culms appear in September or October, probably with the first of the northeast monsoon rains (Gamble 1896). If rains are late in the rainy season, new culms keep appearing up to the end of season (October–November). New culms of *Dendrocalamus strictus* were however noticed during December 1985 in Lucknow after late rains during October in that year (Chaturvedi 1986). Combined effect of appropriate temperature and rainfall usually results in higher rate of clump growth.

In general, the productive months of culm emergence have days of longer photoperiod. Thus, it appears that the period of culm emergence also depend on climatic condition of locality. Heavy shade of upper tree canopy inhibits bamboo regeneration and growth. It has been observed that clumps growing in the open sites produce culms of much better quality and quantity than the clumps growing under heavy shade (Banik 2000). The shoot emergence is earlier in the south slope of a hill than that in the north slope, and annually growing period of the stand is longer than that in the north slope. Partial sunlight was found to be essential for the survival and growth of bamboo seedlings during the initial stage of establishment (Banik 1997b). The seedlings of *B. tulda* produced elongated culms, five to seven times more in height, when grown under the partial shade of shrubby plants like *Cajanus cajan*, *Boga medula* and *Sesbania* sp. during first 2 years of outdoor planting.

In monopodial species, *Phyllostachys*, at the beginning of January, the amount of reserve starch is highest in the upper part of the culm than in the lower part (Uchimura 1980). Most of the sympodial bamboo species of tropical and tropical to subtropical climates show highest amount of carbohydrate reserves in February to March (beginning of spring).

3.3.2 Culm Elongation and Growth

The shoot starts to emerge from the soil mostly at the beginning of rainy season. The height growth of the shoot (culm) results mainly through the elongation of internodes. The internodal elongation begins at the basal portion of the culm and then gradually proceeds to the top. That is, elongation is mainly due to the intercalary meristem present at the node. In intercalary growth, the immature axis increases in length by the elongation of cells in zones of secondary meristems each located just above the node. In the elongating segmented axes of an emerging shoot of *M. baccifera* plant, the locus of each zone of intercalary growth is just above the locus of insertion of a sheath (Fig. 3.6; Banik 2010). There is empirical evidence that the sheath may be the origin of substances that control, or at least influence, the process of intercalary growth and possibly the initiation of roots and branch primordia. When Chinese gardeners wish to dwarf a bamboo, they remove each culm sheath prematurely, beginning with the lowest, before the elongation taking place above its node is completed. Upon the removal of sheath, the elongation above its node ceases. The initiation of branch buds and root primordial on any segmented axis always takes place within a zone of intercalary growth, before the

Fig. 3.6 In the elongating segmented axes of an emerging shoot of *M. baccifera*, the locus of each zone of intercalary growth is just above the locus of insertion of a sheath. Culm elongation is mainly due to the intercalary meristem present at the node

tissue loses their meristematic potential and while the subtending sheath is still living (McClure 1966). The daily growth of culm is the sum total of the daily elongation of the internodes. In the internodes near the apex of young shoot, the cell divides evenly. The cell division in the internodes of the upper part of culm is weaker than that in the lower part. Though the culms do not grow significantly in diameter after sprouting, they continue to change in density and strength properties.

Just after the appearance of culms on the ground, they elongate very slowly up to 1–2 weeks and then gradually gain speed until they attain the optimum size, and thereafter the rate of elongation quickly slows down (Fig. 3.7a). The culm elongation curve of each of the species (*Bambusa balcooa*, *B. vulgaris*, *Melocanna baccifera*) shows two distinct peaks, the first one (P_1) being short and the second one (P_2) tall. These two peaks are due to the two rapid elongation phases exhibited by the culm in its total growth period (Fig. 3.7a; Banik 2000). Such behaviour of culm elongation in bamboos could not be explained and needs further studies. It was also observed that culm did not show any diameter increment during or after the elongation period. The diameter with which it emerges remains unchanged throughout its life. Thus, it seems that the size of the diameter of a culm is determined by the size and vigour of the bud present in the mother rhizome from where it originates.

The time period starting from the sprouting of shoot to the end of height growth (*culm elongation period*) varies with the nature of bamboo species. The culm elongation continues both day and night; the genus *Phyllostachys* in Japan grows more during the day (Ueda 1960), whereas in the tropical regions, bamboos grow preferably during the night. The daily (24 h) extension growth amounts to about 10–30 cm but reaches 58 cm for *Dendrocalamus giganteus* (Osmaston 1918) and up to 121 cm for *Phyllostachys reticulata* (Ueda 1960). Bamboo shoots of *Dendrocalamus asper* can grow 90–120 cm/day under ideal conditions (Subsansenee 1994). The maximum elongation rate of a culm per day was 77 cm in *B. balcooa*, 66 cm in *B. vulgaris* and 44 cm in *M. baccifera* during each of their total elongation period in Chittagong (Fig. 3.7b; Banik 1993a). Such rapid rate of elongation was observed during the second half of the complete culm elongation period. The total culm elongation periods for *Bambusa balcooa* and *B. vulgaris* were observed to be within 75–85 days and 55–60 days in *M. baccifera* (Fig. 3.7a). The culm emergence period, in most of the species, was during May to October (Table 3.5). In later part of the growing season (October), the average total rainfall becomes low with a few numbers of rainy days and gradually decreases the day length and starts inducing the dormancy in plant body. Kadambi (1949) observed that, in general, years of plentiful rainfall produced larger numbers of shoots than the years of deficient and scanty rainfall. Rain also prompted the shoot emergence and culm growth of all Japanese bamboos (Shigematsu 1960).

The action of reserve nutrients must be considered as one of the factors that influences the growth of bamboo culm and rhizome. The reserve starch in 1-year-old culms is highest just before sprouting occurred, and it decreases maximum during the period of growth of the sprouts and increases after the growth is completed; finally, the amount of reserve starch in the culm decreases again when

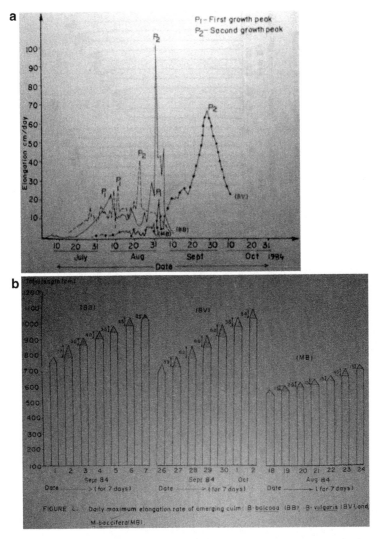

Fig. 3.7 Rate of culm elongation in three bamboo species: (**a**) elongation pattern of emerging culms; (**b**) daily maximum elongation rate of an emerging culm during vigorous growth period

the growth of the rhizome starts and then gradually increases (Uchimura 1980). In other words, the consumption of the reserve starch in 1-year-old culms is higher during the growing period and increased during the period of elongation. It has been further observed that the larger the culm, the higher the amount of reserve starch.

As regards growth, bamboo stem (culms) differs from trees. The periodicity and nature of growth in height and diameter are the most important differences in these two groups of plants.

Table 3.5 Mortality of emerging culm from full-grown clumps in different bamboo species at Chittagong

Species	Year 1978			Year 1979			Year 1980		
	Emerged total X ± SE	Dead nos.	Mortality (%)	Emerged total	Dead nos.	Mortality (%)	Emerged total	Dead nos.	Mortality (%)
Thick-walled species									
B. bal	2.9 ± 0.4	0.9 ± 0.2	31.0	3.8 ± 0.7	1.4 ± 0.3	36.8	3.5 ± 3.2	1.5 ± 3.2	42.9
B. lon	8.2 ± 0.1	2.4 ± 0.7	29.3	6.1 ± 1.8	1.7 ± 0.4	27.9	9.2 ± 1.8	3.0 ± 0.4	33.0
B. nut	8.1 ± 2.1	3.3 ± 0.9	40.7	7.3 ± 1.8	3.0 ± 0.7	41.1	9.1 ± 2.5	4.0 ± 1.0	43.9
B. vul	9.0 ± 1.2	6.1 ± 1.0	68.0	8.9 ± 1.5	6.2 ± 1.3	69.6	8.7 ± 1.5	5.9 ± 1.0	67.8
D. gig	7.0 ± 5.0	4.7 ± 3.7	67.0	11.7 ± 8.7	6.0 ± 4.5	51.3	11.0 ± 2.2	5.0 ± 1.5	45.5
Thin-walled species									
B. tul	12.6 ± 3.8	4.7 ± 1.6	37.3	3.6 ± 0.7	1.1 ± 0.3	30.6	6.8 ± 2.1	2.0 ± 0.9	29.1
D. lon	7.0 ± 2.1	1.5 ± 0.6	21.4	8.9 ± 2.6	2.1 ± 0.6	23.6	4.9 ± 1.6	1.4 ± 1.3	28.6
M. bec	2.9 ± 0.4	2.7 ± 0.7	10.3	25.5 ± 4.0	2.5 ± 0.4	9.8	33.2 ± 6.5	2.9 ± 0.7	8.7

Species: B. bal = *Bambusa balcooa*, B. lon = *B. longispiculata*, B. nut = *B. nutans*, B. vul = *B. vulgaris*, B. tul = *B. tulda*, D. lon = *D. longispathus*, M. bec = *Melocanna baccifera*, D. gig = *D. giganteus*,
Source: Banik (1983, 2000)

Item	Bamboo culm	Timber tree stem
Height growth	Height growth (elongation) mainly by intercalary meristems	Height growth is realised by primary meristem on apex
	Height growth (elongation) begins and ends in different internodes, from base to tip, but not simultaneously, the total elongation is realised by intercalary meristems	Height growth does not take place on secondary growth tissue
	Height growth completes within 2–4 months. The sum of growth in length of all internodes is the total height growth of bamboo stem	Height growth (elongation) lasts in all lifetime of trees; the speed of growth declines with ageing
Diameter growth	The diameter of a culm is primarily determined by the health and size of bud present in the underground rhizome from which it develops. However, while elongating, the diameter of culm and the thickness of stem wall increase slightly	The growth of diameter is realised by cambium
	The diameter does not increase after completion of culm elongation	Diameter growth lasts in all lifetime of tree

During long life of a bamboo culm, the parenchyma cells and the phloem remains physiologically active. No residues of the metabolic processes are known so far. Ageing has no effect on the tissue comparable with the production of polyphenols against biological deterioration. However, culm injuries and ageing induce parenchyma cells to produce phenolic substances in small amounts. Wound reaction induces formation of tyloses in metaxylem and protoxylem, and lignification in sieve tubes and ground parenchyma (Liese 1998). Bamboo completes the maturation of culm tissue as well as its height growth with the full development of twigs and leaves within a year. Although it is quite soft and fragile in the stage of sprout, it becomes increasingly hard in proportion to the elongation of culm. Consequently, it is presumed that the content of lignin as well as that of cellulose must increase with the development of culm. *Lignification* is the formation of a polymer within the cell wall that provides strength to the culm. The lignin content of *D. strictus* decreases with the increasing culm height, whereas in *B. vulgaris*, it is unaffected by changes in culm portion location. In case of *Ph. pubescens*, the lignification within every one internode proceeds downwards from top to bottom, and transverse progress of lignification proceeds inwards from outside to inside. During the growth of new lateral shoots and bud flush, the process proceeds progressively and reaches to its highest level after the full opening of new leaves when full photosynthetic function is ready (Itoh and Shimaji 1981). In this species, full lignification of bamboo culm is completed within one growing season.

In stem (culm) maturing phase, the thickening process of cell wall is accompanied by the accumulation of fundamental tissue matters. The lignification and thickening take place simultaneously. Lignin, cellulose and hemicellulose accumulate, respectively, but their weight ratios to each other remain the same or vary little. A 1-year-old culm normally relies on food supplement from the rhizome of the mother plant, but a 2-year-old bamboo starts to utilise all the carbohydrate contents, particularly sugar, for the development of its culm.

3.3.3   Juvenile Mortality of Elongating Culm

All emerging culms do not always develop into full-grown culms. The natural mortality of emerging culms was found to be higher (28–69 %) in thick-walled and tall species (*B. balcooa, B. nutans, B. longispiculata, B. vulgaris, D. giganteus*), whereas it was low (9–37 %) in thin-walled and small-sized bamboo (*M. baccifera, D. longispathus, B. tulda*) species (Table 3.5; Banik 1983). In all the species, except *M. baccifera*, the death of the emerging culms was observed within the average height of 22.0 cm and within the distance of 27.0 cm to the nearest older culm. The emergence of large number of culms within this short distance created congested clumps condition. The length of rhizome neck and the nature of rhizome movement control the distance of emerging culm from the nearest older culm. Due to elongated rhizome necks (1.0–2.0 m), *M. baccifera* has a more open type of clump than other clump-forming bamboos that have a short-necked sympodial rhizome system (Banik 1980). Clumps having a congested sympodial rhizome system produce most of the culms within a short distance. Therefore, competition for survival among the developing culms becomes intensive. This condition increases mortality percentage among the young developing culms in the species with short-necked sympodial rhizome systems. On the other hand, the mortality percentage is lesser in *M. baccifera* where the clumps are open (Table 3.5). Thus, clump congestion could also be a factor for death of emerging culms.

Generally, the culm mortality was higher in thick-walled than in thin-walled bamboo species (Table 3.5). Average mortality percentage varied from 28 to 69 in the case of thick-walled, large-dimension species, whereas it was 9–37 % in the case of thin-walled, small-dimension species. High juvenile mortality (60 %) was also reported for a thick-walled bamboo species like *Gigantochloa levis* (Sharma 1982).

Elongation rate of an emerging bamboo culm, in general, can be grouped among the fastest growing stems of the plant kingdom (Ueda 1960; McClure 1966). To meet the requirements of such rapid rate of elongation, the mother rhizome has to supply sufficient food material to the developing culms. The amount of reserve food material in the rhizome declines during the culm growing period (Uchimura 1980). Therefore, competition for food among the developing culm is not unlikely. It is likely that the food requirement in the species having larger culm dimension is higher. It is probable that the stored food in the rhizome of mother culm is not enough to meet the requirement of elongation of all the emerging culms in thick-walled bamboo species. Inadequate food supply could be a reason for higher mortality in the elongating shoots of thick-walled bamboo species.

Ueda (1960) and Uchimura (1980) reported that survival rate of emerging culms can be increased by applying the fertiliser. Xianhui and Yuemei (1983) observed that transcription and translation losses of DNA and RNA prevent the synthesis of protein and force bamboo shoots to stop their growth resulting in degradation.

Scanty rainfall is likely to reduce turgescence, a condition essential for cell elongation and finally the culm elongation and growth. When monsoon comes at the abnormal time or if there is a break after the first heavy showers, the production of new culms is unfavourably affected and juvenile mortality of emerging culm takes place (Banik 1983). Moreover, during the later part of the growing season, stored food materials in the mother rhizome may become exhausted by supplying most of its food to the culms produced and developed in the early part of the season. Mortality of emerging culms was higher (50–60 %) in September and October, when average rainfall was low (Banik 1983). Thus, it appears that ecophysiological condition (clump congestion, soil moisture, food storage, etc.) and genetic makeup of the clump in each bamboo species seem to influence the rate of mortality of emerging culms.

3.3.4   The Bud Dormancy, Branching and Leaf Fall

The branch buds primordial are formed in sheath axils as a form of axillary bud during the early stage of shoot growth. The *buds* are developed and arranged alternately on the nodes of emerging culm and branches. Buds on the culm node produce primary branches, and buds on branches may produce both secondary and tertiary branches and leaves. In many bamboos, the branch primordium at each mid-culm node ramifies precociously before it ruptures the prophyllum. However, in *Chusquea*, the branch that arises from the large central bud at each mid-culm node is strongly dominant.

Very rarely buds on the emerging culms show any branch growth. Bud dormancy and the breaking of it vary with the nodal position on the culm (Banik 1980); in *M. baccifera* and *S. dullooa*, the breaking of bud dormancy on the node of a culm was found to be in basipetal order, gradually moving towards the base and being completed in 3–4 years. Most of the time, nodal buds remain dormant on the lower one-third portion (basal 3–7 nodes) of culm with persistent culm sheaths. Although the same pattern was observed in *B. vulgaris*, 5–10 buds around the middle of the culm remained dormant up to 3–4 years. After that, all opened except those in the 2–3 base nodes. However, in *B. balcooa*, breaking of bud dormancy on the culm nodes is from base to the tip of culm (acropetal) and completes within 1 year. In *B. tulda*, all the buds awakened at nearly the same time with the exception of a few at mid-culm or below. Most of the remaining buds started opening about age 1–2 years. In all species of *Phyllostachys*, these buds awaken in acropetal order, that is, each bud breaks as soon as the internode to which it is basal has completed its growth (McClure 1966).

Normally, the monopodial bamboo has a high clear-boled habit, while branching position in sympodial bamboo depends on its position of the branch buds. In some sympodial bamboo species, culms of progressively larger sizes (those approaching mature statures) have a progressively longer series of budless lower aboveground nodes. In *B. cacharensis*, *B. polymorpha*, *B. textilis*, *Thyrsostachys oliveri* and *T. siamensis*, for example, culms from plants of mature size may lack buds and branches in the lower 1/2 to 2/3 or even 3/4 of their length. This is a desirable aesthetic feature in bamboos planted for the purpose of ornamentation. In culms used in handicrafts or in industry, the absence of knots that mark the insertion of branch complements makes for economy and ease in working the material.

In mature bamboo plants, one or more of the lowest aboveground buds on each culm may remain dormant indefinitely. Buds on the nodes of the lower portion of culms of *B. polymorpha*, *B. cacharensis*, *D. asper*, *D. hamiltonii*, *M. baccifera* and *Thyrsostachys oliveri* usually remain dormant and do not produce branches, but in some occasions when sunlight penetrates through the crown to the basal portion of the culm, buds on the lower culm nodes activate and produce small branches and leaves which are usually seen in the clumps of these species grown at the periphery of a compact plantation.

The growing culm is without branches, but branching begins either after the culm has ceased growth or in the following growth season. Simultaneously many thin branches are also produced in assembly from each culm bud. Leaves start developing within 2–4 weeks from the developing branches. There exist numerous and diverse pattern of branching in bamboos. Majority of the species (e.g. *B. balcooa*, *B. bambos*, *B. tulda*, *B. vulgaris*, etc.) produce branches more or less throughout from culm base to tip, while in others (e.g. *B. cacharensis*, *B. polymorpha*, *D. longispathus*, *M. baccifera*, *S. dullooa*) it is confined only to upper-mid part of the culm. However, only in a few cases (*Thyrsostachys oliveri* and *T. siamensis*), branches are confined only at the top portion of the culm.

Most of the clump-forming bamboos partly shed their leaves in winter when it is dry and renew the leaves simultaneously in a short time. Generally, in bamboos of Northeast India and Bangladesh, new leaves and branches start developing during March–April (spring–summer) from the culm buds and branch buds after each defoliation. All forest-grown bamboo species of the subcontinent are evergreen in nature. In most of the bamboo species of the Indian subcontinent, leaf fall occurs at about the same time as bud break but may be a few days later or rarely earlier, depending on the particular weather conditions or on the local climate. The clumps of *B. polymorpha* are ordinarily evergreen in nature at its native places have high to moderate rainfall, but in some drier pockets and severe draught season, clumps of these species may exhibit deciduous behaviour. The clumps of *B. balcooa* in some occasion, especially in drier habitat and in prolonged severe draught period, also exhibited pure deciduous behaviour. Some species, like *D. strictus*, which are usually deciduous, may be evergreen in damp climate or moist habitat.

The life of leaves is about 9–18 months, and leaves of several flushes occur together mainly during spring to rainy summer (March to August). The amount of leaf fall from *M. baccifera*, *G. andamanica* (syn. *Oxytenanthera nigrociliata*), *B. tulda* and *D. giganteus* is 6.0, 5.6, 5.8 and 7.0 t/ha, respectively (Banik 2000). Leaf-decomposition starts in June–July and humification reaches its peak in October–November. Each of the bamboo species contributes about 20 kg/ton of organic matter to the soil more or less similar to other broad-leaved tree species (*Hevea brasiliensis*, *Syzygium grande*, etc.). Due to the scarcity of fuel wood nowadays, people collect dry bamboo leaves every year from the ground and use these as fuel. This practice gradually depletes the organic matter and reduces moisture content of the soil, resulting in poor annual yield and growth of bamboo culms per clump. As collection of leaves exposes the ground, the ensuing monsoon shower directly hits the soil, accelerating the erosion, especially, on the slopes. Thus, it is advisable to allow leaves to decompose on the ground for controlling erosion and maintaining the soil nutrition level.

The blanket of fallen leaves is an effective mulch to keep the moisture in and an organic fertiliser to rejuvenate the soil. Bamboos growing in flatland site usually accumulate litter fall twice as much as the hill side. The site conditions significantly affected the decomposition rate. The constant temperature and high humidity favour the development of certain kinds of microbes so that the resulting population can decompose the material faster than if the temperature fluctuates.

3.3.5 Culm Production and Clump Expansion

The number of new culms that develop from a clump varies by species, soil and climatic conditions, harvesting method, size of clump, overhead cover, etc. If the monsoon comes in the normal time and the rainfall is well distributed and adequate in amount, the production of new culm is good. If the monsoon comes at the abnormal time or if there is a break after the first heavy showers, the production of new culms is unfavourably affected (Ueda 1960). The diameter of the successively emerged culms increases with the age of the clumps. Banik (1988) studied culm production, growth and clump expansion behaviour of five priority bamboo species at Chittagong and reported that only one to three full-grown culms (FG) were produced per clump in the first year after plantation of offset or cutting (Table 3.6).

The number of FG culms gradually increased in the subsequent years and became a maximum (3.2–8.8) in the fifth year of clump age in the case of all *Bambusa* species studied. After that culm production gradually decreases if there is no felling, but the growth in diameter and height is more or less similar to those of previous years. Cumulative total of FG culms produced in each of the clumps of *Bambusa balcooa*, *B. longispiculata*, *B. tulda* and *B. vulgaris* up to the fifth and sixth year of age made for a crowded condition due to the short (7–12 cm) neck rhizome system. If culms are not harvested, this crowded and congested condition would result in scarcity of room for the new emerging culms and also increase competition among them for survival. Probably due to this reason, culm production subsequently decreased from the sixth year in the clumps of these species. Similarly Kondas (1981) observed in a clump of *B. vulgaris* planted in Dandeli, India, that the culm production was higher up to the fourth year after planting and then decreased in the subsequent years. Decrease in culm production might have slowed down the

Table 3.6 Annual production of full-grown (FG) culm and gradual expansion of clump girth up to 10 years of clump age in five bamboo species (Banik 1988)

Clump age (year)	B. balcooa		B. longispiculata		B. ʰulda		B. vulgaris		M. baccifera	
	FG Culms (nos.)	Clump girth (cm)	FG Culms (nos.)	Clump girth (cm)	FG Culms (ncs.)	Clump girth (cm)	FG Culms (nos.)	Clump girth (cm)	FG Culms (nos.)	Clump girth (cm)
1 (1973)	1.2 ± 0.2	28.5 ± 2.0	1.8 ± 0.4	30.0 ± 5.2	3.0 ± 0.5	87.0 ± 13.2	1.6 ± 0.2	50.0 ± 4.5	2.5 ± 0.5	92.2 ± 20.5
2 (1974)	2.3 ± 0.9	66.2 ± 13.5	2.7 ± 0.6	66.5 ± 13.5	3.8 ± 0.4	153.0 ± 16.6	2.8 ± 0.3	105.0 ± 10.0	5.6 ± 1.7	145.9 ± 23.2
3 (1975)	1.7 ± 0.2	102.0 ± 14.0	4.1 ± 0.6	140.2 ± 16.5	8.2 ± 1.3	243.0 ± 28.3	3.2 ± 0.7	162.6 ± 22.0	12.5 ± 2.5	301.0 ± 42.0
4 (1976)	2.7 ± 0.4	177.0 ± 22.6	6.2 ± 0.5	222.0 ± 24.7	5.7 ± 0.9	307.0 ± 36.5	5.3 ± 1.2	273.4 ± 35.5	16.4 ± 4.2	395.4 ± 53.3
5 (1977)	3.2 ± 0.8	232.0 ± 38.0	7.0 ± 1.3	286.4 ± 20.6	8.8 ± 2.3	439.3 ± 56.3	5.2 ± 1.7	378.0 ± 50.3	23.6 ± 7.2	529.3 ± 105.6
6 (1978)	2.2 ± 0.4	250.0 ± 29.4	6.7 ± 1.3	372.0 ± 28.2	4.5 ± 1.0	464.0 ± 51.0	3.6 ± 0.8	456.3 ± 62.1	25.3 ± 4.6	624.3 ± 115.2
7 (1979)	2.4 ± 0.5	312.5 ± 32.5	5.0 ± 1.0	430.0 ± 35.0	4.9 ± 1.2	520.1 ± 42.0	3.2 ± 0.6	508.3 ± 64.4	26.6 ± 8.7	805.4 ± 146.1
8 (1980)	1.8 ± 0.5	320.5 ± 24.0	6.6 ± 1.4	507.6 ± 66.6	3.7 ± 0.3	527.2 ± 61.6	3.2 ± 0.6	555.0 ± 75.8	25.1 ± 6.2	980.6 ± 170.5
9 (1981)	2.2 ± 0.5	376.0 ± 53.7	5.8 ± 1.0	648.3 ± 109.6	4.8 ± 1.8	571.0 ± 57.6	2.6 ± 0.6	603.0 ± 58.0	31.1 ± 8.0	1268.0 ± 247.0
10 (1981)	2.2 ± 0.5	414.2 ± 58.3	5.7 ± 1.6	690.0 ± 132.3	2.7 ± 0.4	586.7 ± 207.8	2.8 ± 0.7	705.0 ± 114.5	35.7 ± 13.3	1432.3 ± 281.0

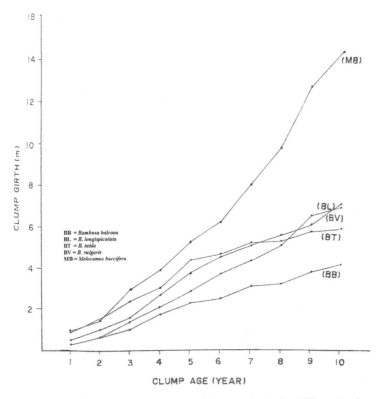

Fig. 3.8 Influence of age on the pattern of clump-girth expansion in five different bamboo species

rate of clump-girth expansion after the sixth and seventh year of age in all these *Bambusa* species (Fig. 3.8). It has been observed that in all the five bamboo species studied, the culms that emerged in the first year after plantation were short (1.5–4.0 m) in length and narrow (0.7–1.6 cm) in diameter at breast height (dbh). Culms produced in subsequent years were distinctly taller and wider in diameter than those produced in the past years, and such trend continued up to 5–6 years in *B. longispiculata* and *B. tulda* and 7 years in the case of *B. vulgaris* and *B. balcooa*. After that period, the increment is very little in the subsequent years and remains more or less static (Banik 1988). The study also revealed that the number of full-grown culms emerged in the clumps of all the four *Bambusa* species increased up to the fifth year of age and then gradually decreased, if not felled.

Consequently to provide room for new culms, clump girth also expanded rapidly up to the fifth to sixth year of plantation (Fig. 3.8). On average, clump girth varied between 232 and 456 cm within this period depending on the species. Considering the upper limit of clump-girth expansion, spacing between the planting stock may be kept not less than 5.0 m in the case of these four *Bambusa* species and other clump-forming caespitose bamboos. In these species, both height and dbh of FG culm produced during 5–7 years of clump age were maximum, and after that period, these were more or less static. As the clumps were left undisturbed for a number of years, a crowded and congested condition developed due to the cumulative production of culms, leading to a decreased yield. Similarly the expansion of clump girth is rapid up to the fifth and sixth year after planting and thereafter becomes slow (Banik 1988). So the selective felling of older culms can be started from a clump of all the clump-forming bamboo species after 4–6 years from planting to minimise the competition for space among culms, and this helps in rejuvenating the clumps, and although bigger clumps produce more new culms, the ratio between new and old culms decreases with the extension of the clump (Banik 2000).

By nature a bamboo clump expands in centrifugal manner in which most of the younger culms are produced on the periphery enclosing the older ones.

On the contrary, in *M. baccifera*, increase of culm production and clump expansion continues even after the age of 10 years of planting (Fig. 3.8). In this species, the average number of FG culms produced in the first year was 2.5, and from third year, the number increased rapidly to 12.5. The increase of FG culm production further accelerated in the subsequent years, reaching 35.7 after the tenth year of clump age (Table 3.6). Unlike other caespitose bamboos (e.g. *Bambusa*, *Dendrocalamus* sp., etc.), a clump of *M. baccifera* starts producing very strongly elongated rhizome necks from 3 to 5 years of age. Probably, a clump of *Melocanna* requires 4–5 years of age to accumulate a sufficient amount of photosynthetic reserves for producing elongated rhizome necks. Therefore, the rate and pattern of clump expansion in *M. baccifera* are more or less similar to those of other clump-forming bamboo species up to 4–5 years of clump age; after that period, the expansion of clump girth rapidly increases every year. Owing to the presence of long rhizome neck (1.0–2.0 m), the species produces culms at varying intervals in all directions forming a diffuse and open type of clump which can accommodate the space required for the increased number of FG culms, and competition among them is likely to be minimum (Banik 2000). With such rapid clump expansion ability, many different neighbouring adult clumps of *M. baccifera* overlap and intermingle with each other, forming a pure bamboo vegetation with an extended rhizome network below the ground, and make impossible to demarcate the boundary of each clump. Due to the aggressive spreading nature, *M. baccifera* bamboo constitutes 70–90 % of the total bamboo resources present in the different hill forests of northeast and eastern part of India, Sylhet and Chittagong Hill Tracts of Bangladesh and Arakan Range of Myanmar. It is interesting to note that while *M. baccifera* clump is expanding, it can also move even towards the upper direction of the hill slope by producing new culms. Every year this vast region of the globe has been experiencing very high rainfall (3,000–6,350 mm) with 5–6-month draught period (commonly from November to mid-April). All these regions of the world have sandy to clay loam alluvial soils on the hilly terrain with rough slopes and thus are extremely vulnerable to soil erosion during monsoon. During draught period, the topsoil gets dried and dusted. With the onset of such monsoonal heavy rainfall, the big raindrops hit hardly on the ground, and as a result, the hilly region becomes extremely vulnerable to soil erosion and landslide. The vastly spread, branched and interconnected ramifying underground rhizome networks of *M. baccifera* vegetation serve as a bamboo-reinforced concrete to bind the soil and thus prevent the soil erosion or landslide in the hills of these regions, resulting in minimum siltation in the downstream and controlled flash floods in the valleys and plains (Banik 1997c, 2010).

3.3.6 Leaf Production in Relation to Culm Age

In general there are fewer leaves on 1-year-old culms than on the older ones. Ueda (1960) measured the fresh weight of leaves collected from 1-, 2- and 3-year-old culms of a clump of *M. baccifera* in Assam and observed higher production of leaf biomass (1.10 kg/culm) in 2-year-old culms than those of 1-year-old (0.70 kg) and 3-year-old (0.60 kg) culms. A study was also conducted at Chittagong on leaf biomass production by different age group culms in the clumps of *D. longispathus*, and it was observed that fresh weight of leaf biomass was maximum in the 2-year-old culms, and then it decreased about 30 % and remained somewhat static in the third and fourth years (Banik 2000; Banik and Islam 2005). In the fifth and sixth year, the leaf production was drastically decreased (Table 3.7).

As a 5-year-old culm contains fewer amounts of leaves, therefore, it is likely that it has little contribution in photosynthesis and overall health of the clump. Thus, it appears that both in *Melocanna baccifera* and *D. longispathus* clumps felling of culms may be started after the third to fourth year of age. It is further observed that irrespective of culm age, leaf contains about 50 % moisture to its fresh weight. Different age groups of culms in the clumps of *Bambusa* spp. may have a similar pattern of leaf production. The species with more leaves and larger leaf area absorb more water, and they can grow well even in wet soil. Majority of bamboo species growing naturally in the moist habitat of high rainfall zone in Northeast India and Chittagong Hill Tracts commonly possess comparatively bigger size of leaves.

Generally, growth of new leaves and culms commences after the winter or dry season followed by growth of new rhizomes and subsequently new roots. One basic difference between monopodial and sympodial bamboo is the growth cycle of their leaves: monopodial bamboo replaces shed leaves in a short time during early spring

Table 3.7 Influence of culm age on the average production of leaf in the clumps of *Dendrocalamus longispathus*

| Culm age (year) | Number of leaves per culm | Total fresh weight (kg) | |
		Fresh weight	Oven-dry weight
1	1,406	0.67	0.31
2	6,975	2.09	1.15
3	4,790	1.44	0.75
4	4,877	1.16	0.62
5	1,755	0.48	0.28
6	907	0.29	0.14

(February and March), whereas shedding of old leaves and growth of new leaves in sympodial bamboo occur during a much longer period from mid-winter to mid-summer (December–July with a peak in March/April) (Kleinhenz and Midmore 2001). Ageing particularly affects the transport tissues of bamboo culms, which lose their conductivity for water, nutrients and photosynthates over the years. The photosynthetic capacity of bamboo leaves decreases with age as well, but this has different consequences in monopodial and sympodial bamboos. The reason for this is the difference in life span of leaves in the two groups. Life span of leaves of monopodial bamboos is not more than 2 years, whereas that of sympodial bamboos is up to 6 years (Kleinhenz and Midmore 2001). Age of all leaves on a culm of a particular age is the same (either 1 year or 2 years) in monopodial bamboos, whereas that in sympodial bamboos is variable (on average 2–5 years). Monopodial bamboos shed only their 2-year-old leaves every year, which can cause a dramatic decrease in leaf area, photosynthesis and consequently shoot/culm production. 'Alternation', when the stand is composed of relatively more culms with 2-year-old than 1-year-old leaves, causes marked biennial variation in growth of monopodial bamboos. This does not happen in sympodial bamboos where the photosynthetic capacity of foliage of culms aged 2 years and over is much reduced due to the older age structure of leaves on them. This is in contrast to culms of monopodial bamboos, which 'refresh' their leaves every 2 years and retain culm productivity longer (Kleinhenz and Midmore 2001). Therefore, despite the fact that felling age of culms also depends on their use (i.e. <1 year for edible shoots, 1 year for pulping and papermaking and >2–3 years for timber), culms of monopodial bamboos can be harvested much later than those of sympodial bamboo. The latter do not contribute much to the new growth after about 2–3 years of age. This is indeed reflected in the average felling age of culms, which is 5–6 years for monopodial bamboo and only 4 years for sympodial bamboo.

3.3.7   Senescence of Culms

It has been seen that the longevity of culms in the clump varied from species to species. Culms in some bamboo species are very short-lived and more commonly persist for 5–10 years (McClure 1966). In eastern region of the subcontinent, the life of culms in the clumps of D. longispathus and Gigantochloa andamanica was 5–6, in B. tulda 5–10, in B. polymorpha and M. baccifera 5–11, in B. longispiculata 6–9, in D. giganteus 10–13 and in B. balcooa 11–14 years. Culms started dying at 5 years of age in B. polymorpha, B. tulda, D. longispathus, M. baccifera and G. andamanica, whereas death started at 10 and 11 years of age in D. giganteus and B. balcooa respectively (Banik 1991). It appeared that culm life in the bamboo clumps of somewhat thin-walled species of forest was comparatively shorter than that in the clumps of thick-walled cultivated species. In B. longispiculata death started when culms were 6 years of age. So live culms of different age groups, from 1 to 6 or even 14 years old, can be found in a bamboo clump of different species, if left unharvested (Banik 2000). It was observed in Karnataka (south India) that the individual culms of *B. bambos* started decaying and dying after 7 years with simultaneous death of roots and rhizomes (Lakshmana 1994). So any bamboo left beyond 7 years will not add any incremental growth nor will it produce any new rhizome. The culms of Japanese bamboo become mature in 3–4 years and then gradually grow old and weak to death in 10 years or so (Ueda and Numata 1961).

However, in a clump, all culms of the same age group do not die at a time; rather it continues for several years (the death years). Banik (1991) reported that depending on the species, 4–33 % culms started dying at the beginning of death years, and the remaining culms died gradually within next 4–7 years. That means in *D. longispathus* and *G. andamanica* from the same age group, some culms died at 5 years of age and gradually the remaining all died within the next 4 years (period of death years is 5 years). In *B. tulda*, some culms from the same age group died at 5 years of age, and the remaining all died gradually within the next 5 years (period of death years is 6 years). In *M. baccifera* and *B. polymorpha*, some culms of the same age group died at 5 years of age and the rest died within the next 6 years (period of death years is 7 years). More often culms among the same age group that are at the centre of a clump die earlier than those which emerge on the periphery. Death symptom starts from the upper portion of the culm by drying up of tip, branches and leaf buds. When most of the leaves on the culm shed, the symptoms of ageing rapidly accelerate by complete yellowing of the culms from tip towards the base. Such natural deterioration due to ageing that accompanies senescence ultimately leads to the death of culm in a clump. Thus, it appears that the reduction in photosynthetic surface on the older culms limits the supply of energy to the different metabolic activities necessary for the life process. Moreover, when a culm attains the age of 6–7 years, its fibrous roots and root hairs through which the essential nutrients are absorbed gradually get reduced in number. Therefore, it is likely that the food reserve in the older culms gradually gets reduced, leading to the loss of culm vitality. It is evident that culms in the clump should not be left unharvested for a longer time (Banik 1991). Life period of individual culms of a bamboo species might be an important criterion for fixing the upper limit for the felling cycle (Seth and Mathauda 1959).

3.3.8 Coppicing and Culm Congestion

The formation of congested clumps, in which the culms are packed tightly together and are often much bent and twisted, is frequently seen under certain conditions, especially in village lands and on the outskirts of the forest where clumps suffer injury and in places where the soil is poor, dry or hardened, in degraded forests. Congestion is one of the most serious problems in proper growth of bamboos. It is particularly common in the case of *B. bambos*, *B. balcooa*, *B. polymorpha*, *D. strictus*, *G. andamanica*, etc., growing in dry and degraded forests. In localities where heavy and irregular cuttings take place, the frequency of congested clumps is high. Such congested clumps of *D. hamiltonii* are available in biotically disturbed tea gardens of Khadimnagar Sylhet forest. When the young culms are injured or felled, many coppice shoots develop and clump congestion starts. This constant injury to the periphery of the culm causes the death of the rhizomes, and, in time, a dense mass of dead rhizomes prevents the living rhizomes from spreading outwards. The latter accordingly develop within the clump, where also the new culms are produced year after year, with the result that congestion takes place and the new culms may bend in all directions in their efforts to penetrate the dense mass of older culms. Congested clump shoots are whippy and thinner than the normal culms. In extreme cases, if mature bamboo culms are not harvested in time, they prevent new shoot growth, and as a result, clumps become congested with degenerated culms as observed in the farms and homesteads of absentee land lords. If bamboo is not cut, the culms get congested and become thinner in size and the forest presents an unhealthy appearance. In order to maintain the bamboo forest in a healthy state, it is necessary to cut the bamboos, but under regulation. Thinning/cleaning operations relieve congestion in clumps and dense stands, promote regeneration and also facilitate harvesting. Thin dead and weak culms are to be harvested to make room for the emerging culms.

3.3.9   Culm Age Determination

Culm age is the important parameter that has been considered in the management of a bamboo clump and bamboo forest. The fixation of felling cycle of a bamboo species depends mainly on the maturity of culms. Generally, 3–5-year-old culms are treated as 'mature'. Usually the felling cycle of culms is fixed at the age when the culms become exploitable and the age at which it dies.

According to Liese (1985), the evidence from chemical and technological tests regarding the beginning of 'maturity' is still contradictory. Bamboo, being a member of monocotyledonous group, does not possess any growth rings in the culms. Unlike dicotyledonous plant species, the age of a bamboo plant, thus, cannot be determined by counting the growth rings. The size of a culm is no criterion of its age. Therefore, the *age of a culm* is difficult to determine. The 1- or 2-year-old culms are easily recognisable by colour of culm or that of sheath at the base of a culm and the white waxy powder on the internodes (e.g. *B. tulda*). In South America, mature culms of *Guadua angustifolia* are recognised by the formation of *white spots* on the culm and lichens (fungi) at the nodes. Experienced bamboo growers can even recognise mature bamboo by the sound in the stem when struck with a stone or the back of a machete. If bamboo stems are covered with fungi and mosses in their entirety and nodes appear whitish grey or even dry, it is a sign that the culms are overmature.

A farmer usually distinguishes between 'mature' and 'immature' culm by flipping or tapping with fingers. From the sound, he can identify the group. Sometimes the colour of the culm is also considered. This is a very skilful job and depends on years of experience. However, Banik (1993b) has identified some morphological field characteristics for determining the age of culms in bamboo clumps. The presence or absence of culm sheath and nodal root rings, bud break and branching pattern and colour were found to be diagnostic morphological characteristics (Table 3.8).

Table 3.8 Morphological characteristics for culm age determination of some bamboo species

Bambusa balcooa

Age up to	Morphological description
First year	Culm sheath: May be present at 2–3 basal nodes of the culm
	Bud break and branches: The large branch bud on the culm node breaks and produces a thick stout branch except on the basal 3–4 nodes. Comparatively small and thin 1–2 branches are also produced from both sides of the main stout branch. Branch bases and some basal nodes of the culms are generally covered with straw-coloured papery sheaths
	Culm: Brown to whitish pubescent ring is present on the basal 6–9 nodes, and usually basal 4–5 nodes also possess a ring of adventitious roots. Nodes are prominent and ridged. Dark glossy green. Basal 6–8 internodes are lightly covered with minute brownish hairs arranged in many vertical lines closely parallel to each other
Second year	Culm sheath: Usually not present on the culms, may be present on the basal 1–2 nodes
	Bud break and branches: Except 4–5 basal nodes, almost all buds of the culm nodes develop thick stout branches. The auxiliary branches up to basal 5–6 nodes transform into curved thornlike structure. Base of the branches may be covered with thin persistent sheath
	Culm: Brown to whitish ring along with adventitious roots may be present on the basal 3–4 nodes. Not so deep green. Basal internodes are slightly covered with minute brownish hairs
Third year	Culm sheath: Absent
	Bud break and branches: Usually buds on the basal 3–5 culm nodes are dead and rotten. Death is more on the congested clumps. Thin wiry auxiliary branches up to 9–12 basal nodes shed their leaves and transform into curved thornlike structure
	Culm: Dead adventitious root rings may be present on the basal 2–3 nodes of the culm. Deep bottle green (hookers green) and internodal surface is smooth
Fourth year	Culm sheath: Absent
	Bud break and branches: Branches have less number of leaves. Most of the auxiliary and secondary branches transform into curve thornlike structure. Stout thick branches on the basal nodes (up to 6) of the culm are usually dead, are shed and keep dead scar on the culm nodes
	Culm: Smooth surface having deep bottle green colour. Black rotten or dried adventitious root ring may be present on the basal 1–2 nodes

Bambusa longispiculata

Age up to	Morphological description
First year	Culm sheath: Generally absent. However, in some cases, present on the nodes of upper portion of the culm and at the 1–2 basal nodes
	Bud break and branches: Most of the branch buds on the upper and basal nodes break and produce branches. The buds on the lower mid-culm zone (5–12 nodes) remain dormant and do not produce branches
	Culm: Deep glossy green. Vertical yellow striations are prominent on the 1–5 basal internodes. White bloom is present in the internode

(continued)

Table 3.8 (continued)

Second year	Culm sheath: Absent
	Bud break and branches: Except for a few buds, 2–4, at the mid-culm zone, all branch buds break and produce branches
	Culm: dark green. Vertical yellow striations are also prominent. Amount of white bloom conspicuously reduced and comes off on fingers
Third year	Culm sheath: Absent
	Bud break and branches: Almost all the branch buds in the culm nodes break and produce branches
	Culm: Dark green, but yellow colour striations are comparatively faded. White bloom is either absent or scanty and turns greyish white
Fourth year	Culm sheath: Absent
	Bud break and branches: Same as previous years, but some main and thin branches at the base and upper portion of culm die and drop keeping black scars on nodes
	Culm: Uniformly dark green with no bloom on the internodes. Yellow colour striations are mostly faded

Bambusa tulda

Age up to	Morphological description
First year	Culm sheath: Usually not present on the culm, may be present on the basal 1–2 nodes and tightly fitted
	Bud break and branches: The branch bud on the culm node breaks acropetally producing stout branch, while 3–8 buds in lower mid-culm zone remain dormant. Sometimes 3–5 branch buds may also remain dormant on the upper portion of the culm. Leafy branches are mostly confined to the top portion of the culm; smaller thin branches are present on the lower nodes. Branch bases and internodes cover with the straw colour sheaths
	Culm: Dark green colour with slight whitish bloom on the internode and comes off easily with finger. Basal 1–2 nodes may have a ring of small adventitious roots. A brown to dark colour woolly band present just above the 1–3 basal nodes
Second year	Culm sheath: Absent
	Bud break and branches: Almost all the branch buds including dormant one of the first year on the lower mid-zone and upper portion of the culms become active and produce branches
	Culm: Whitish bloom may slightly be present, comes off with rubbing
Third year	Culm sheath: Absent
	Bud break and branches: Generally branchy sprouts in a branch complement on the basal 1–4 nodes of culm become dead and shed off, keeping the black scars. As a result, the number of thin branches at the basal portion of the culm gets reduced
	Culm: Green throughout, except yellowish stains present at the lower part of the internodes. Basal 1–3 internodes may not have this yellowish stain
Fourth year	Culm sheath: Absent
	Bud break and branches: Branches on the basal 3–7 nodes and in some cases on the upper portion the culms (2–4 nodes) become dead and shed off, keeping the black scars. More numbers of thin branches are dead
	Culm: Dark green. Yellowish stain at the lower part of the internode is comparatively prominent

(continued)

Table 3.8 (continued)

Bambusa vulgaris	
Age up to	Morphological description
First year	Culm sheath: Except basal 2–3 nodes, no sheaths are present
	Bud break and branches: One-third of culm top has leafy branches. Buds break producing central stout branches with small auxiliary branches throughout culm except 3–4 basal nodes
	Culm: Bright glossy green, basal 3–4 nodes have adventitious white root rings
Second year	Culm sheath: Light dark colour sheath may be present at 1–2 basal nodes of culm
	Bud break and branches: More than half of the upper portion of the culm has thick stout leafy branches. Small auxiliary branches disappear
	Culm: Dull grassy green. Adventitious root rings on the basal nodes are drying out and turning black
Third year	Culm sheath: Absent
	Bud break and branches: Branches are more thick and stout with light green colour; leaves are few. Bud breaks start at basal 3–4 nodes but forms small thin wiry leafy branches
	Culm: Slightly yellowish green, no adventitious root rings on the nodes
Fourth year	Culm sheath: Absent
	Bud break and branches: Thick stout branches are only on the upper one-third of the culm, turning yellowish with few leaves. Buds and branches on the basal nodes mostly dead leaving black scar
	Culm: Turning yellowish, smooth
Melocanna baccifera	
First year	Culm sheath: All nodes of the culm covered with culm sheath except thin tip
	Bud break and branches: No bud break on the culm node, 2–4 tip buds on the culm produce drooping large leaves directly on the nodes, no branches
	Culm: Green branchless straight culm with drooping tips having 2–4 leaves
Second year	Culm sheath: Sheath persist on the lower two-third culm, blade of the sheath is loosely fitted
	Bud break and branches: Buds on the one-third upper portion of the culm nodes break and produce many thin branches in assembly
	Culm: Green
Third year	Culm sheath: Persist loosely only on basal one-third portion of culm (3–5 internodes), mostly black colour, blade of the culm sheath shed off
	Bud break and branches: Further buds break up to two-thirds of the culm, and many thin branches are produced in assembly with comparatively less number of leaves
	Culm: Dull dark green
Fourth year	Culm sheath: Absent
	Bud break and branches: Buds mostly dead on the basal culm nodes. Some branches in the branch assembly die mostly on the upper portion of the culm. Leafy branches are less, dead leaves scars present on the branches
	Culm: Light green to yellowish straw colour

Source: Banik (1993b, 2000)

Furthermore, the age of a bamboo can be determined by *counting the leaf scar* (base node) on leaf sheath present in the twig of a culm of a bamboo species, such as *B. balcooa*, *B. vulgaris*, *D. longispathus*, etc., especially in dry winter (Fig. 3.9; Banik 2000). Bamboo leaves fall in a year or a year and a half, and soon new leaves are developed from the near node of leaf-fall portion, keeping marks of leaf scar on the twig. So within 12–15 months of culm age, mark of one leaf scar is formed, and again in the next 24–30 months of age, another mark totalling two marks of leaf scar is formed. Thus, in the third year of age, three marks of leaf scar are found on the

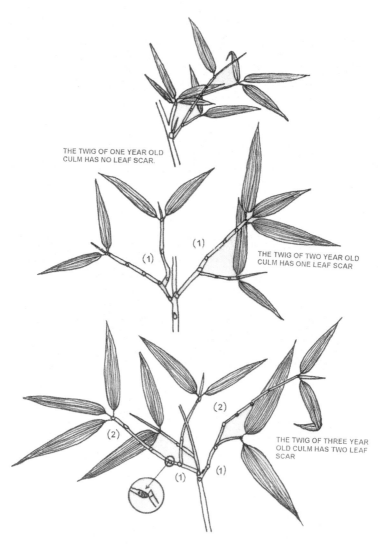

THE TWIG OF ONE YEAR OLD
CULM HAS NO LEAF SCAR.

THE TWIG OF TWO YEAR OLD
CULM HAS ONE LEAF SCAR

THE TWIG OF THREE YEAR
OLD CULM HAS TWO LEAF
SCAR

Fig. 3.9 Diagrammatic presentation of the method of determination of culm age in bamboo by counting the number of leaf scar on twigs

twig. Earlier the same method was also used for counting the culm age in *Phyllostachys edulis* (Ueda 1960).

3.4 Rhizome Growth

3.4.1 Rhizome Initiation, Development and Other Growth in Seedling

The nature of a given bamboo plant becomes clear only after the form and growth habit of its rhizome systems are fully understood. The rhizome system constitutes the structural foundation of the plant. In caespitose, types of bamboos (e.g. *Bambusa tulda*) after seed germination plumule emerge in the form of a pointed conical bud with sheathing scale-like leaves, which rapidly develop into a thin, wiry stem bearing single foliage leaves arising alternately at the nodes. Meanwhile fibrous roots develop from the base of the young shoot, and 1–2 shoots also develop within 4–6 weeks from the base of the main young shoot. The seedling of *B. tulda* attains 4–5 shoots (culms) at 9 months of age, and a miniature clump of the young plant is developed (Banik 1987). This is affected by the production on the rhizome of successive pointed buds, from which short rhizomes developed below the ground move geotropically that curve upwards and form shoots (culms). The buds and rhizomes, and the shoots arising from them, become successively larger and larger as the time passes (Fig. 3.10a).

In *M. baccifera*, a non-caespitose type of bamboo, soon after 15–21 days of seed germination, fibrous roots develop from the base of the young shoot of a seedling. Rhizome formation started after 30–40 days of seedling age (Banik 1994). The rhizome neck was more elongated and development was strongly geotropic. It carried a bud at the tip (Fig. 3.10b-i). Both neck and the rhizome buds were covered with comparatively thick tough sheaths fitted imbricately one above the other. After about 5–6 months of seedling age, the rhizome proper took a horizontal course for a very short distance deep inside the ground and then moved with negative geotropism turning upwards piercing the soil and producing a new culm on the ground (Fig. 3.10b-ii). Within 11–12 months, a tufted young plant is formed by the production of successive pointed shoot buds on the aggressively developing young rhizome system.

Both green and oven-dry weight of different organs of seedlings of *M. baccifera* were found higher than those of *B. tulda* at the initial stage of growth and development (Table 3.9). Long rhizome necks develop from the seedling of *M. baccifera*, and as a result, allocation of biomass to rhizome was higher than that of root. On the other hand, in case of *B. tulda*, the rhizome necks were very short, and most of the roots come out from the rhizome itself. Therefore, the rhizome biomass was low than that of root in *B. tulda*.

The biomass growth of culm with leaves, rhizome and roots gradually increased as the *Melocanna* seedling became older (Table 3.9).

Fig. 3.10 Rhizome development in bamboo seedling. (**a**) *Bambusa tulda*: a seedling attained five shoots (culms) stage at 9 months of age, and the buds and rhizomes, and the shoots arising from them, become successively larger and larger as the time passes. (**b**-i) *Melocanna baccifera*: the rhizome neck is strongly geotropic, more elongated neck and carried a bud at the tip; both are covered with comparatively thick tough sheaths fitted imbricately one above the other; (**b**-ii) at 5–6 months of seedling age, the rhizome proper took a horizontal course for a very short distance and then moved with negative geotropism turning upwards and producing a new culm on the ground

In monopodial bamboo having leptomorph rhizome (*Phyllostachys* sp.), the 1-year-old seedlings tiller and grow thickly, and the 2-year-old ones have side buds, some of which sprout out or grow down into soil and develop horizontally

Table 3.9 A comparative biomass measurement (average value) *Melocanna baccifera* and *Bambusa tulda* seedling at different age and season in the nursery

Species	Age in months (observed month)	Green weight (g)			Oven-dry weight (g)		
		Culm and leaves	Rhizome	Root	Culm and leaves	Rhizome	Root
Melocanna baccifera	3 (Nov.)	60.14 ± 9.98	11.11 ± 5.32	12.63 ± 5.49	16.75 ± 3.36	2.81 ± 1.09	5.09 ± 1.80
B. tulda	3 (Sep.)	10.71 ± 2.11	1.27 ± 0.43	6.78 ± 1.76	4.01 ± 0.71	0.32 ± 0.09	1.95 ± 0.44
M. baccifera	6 (Feb.)	43.59 ± 10.06	35.40 ± 10.39	13.28 ± 6.96	19.87 ± 5.22	12.38 ± 5.24	7.90 ± 5.14
B. tulda	6 (Dec.)	44.20 ± 6.50	6.83 ± 1.42	49.40 ± 30.10	18.52 ± 3.70	1.72 ± 0.21	14.90 ± 3.42
M. baccifera	10 (May)	86.76 ± 31.40	38.05 ± 8.17	14.00 ± 6.80	43.59 ± 16.58	16.42 ± 10.01	5.05 ± 2.16
B. tulda	10 (Mar.)	87.05 ± 10.50	18.70 ± 9.65	49.40 ± 25.01	42.08 ± 6.28	10.63 ± 5.31	27.47 ± 17.39

Source: Banik (2010)

Table 3.10 Comparative average leaf size (length, mid-width, cm) in the seedlings and 15-year-old adult plants of *Melocanna baccifera* and *Bambusa tulda*

	M. baccifera				B. tulda			
	Seedling leaf		Adult plant leaf		Seedling leaf		Adult plant leaf	
Age	Length	Mid-width	Length	Mid-width	Length	Mid-width	Length	Mid-width
3 months	32.31	8.76	–	–	11.18	2.58	–	–
6 months	28.68	6.57	–	–	17.91	3.24	–	–
10 months	21.30	3.94	–	–	19.5	3.61	–	–
About 15 years old			23.74	3.26			21.11	3.05

Source: Banik (2010)

there as short rhizomes (Naixun and Wenyan 1994), which is distinctly late than that of bamboos with pachymorph rhizome (as mentioned above).

Leaves on the *M. baccifera* seedlings are bigger, one and a half times more in size than those of adult clumps (Fig. 3.5, Table 3.10). The seedling leaves of *M. baccifera* are also bigger than those of *B. tulda*, probably because of trapping more sunlight to produce higher amount of food material required for rapid production of rhizome system with elongated necks. The leaves chronologically produced in the seedling become smaller with the age. After about 10–12 months, the leaves produced by the seedlings have attained more or less similar size to that of adult clumps. However, at about 15 years of age, the leaf size becomes elongated by 2.0 cm and slightly less in mid-width than that of 12-month-old plant.

3.4.2 Rhizome Growth Periodicity in Adult Clump of M. baccifera

Most of the tropical clump-forming bamboo species exhibit seasonal movement and growth of rhizomes only in the spring. The development of daughter rhizome starts from the base of a culm only after it completes elongation and becomes at least 12–15 months old. In contrast to that, a clump of *M. baccifera* while producing a new culm simultaneously also start developing the rhizome neck(s) below the ground from the rhizome of the emerging culms (Banik 1999, 2010). Study shows that the movement and elongation of rhizome neck of *M. baccifera* go on, either actively or slowly throughout the year, even in winter and dry season (Banik 1999). However, the rhizome neck elongation slows only when mother culm exhibits growth and produces branches and leaves. In contrast to *M. baccifera*, most of the tropical clump-forming species of *Bambusa* and *Dendrocalamus*, with compact pachymorph rhizome systems (short 3–10 cm rhizome neck; Plate 3.1b, c), exhibit only seasonal, primarily during spring–summer, movement and growth of rhizomes in a year. In these genera (clump-forming bamboo), development of daughter rhizome starts from the base of culm only when it is at least of 11–15 months of age. The rhizome apex grows faster, and the yearly elongation is more in loose and fertile soil with moderate moisture content and less mechanical obstacles.

In monopodial bamboo species, the mature lateral buds in the alternate nodes of mother rhizomes may sprout to new rhizomes. About ten internodes in the forepart of a growing rhizome are regarded as elongation zone where each internode elongates orderly and successively from the rear to the fore tip to achieve rhizome growth. The internodes in the rear part far from the tip of elongation zone complete its growth in length and diameter earlier than those in the fore part near the apex (Jianghua 1994).

3.4.3 Identification of Old and New Rhizome

A 1-year-old rhizome is partly covered with sheaths and its roots have few fibrous roots at the nodes. In determining the age of over 2-year-old rhizomes, it is possible to distinguish them from all other rhizomes excavated, belonging to one system of a bamboo clump. But the correct age of a single part of a rhizome can neither be distinguished nor determined, although one may be able to decide whether it is young or old. In general the young rhizomes are yellowish and have vigorous buds; furthermore, they bear roots with many fibrous roots at each node. The rhizome of over about 5 years becomes brownish, and its nodes have but few vigorous buds and fibrous roots.

3.4.4 Growth of Roots in Rhizome and Culms

Many new roots are generally growing in the underground basal part of an emerging sprout. These roots continue growing gradually and complete their growth within a year. The elongation of roots is usually 40–100 cm and thereafter they neither grow nor thicken in diameter. When a culm becomes older than 6–7 years, its fibrous roots or root hairs, which are the organs of nutrient absorption, markedly reduce in number, and accordingly the absorption of minerals and productive power of the culms also deteriorates.

3.5 Variation in Below- and Aboveground Clump Growth Ratio in Relation to Clump Age

The regenerating M. baccifera seedlings in the first year produce thick, tender, soft (not woody) culms which rapidly elongate up to 150–180 cm within 3 months of age. Therefore, the crown structure of M. baccifera is predominantly herbaceous at the first 12–18 months of naturally regenerated vegetation after gregarious flowering. During this phase of development, the aboveground biomass is more

Table 3.11 Aboveground (culm, leaves, branches) and belowground (rhizomes, roots) oven-dry biomass and ratio of *Melocanna baccifera* bamboo at 3 months to 25 years of clump age

Age of plant (clump) (months/years)	Biomass (kg)			AG/BG ratio of biomass
	Aboveground (AG)	Belowground (BG)	Total	
3 months (Nov.) oven dry at 102 °C	4 culms = 0.170 kg	0.08 kg		2.13
0 months (May) oven dry at 102 °C	6 culms = 0.436 kg	0.215 kg		2.03
24 months (2 years) oven dry at 102 °C	6 old + 35 new culm = 41 (0.44 + 7.38 kg = 7.82)	3.32 kg		2.35
3 years, a whole of clump (air-dry value)	41 old + 79 new culms = 120 (7.82 + 27.26 kg = 35.47)	14.73 kg		2.41
4 years, a whole of clump (air-dry value)	120 old + 121new culms = 289 (35.47 + 83.25 kg = 118.72 kg)	48.13 kg		2.46
5 years, a whole of clump (air-dry value)	289 old + 187 new culms = 466 (118.72 + 128.67 kg = 247.39 kg) (71.52 %)	98.53 kg (28.48 %)	345.92	2.51
17 years (approx.), from 5x5m plot at CHT forest flowered in 1957–1961. Data collected in 1978 (air-dry value)	57 old + 34 new culms = 91 (150.15 kg) (73.28 %)	54.76 kg (26.72 %)	204.91	2.74
25 years (approx.), from 5 × 5 m plot at CHT forest flowered in 1957–1961. Data collected in 1986 (air-dry value)	63 old + 24 new culms = 87 (137.46 kg) (67.48 %)	66.25 kg (32.52 %)	203.71	2.07

Note: CHT = Chittagong Hill Tract
Source: Banik (2010)

than double to that of belowground part of the clump (Table 3.9). As the age increased, the canopy starts closing due to the production of more number of culms. More amount of underground biomass starts producing after three years of clump age due to the formation of comparatively elongated rhizome necks. The aboveground biomass was found to be maximum (73.28 %) in 17 years to any time below 25 years of clump age and showed highest value of aboveground/belowground ratio (2.74) of biomass (Table 3.11).

Thus, it seems the canopy architecture of *M. baccifera* showed maximum plasticity during this period of clump age (17–24 years). Another fairly common clump-forming bamboo species, *Schizostachyum dullooa* (syn. *Neohouzeua dullooa*), growing in the forest of Northeast India and CHT has somewhat similar (height and diameter)-looking culms to that of *M. baccifera*, which also showed maximum total standing biomass per clump in a 15-year-old fallow (Rao and Ramakrishnan 1988). At the end of 25 years, *M. baccifera* clumps became thickly populated with more number of culms and due to the reduction in space and light canopy assume a cylindrical form with culms having shorter branches mostly in the upper part. As a result, the allocation of biomass started building up (32.52 %) at belowground portion and also due to the production of optimally elongated solid rhizome necks. Therefore, at 25 years of clump age, the aboveground/belowground biomass ratio started declining to 2.07 (Banik 2010).

References

Arber A (1934) The gramineae, a study of Cereal, bamboo and grass. Cambridge University Press, Cambridge, pp 1–480

Banik RL (1980) Propagation of bamboos by clonal methods and by seeds. In: Lessard G, Chouinard A (eds) Bamboo research in Asia. Proceedings of a bamboo workshop, Singapore, 28–30 May 1980, IDRC Ottawa, Canada. IUFRO, Vienna, pp 139–150

Banik RL (1983) Emerging culm mortality at early developing stage in bamboos. Bano Biggyan Patrika 12(1–2):47–52

Banik RL (1987) Techniques of bamboo propagation with special reference to prerooted and prerhizomed branch cuttings and tissue culture. In: Rao AN, Dhanarajan G, Sastry CB (eds) Recent research on bamboos. Proceedings of the international bamboo workshop, 6–14 Oct 1985, Hangzhou, China. The Chinese Acad of Forest; IDRC, Canada, pp 160–169

Banik RL (1988) Investigation on the culm production and clump expansion behaviour of five bamboo species of Bangladesh. Indian For 114(9):576–583

Banik RL (1991) Biology and propagation of bamboos of Bangladesh. Ph.D. thesis. University of Dhaka, pp 1–321

Banik RL (1993a) Periodicity of culm emergence in different bamboo species of Bangladesh. Ann For 1(1):13–17

Banik RL (1993b) Morphological characters for culm age determination of different bamboo species of Bangladesh. Bang J For Sci 22(1–2):18–22

Banik RL (1994) Studies on seed germination, seedling growth and nursery management of *Melocanna baccifera* (Roxb.) Kurz. In: Proceedings of the 4th international bamboo workshop on bamboo in Asia and the Pacific, 27–30 Nov 1991, Chiangmai, Thailand. FORSPA Publication No. 6, IDRC, FAO-UNDP, pp 113–119

Banik RL (1997a) The edibility of shoots of Bangladesh bamboos and their continuous harvesting effect on productivity. Bang J For Sci 26(1):1–10

Banik RL (1997b) Growth response of bamboo seedlings under different light conditions at nursery stage. Bang J For Sci 26(2):13–18

Banik RL (1997c) *Melocanna baccifera* (Roxb.) Kurz – a priority bamboo resource for denuded hills of high rainfall zones in South Asia. In: Karki M, Rao AN, Rao VR et al (eds) The role of bamboo, rattan and medicinal plants in mountain development. Proceedings of a workshop, Institute of Forestry, Pokhara Nepal, 13–17 May 1996. INBAR technical report No. 15. INBAR/IPGRI/ICIMOD/IDRC, pp 79–86

Banik RL (1999) Annual growth periodicity of culm and rhizome in adult clumps of *Melocanna baccifera* (*Bambusoideae*: *Gramineae*). Bang J For Sci 28(1):7–12

Banik RL (2000) Silviculture and field-guide to priority bamboos of Bangladesh and South Asia. BFRI, Chittagong, pp 1–187

Banik RL (2010a) Biology and silviculture of muli (*Melocanna baccifera*) bamboo. NMBA (National Mission on Bamboo Applications), TIFAC, Department of Science and Technology, New Delhi, pp 1–237

Banik RL, Islam SAMN (2005) Leaf dynamics and above ground biomass growth in *Dendrocalamus longispathus* Kurz. J Bamb Rattan 4(2):143–150

Chatterjee RN, Raizada MB (1963) Culm-sheaths as aid to identification of bamboos. Indian For 89(11):744–756

Chaturvedi AN (1986) Bamboos for farming. UP Forest Bulletin No. 52, Lucknow, pp 1–36

Gamble JS (1896) The Bambuseae of British India. Annals of the Royal Botanic Garden, Calcutta, vol 7. Printed at the Bengal Secretariat Press, Calcutta, London, pp 1–133

Holttum RE (1958) The bamboos of the Malay Peninsula. The Garden Bulletin 16, Singapore, pp 1–135

Itoh T, Shimaji K (1981) Lignification of bamboo culm (*Phyllostachys pubescens*) during its growth and maturation. In: Higuchi T (ed) Bamboo production and utilization. Proceedings of the congress group 5.3A, 17th IUFRO world congress, Kyoto, Japan, 6–17 Sept, pp 104–110

Jianghua X (1994) Biology and ecology of bamboo. Part I. Bamboo growth and its environment. In: Fu M, Jianghua X (eds) Cultivation and utilization on bamboos. The Research Institute of Subtropical Forestry, The Chinese Academy of Forestry, pp 9–31

Kadambi K (1949) On the ecology and silviculture of *Dendrocalamus strictus* in the bamboo forests of Bhadravati Division, Mysore state, and comparative notes on the species *Bambusa arundinacea*, *Ochlandra travancorica*, *Oxytenanthera monostigmata* and *O. stocksii*. Indian For 75:289–299, 334–349, 398–426

Kleinhenz V, Midmore DJ (2001) Aspects of bamboo agronomy. In: Sparks DL (ed) Advances in agronomy, vol 74. Academic, New York, pp 99–153

Kondas S (1981) Bamboo biology, culm potential and problem of cultivation. In: Higuchi T (ed) Bamboo production and utilization. Proceedings of the 17th IUFRO world congress, Kyoto, Japan, pp 184–190

Lakshmana AC (1994) Culm production of *Bambusa arundinacea* in natural forests of Karnataka, India. In: Proceedings of the 4th international bamboo workshop on bamboo in Asia and the Pacific, 27–30 Nov 1991. FORSPA Publication No. 6. IDRC FAO-UNDP, Chiangmai, Thailand, pp 100–103

Liese W (1985) Bamboos – biology, silvics, properties, utilization. Eschborn, Germany, pp p1–132

Liese W (1998) The anatomy of bamboo culms. INBAR Tech Report-18. New Delhi, pp 1–208

McClure FA (1966) The bamboos: a fresh perspective. Harvard University Press, Cambridge, MA, pp 1–347

Naixun M, Wenyan Z (1994) Bamboo silviculture. In: Fu M, Jianghua X (eds) Cultivation and utilization on bamboos. The Research Institute of Subtropical Forestry, The Chinese Academy of Forestry, pp 59–80

Osmaston BB (1918) Rate of growth of bamboos. Indian For 44(2):52–57

Rao KS, Ramakrishnan PS (1988) Architectural plasticity of two bamboo species (*Nehouzeua dulloa* A Camus and *Dendrocalamus hamiltonii* Nees and Arn.) in successional environments in north-east India. Proc Indian Acad Sci (Plant Sci) 98(2):121–133

Seth SK, Mathauda GS (1959) Bamboo experiments. Indian For 85(2):699–709

Sharma YML (1982) Some aspects of bamboos in Asia and the Pacific. FAO regular programme No. Rapa 57, Bangkok, pp 1–56

Shigematsu Y (1960) Studies on the growth types of Japanese bamboo. Bull Fac Agric Univ Miyazaki 6(1):14–105

Subsansenee W (1994) Thailand. In: Patrick BD, Ward U, Kashio M (eds) Non-wood forest products in Asia. Regional Office for Asia and The Pacific (RAPA Publication 1994/28)/FAO. Bangkok, pp 127–150

Uchimura E (1980) Bamboo cultivation. In: Lessard G, Chouinard A (eds) Bamboo research in Asia. Proceedings of a bamboo workshop, Singapore, 28–30 May 1980. IDRC, Ottawa, Canada; IUFRO, Vienna, pp 151–160

Ueda K (1960) Study on the physiology of bamboo with reference to practical application. Prime Minister's office, Resources Bureau, Science and Techniques Agency. Tokyo, Japan, pp 1–167

Ueda K, Numata N (1961) Silvicultural and ecological studies of a natural bamboo forest in Japan. Bull Kyoto Univ For 33:27–54

Watanabe M (1986) A proposal on the life form of bamboos and the ecological typification of bamboo forests. In: Higuchi T (ed) Bamboo production and utilization. Proceedings of the 18th IUFRO world congress, Ljubljana, Yugoslavia, pp 94–98

White DG, Childers NF (1945) Bamboo for controlling soil erosion. J Am Soc Agron 37:839–847

Xianhui L, Yuemei L (1983) Change in nucleic acid content and nuclease activity in degraded bamboo shoots. Bamb Res 2(1):54–56

Chapter 4
Resources, Yield, and Volume of Bamboos

Yannick Kuehl

Abstract This chapter assesses global bamboo resources and their characteristics. Based on available data, there are 31.5 M ha of bamboo forests in the world, representing around 1 % of total global forest area. While many countries report bamboo resource data, the need for widened and uniform global reporting becomes apparent. In addition, this chapter describes the volume and yield of selected bamboo species. The description of the size and volume of bamboo is also linked to the harvestable yield of several bamboo species. Also this data show large variation and emphasize the need for uniform/standardized reporting and measurements. This chapter also looks at the special role of the root and rhizome system—especially with regard to biomass stored belowground. Finally, this chapter surveys the potential role of carbon sequestration with bamboo in order to mitigate climate change. In order to utilize bamboo's characteristics of fast growth and high renewability, it is prerequisite to manage and sustainably harvest bamboo stands.

Keywords Global bamboo resources • Volume growth • Harvestable yield • Root and rhizome system • Biomass • Carbon sequestration

4.1 Introduction

Bamboos are fast-growing woody grasses that grow in the tropics and subtropics in mixed forests or as pure stands. They can be cultivated in plantations, on homesteads, and on farms—within forestry of agroforestry systems. Bamboos are grown for their long, usually hollow, culms that can be used as whole or sectioned poles and that yield softwood and fiber for processing; shoots of several species are also edible (Mohanan 2002). Millions of the world's poor people live with and rely on bamboo for their livelihoods, and it can be a significant pathway out of poverty (Belcher 1995; Hogarth and Belcher 2013). However, "remarkably little is known about this entire subfamily of tall graminaceous plants, despite its everyday utilization, by about 2.5 billion people" (Scurlock et al. 2000). Besides limited research and data, bamboo has been characterized as an "institutional orphan" (Buckingham

Y. Kuehl (✉)
TRAFFIC (former INBAR, Beijing), Hong Kong, People's Republic of China
e-mail: yannick.kuehl@traffic.org

© Springer International Publishing Switzerland 2015   91
W. Liese, M. Köhl (eds.), *Bamboo*, Tropical Forestry 10,
DOI 10.1007/978-3-319-14133-6_4

et al. 2011)—indicating needs for increased research activities about bamboo and its utilization.

Bamboo is a main forest product—it represents an economically important non-timber forest product (NTFP). Bamboo is popularly known as "poor man's timber" denoting its popularity among poor people as a good substitute for expensive wood from trees (Lobovikov et al. 2012). It has been argued that promoting the use of bamboo as a renewable and sustainable substitute of wood from trees (in the form of fuel wood, charcoal, wood tiles, walls, beams and columns, furniture, etc.) may reduce pressure on tree-dominated forests thereby helping to avoid further defor-estation and forest degradation (Lobovikov et al. 2009). In order to assess bamboo's potential as a substitute or for ecosystem services (such as climate change mitiga-tion or adaptation) or productive services, it is essential to estimate the current global resource base, as well as growth characteristics of important species.

Therefore, this chapter first looks at existing bamboo forest resources in the world. Further, it assesses the yield, height, and volume of selected bamboo species. Finally, based on bamboo's biomass generation characteristics, this chapter dis-cusses the potential role of bamboo biomass in climate change mitigation.

4.2 Bamboo Resources in the World

Figure 4.1 displays that bamboos are distributed about the tropical and subtropical parts of the world. Some species can even in temperate zones of Europe and North America (FAO 2007)—even though bamboos are not indigenous in these areas. Indigenous bamboo species exists on all regions which grow bamboo: the Asia-Pacific region, the Americas region, and the African region (I, II, and III, respec-tively, in Fig. 4.1).

Globally, bamboos make up around 1 % of the total global forest area (FAO 2007). A recent global survey estimated the total global bamboo area to be 31.5 mio ha—see Table 4.1. Table 4.1 presents data on bamboo resources in Africa.

Table 4.1 shows little variation in African bamboo resources from 1990 to 2010. Nevertheless, a slight decrease of African bamboo resources over the last decades can be observed. According to the reviewed data, Nigeria has the largest bamboo resources in Africa. It should be noted that not all countries which grow bamboo in Africa reported respective data. Table 4.2 presents data on the bamboo resources of Asian countries.

Many Asian countries reported dynamic changes of bamboo resources during the last decades. Overall, a trend of increasing bamboo resources in Asia from 1990 to 2010 can be observed. During that period, bamboo resources in China increased the most: from 3.8 million ha in 1990 to 5.7 million in 2010. India's bamboo resources increased by 0.3 million ha in the same time. At the same time, Sri Lanka's bamboo resources decreased by 0.5 million ha between 1990 and 2010. It should be noted that not all Asian countries which grow bamboo reported respective

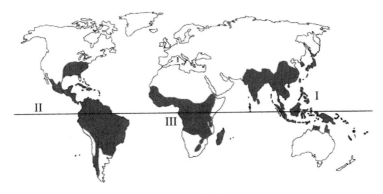

Fig. 4.1 Global distribution of bamboo (INBAR 2010)

Table 4.1 Bamboo resources in Africa (FAO 2010)

Country	Area of bamboo (1,000 ha)			
	1990	2000	2005	2010
Ethiopia[a]	1,000	1,000	1,000	1,000
Kenya	150	150	150	150
Mauritius	n.s.	n.s.	n.s.	n.s.
Nigeria[b]	1,590	1,590	1,590	1,590
Senegal	723	691	675	661
Sudan[a]	30	30	30	31
Uganda[b]	67	67	67	67
United Republic of Tanzania[b]	128	128	128	128
Total Africa	3,688	3,656	3,640	3,627

n.s. not significant

[a]Data for 1990, 2000, and 2005 from FAO (2007). Data from 2005 also used for 2010. For countries in South America, the figures for 2000 were also used for 1990

[b]Gaps in data series filled by FAO estimates

data. Table 4.3 presents data on bamboo resources of countries in North, Central, and South America, as well as in Oceania.

Little variation of reported bamboo resource areas can be observed for the countries in Table 4.3. Only a few countries in North, Central, and South America—as well as Oceania—reported bamboo resource data. According to the available data, Brazil represents the country with the largest bamboo resources in the world—with 9.3 million ha. To summarize global data, Table 4.4 displays bamboo resources of the regions of the world.

As Table 4.4 shows, globally an increase of bamboo resources can be observed from 1990 to 2010. The strongest increase of bamboo resources can be observed in Asia. It should be noted that "the area of bamboo is difficult to assess, as these species often occur as patches within forests or as clusters outside them. Nevertheless, preliminary findings based on information from 33 of the main bamboo-rich countries indicate that the total area is about 31.5 million hectares" (FAO 2010).

Table 4.2 Bamboo resources in Asia (FAO 2010)

Country/region	Area of bamboo (1,000 ha)			
	1990	2000	2005	2010
Bangladesh	90	86	83	186
Cambodia	31	31	36	37
China	3,856	4,869	5,426	5,712
India	5,116	5,232	5,418	5,476
Indonesia[a]	1	1	1	1
Japan	149	153	155	156
Lao People's Republic[b]	1,612	1,612	1,612	1,612
Malaysia[b]	422	592	677	677
Myanmar	963	895	859	859
Pakistan[b]	9	14	20	20
Philippines	127	156	172	188
Republic of Korea	8	6	7	8
Sri Lanka	1,221	989	742	742
Thailand[b]	261	261	261	261
Vietnam	1,547	1,415	1,475	1,425
Total Asia	15,412	16,311	16,943	17,360

n.s. not significant
[a]Data for 1990, 2000, and 2005 from FAO (2007). Data from 2005 also used for 2010. For countries in South America, the figures for 2000 were also used for 1990
[b]Gaps in data series filled by FAO estimates

Table 4.3 Bamboo resources in North, Central, and South America and Oceania (FAO 2010)

Country/region	Area of bamboo (1,000 ha)			
	1990	2000	2005	2010
Cuba[a]	n.s.	n.s.	n.s.	2
El Salvador	n.s.	n.s.	n.s.	n.s.
Jamaica	34	34	34	34
Martinique	2	2	2	2
Trinidad and Tabogo	1	1	1	1
Total North and Central America	**37**	**37**	**37**	**39**
Brazil[b]	9,300	9,300	9,300	9,300
Chile[b]	900	900	900	900
Ecuador[b]	9	9	9	9
Peru[b]	190	190	190	190
Total South America	**10,399**	**10,399**	**10,399**	**10,399**
Papua New Guinea[b]	23	38	45	45
Total Oceania	**23**	**38**	**45**	**45**

n.s. not significant
[a]Data for 1990, 2000, and 2005 from FAO (2007). Data from 2005 also used for 2010. For countries in South America, the figures for 2000 were also used for 1990
[b]Gaps in data series filled by FAO estimates

Table 4.4 Bamboo resources in the world (FAO 2010)

Region	Area of bamboo (1,000 ha)			
	1990	2000	2005	2010
Total Africa	3,688	3,656	3,640	3,627
Total Asia	15,412	16,311	16,943	17,360
Total Europe	0	0	0	0
Total North and Central America	37	37	37	39
Total Oceania	23	38	45	45
Total South America	10,399	10,399	10,399	10,399
World	29,560	30,442	31,065	31,470

n.s. not significant

Moreover, not all countries which have bamboo resources are covered in above data. Therefore, it can be expected that global bamboo resources are considerably larger than the reported figures. In addition, other studies state different data for bamboo resources, e.g., Jiang et al. (2011) estimate China's bamboo areas to be 7.2 million ha (compared to 5.7 million ha in reported data in Table 4.4)—of which more than 2/3 are *Phyllostachys pubescens*. Nevertheless, a trend of increasing bamboo resources can be observed: global resources increased by nearly 2 million ha between 1990 and 2010. "However, there is clearly a need for better internal communication and more accurate assessments of the area of bamboo in many countries" (FAO 2010). This also links to the need for standardized and uniform monitoring of global bamboo resources. Related guidelines and clear definitions for "bamboo forests" and other land-use forms with bamboo would be useful to support such processes. In addition, data collection should be widened to cover more global bamboo resources and all bamboo-growing countries.

The above tables simply state the amount of bamboo resources in the respective countries. It should be noted that they do not assess the utilization, condition, and productivity of the bamboo stands. Moreover, they do not indicate the share of plantations, mixed forests, natural forests, and agroforestry systems. Such data would be useful to assess and compare the state of the world's bamboo resources.

When assessing bamboo resources, it is important to determine the share of monopodial and sympodial bamboo species. Therefore, Fig. 4.2 displays the composition of bamboo types within continents which grow bamboo.

Figure 4.2 shows that only monopodial bamboo species are grown in Latin America. Africa is dominated by sympodial bamboo species, and only a small share of the resources are not specified. However, more than 35 % of bamboo resources in Asia are not specified; within the specified resources, sympodials make up the larger share in Asia. The data indicate that, globally, there seem to be more monopodial bamboo resources. Figure 4.3 assesses the share of planted and natural bamboo forests in Asia.

In Asia, most of the bamboo resources occur naturally. Generally, bamboo areas in Asia have been growing in the last decades, both plantations and natural bamboo forests. It should be noted that not all bamboo systems are covered by the

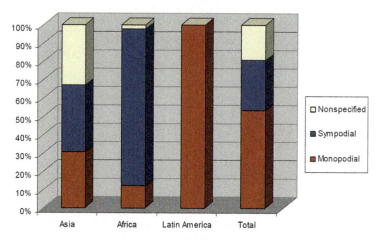

Fig. 4.2 Composition of bamboo types by continent (FAO 2007)

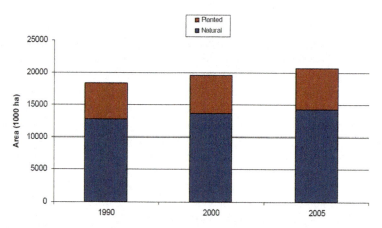

Fig. 4.3 Areas of planted and natural bamboo in Asia (FAO 2007)

classifications "natural" and "planted" bamboo forests, e.g., agroforestry systems might not fall in these categories, as they are often grown on land which is classified as "agricultural." Again, this highlights the need for uniform definitions and standardized classifications of "bamboo forests" and other land-use forms.

Bamboo resources in the major Asian bamboo countries have been growing in the past decades. Therefore, the total growth of bamboo resources displayed in Fig. 4.3 is in line with other studies which report for China a "3 % increase every year since 1980" (Cao et al. 2011).

4.3 Yield, Height, and Volume of Bamboo

Bamboos occur in pure stands, mixed forests, and agroforestry systems. Yield, volume, and height of bamboo stands vary considerably between species and are related to management practices as well as to ecological and site conditions.

"Bamboo shoots and culms grow from the dense root rhizome system. There are two main categories of rhizomes: monopodial and sympodial. Monopodial rhizomes grow horizontally, often at a surprising rate, and thus their nickname 'runners.' The rhizome buds develop either upward, generating a culm, or horizontally, with a new tract of the rhizomal net. Monopodial bamboos generate an open clump with culms distant from each other and can be invasive. They are usually found in temperate regions and include the genera Phyllostachys and Pleioblastus. Sympodial rhizomes are short and thick, and the culms aboveground are close together in a compact clump, which expands evenly around its circumference. Their natural habitat is tropical regions and they are not invasive. The main genus is Bambusa" (FAO 2007).

Generally, "dwarf bamboo species grow to only a few centimeters (cm), while medium-sized bamboo species may reach a few metres (m) and giant bamboo species grow to about 30 m, with a diameter of up to 30 cm" (FAO 2007). Diameter at Breast Height (DBH) "is a standard and the most common method of measuring tree dimension apart from tree height. It can be applied to monitor the growth of trees and compare the dimensions of different trees" (AFCD 2006). DBH is measured 1.3 m aboveground and can also be applied to bamboo. Table 4.5 presents diameter and height data of selected monopodial bamboo species.

The data in Table 4.5 reveals large ranges for diameter and height data of respective monopodial bamboo species. Amongst the presented species, *P. pubescens* is the largest in terms of diameter and height (30 cm and 35 m, respectively). *P. pubescens* is very popular and widespread across sub-tropical regions of East Asia. Table 4.6 presents diameter and height data of selected sympodial bamboo species.

Table 4.6 shows that *Dendrocalamus brandisii* has the biggest height of all presented sympodial bamboo species (36 m). The diameter of *Dendrocalamus giganteus* can reach up to 35 cm. It should be noted that not all presented data is DBH data, which indicates that data collection is not uniform and standardized. This makes it difficult to compare and assess presented data—as measurements might have been taken at varying heights.

Moreover, large variation in diameter and height within presented species can be observed. The presented data also has no reference to the age of plantation, management practices, site, and climatic conditions. The geographic locations where the measurement were taken (including elevation) should also be indicated to facilitate data comparison. Additionally, no information regarding at which height diameter was measured is provided. All of these observations indicate that no standardized or uniform measurement and reporting systems were applied. In order to allow comparison and better analysis of global data, it is recommended to

Table 4.5 Diameter and height of monopodial bamboos

Species	Diameter (cm)	Diameter measurement	Height (m)	Reference
Arundinaria alpina	8–10	n.a.	10–20	Liese (1985)
Arundinaria japonica	4–5	n.a.	2–5	Liese (1985)
Indosasa sinica	n.a.	n.a.	8–10	INBAR (2010)
Phyllostachys aurea	5	n.a.	7–9	Liese (1985)
Phyllostachys bambusoides	12–20	n.a.	15–25	Liese (1985)
Phyllostachys edulis	11.60 (avg.) 20.00 (max.)	DBH	18.00 (avg.) 22.50 (max.)	INBAR (2010)
Phyllostachys nigra	2–3	n.a.	5–7	Liese (1985)
Phyllostachys nigra f. henonis	6–10	DBH	12–15	INBAR (2010)
Phyllostachys pubescens	10–30	n.a.	10–35	Liese (1985)
Phyllostachys pubescens	8.75	DBH	n.a.	Yen and Lee (2011)

carry out standardized research—with reporting of environmental conditions—to generate uniform data.

Studies show that lower culm densities result in increased diameter-at-breast-height (DBH), but decrease total biomass (per unit area), whereas higher culm densities are related to reduced DBH and higher total biomass (Kleinhenz and Midmore 2001). Patil et al. (1994) also demonstrated that rising culm density results in thinner culms. This indicates that the culm density is a fundamental factor when trying to optimize biomass productivity of individual culms per unit area. Chen et al. (2004) studied the impact of changing DBH on the proportional distribution of biomass between the parts of bamboo and concluded that changes in DBH do not significantly influence the following aboveground biomass proportion culm > branch > leaves. These results also indicate the needs and benefits of further comprehensive and standardized measurement and reporting of bamboo diameter and height data—including parameter such as culm density.

Research has demonstrated that fertilization can have "dramatic" impacts on bamboo productivity—especially under poor soil conditions (Kleinhenz and Midmore 2001). Optimal fertilization is also related to the purpose of cultivation (i.e., for food or for timber). However, besides several Chinese studies, specific fertilization regimes for bamboo are rare and, e.g., knowledge on the impacts of cultivation methods on the productivity of bamboo is still limited. However, the impacts of intra-plant competition for nutrients and water need to be better known in order to be able to optimize bamboo management. A comprehensive Chinese standard for Moso bamboo (*P. pubescens*) plantations exists (Chinese National Standard 2006), but for other species this kind of data are rare. Moreover, the few

Table 4.6 Diameter and height of sympodial bamboos

Species	Diameter (cm)	Diameter measurement	Height (m)	Reference
Bambusa arundinacea	15–18	n.a.	26–30	Liese (1985)
Bambusa bambos	8.30	DBH	28.50	Shanmughavel and Francis (1996)
Bambusa cacharensis	5.10	DBH	11.35	Nath et al. (2009)
Bambusa chungii	4–7	n.a.	8–12	INBAR (2010)
Bambusa emeiensis	4.06 (avg.)	DBH	9 (avg.)	INBAR (2010)
Bambusa longispiculata	6–8	n.a.	15	Liese (1985)
Bambusa multiplex	1.0–3.5	n.a.	3–7	Liese (1985)
Bambusa tulda	10–15	n.a.	6–9	Liese (1985)
Bambusa vulgaris	5–10	n.a.	8–18	Liese (1985)
Bambusa vulgaris	7.35	DBH	14.52	Nath et al. (2009)
Bambusa balcooa	7.39	DBH	14.25	Nath et al. (2009)
Cephalostachyum pergracile	5–8	n.a.	10–15	Liese (1985)
Cephalostachyum pergracile	n.a.	<3.00 (avg.) >7.50 (max.)	<12 (avg.)	INBAR (2010)
Dendrocalamus barbatus	4.50–12.50	At 15 cm of basal area	n.a.	Ly et al. (2012)
Dendrocalamus brandisii	10–20	n.a.	24–36	Liese (1985)
Dendrocalamus giganteus	30–35	n.a.	30–35	Liese (1985)
Dendrocalamus hamiltonii	10–15	n.a.	20–25	Liese (1985)
Dendrocalamus longispathus	8–15	n.a.	20	Liese (1985)
Dendrocalamus membranaceus	5–12	n.a.	18–24	Liese (1985)
Dendrocalamus strictus	5–12	n.a.	8–18	Liese (1985)
Gigantochloa apus	5–12	n.a.	12–20	Liese (1985)
Gigantochloa verticillata	8–15	n.a.	25–30	Liese (1985)
Guadua angustifolia	10–15	n.a.	10–18	Liese (1985)

(continued)

Table 4.6 (continued)

Species	Diameter (cm)	Diameter measurement	Height (m)	Reference
Guadua angustifolia	8.41, 7.62, 5.25	Lower, middle and upper diameter	16.70	Riano et al. (2002)
Melocanna bambusoides	5–15	n.a.	13–23	Liese (1985)
Ochlandra travancorica	2–5	n.a.	2–6	Liese (1985)
Oxytenanthera abyssinica	5–15	n.a.	5–15	Liese (1985)
Oxytenanthera albociliata	1–3	n.a.	7–10	Liese (1985)
Oxytenanthera nigro-ciliata	6–10	n.a.	10–15	Liese (1985)
Schizostachyum brachycladum	2–7	n.a.	10–13	Liese (1985)
Shizostachyum funghomii	7–10	DBH	15–20	INBAR (2010)
Teinostachym dulloa	5–10	n.a.	9–23	Liese (1985)
Thyrsostachys oliveri	5–8	n.a.	12–25	Liese (1985)
Thyrsostachys siamensis	3–6	n.a.	7–13	Liese (1985)

existing studies focus on optimizing productivity (i.e., for food or for timber), but do not aim to optimize carbon storage or biomass within a bamboo ecosystem. For example, it is known that precipitation and temperature are significant growth limiting factors (Biswas 1988), but knowledge on total water usage and water requirements during specific growth stages is still limited (Kleinhenz and Midmore 2001). Such data, however, are needed to compare and assess yield, volume, and height reports of bamboo—further indicating the need for comprehensive and standardized reporting.

Tables 4.5 and 4.6 presented diameter and height of bamboo species at a given point in time. It is, however, also important to assess how the size and volume of bamboo culms are developing with the age of a plantation. Therefore, Table 4.7 presents the diameter of new *P. pubescens* culms from planting to maturity.

The data in Table 4.7 shows that during formation of a bamboo plantation, the DBH of new bamboo culms is increasing. It should be noted here that the diameter of newly emerging culms is growing—the diameter of existing culms does not increase once they get older. Diameter of individual culms does not change significantly throughout their lifetime. Once a bamboo plantation enters maturity, the diameter of new culms remains stable (no significant change in diameter). These

Table 4.7 Mean DBH of new culms in a newly established *Phyllostachys pubescens* plantation (Chen et al. 2004; Kuehl et al. 2013)

Years after planting	DBH (cm)
1	3.8
2	4.5
3	5.2
4	5.8
5	6.5
6	7.1
7	7.5
8	7.9
9	8.2
Mature	8.4

Table 4.8 Average number of culms per ton air-dried biomass (Liese 1985)

Species	Number of culms per ton (air dried)
Dendrocalamus strictus	400–700
Melocanna bambusoides	350–500
Bambusa tulda	30–200
Dendrocalamus longispathus	50–150

results show that it is important to indicate age of plantations or bamboo forest—in combination with the age of the culm—when reporting DBH data of bamboo.

Not only diameter and height determine the biomass generation of bamboo, but also the wall thickness of culms or water content are important. Table 4.8 compares the average number of culms per ton of air-dried biomass of selected bamboo species.

The number of culms per ton is determined by the average size of a species, but—of course—depends on a range of factors. The above table indicates that there are considerable differences in weight of culms of different bamboo species—depending on the age of the harvested culms or on the management of the bamboo stand. Besides, the table also indicates that volume and height of bamboos differentiate between species.

Yield refers to the productivity of a bamboo stand, i.e., shoots or culms which are regularly removed. It should be noted that management practices can be adjusted according to the objectives of a stand (such as timber, erosion control, or carbon sequestration) and is also influenced by climatic and site conditions. Table 4.9 displays the yield of different products in stands of monodpodial bamboo—depending on the usage as food (i.e., shoots) or timber.

Table 4.10 presents recorded yields of sympodial bamboo stands—depending on the usage of the culms as food (i.e., shoots) or timber.

The data in Tables 4.9 and 4.10 indicate that, generally, yield of shoots (in weight) is lower than that of timber—but bamboo shoots can usually be sold at a higher price (per weight). Tables 4.9 and 4.10 also revealed a high variation in yields. Bamboo yields depend on a range of factors, including management

Table 4.9 Yields of monopodial bamboo species[a]

Species	Product	Standing culm-density (culms/ha)	Annual yield (t/ha)	Reference
Acidosasa notate	Shoots	9,000–12,000	9–11	Zheng et al. (1996)
Bashania fargesii	Timber	7,400–11,000	10–11	Tang and Wei (1984)
Phyllostachys fimbriligula	Shoots	9,600–10,200	31	Cai and Wang (1985)
Phyllostachys makinoi	Timber	4,900	30	Hwang (1975)
Phyllostachys nidularia	Timber	52,500	11	Zhang et al. (1997)
Phyllostachys nidularia	Shoots or timber	50,000	20 or 14	Shen et al. (1993)
Phyllostachys pubescens	Shoots	1,600	7–28	Oshima (1931)
Phyllostachys pubescens	Shoots	1,700	7	Hu and Pan (1983)
Phyllostachys pubescens	Shoots	2,200	10–20	Fu and Banik (1995)
Phyllostachys pubescens	Timber	1,500–2,300	10	Fu et al. (1991)
Phyllostachys pubescens	Timber	3,000	7–10	Fu and Banik (1995)
Phyllostachys pubescens	Timber	3,200	9	Fang et al. (1997)
Phyllostachys pubescens	Timber	3,300	24	Cheng (1983)

[a]This table draws heavily from Kleinhenz and Midmore (2001), data were modified and added

practices, age, and site conditions. The presented variation could be a result of differences in applied measurement and reporting approaches. Yield data can be based on green, air-dried, or over-dried weight of bamboo culms. Bamboo culms can have a moisture content of 60–80 % which varies considerably during the year (Liese 1985). Also the age of the stand, as well as soil and climatic conditions, is not reported and could be a reason for the large variation in presented data. That is why research on bamboo's growth and development should be intensified. With such findings and sufficient scientific evidence, bamboo can be increasingly used for purposes such as afforestation, climate change mitigation, or the production of highly renewable, fast-growing tree-wood alternatives. However, for many of these potential uses, comparable and evidence-based scientific data is needed in order to forecast or simulate the performance of bamboo plantations. Therefore, more research on bamboo's productivity and performance is necessary.

For a limited number of species, calculations for the volume of bamboo—so-called allometric equations—exist. These functions are the result of individual

Table 4.10 Yields of sympodial bamboo species[a]

Species	Product	Standing culm-density (culms/ha)	Annual yield (t/ha)	Reference
Bambusa chungjii	Timber	n.a.	10–12	INBAR (2010)
Bambusa oldhamii	Shoots	9,600–18,000	12–15	Lin (1995)
Dendrocalamus barbatus	Timber	1,580–5,600	39	Ly et al. (2012)
Dendrocalamus latiflorus	Shoots	900	13	Chen (1993)
Dendrocalamus latiflorus	Shoots	700–14,000	10–30	Fu and Banik (1995)
Oxytenanthera abyssinica	Timber	43,000	37	Kigomo and Kamiri (1985)

[a]This table draws heavily from Kleinhenz and Midmore (2001); data were modified and added

studies on the growth of specific bamboo species. As mentioned before, bamboo's growth is impacted from management practices, as well as climatic and site conditions. Therefore, it is recommended that the applied management practices, age of a plantation, climatic conditions, site conditions, and other factors will be reported—for reference and comparison purposes. Standardized comprehensive reporting will allow users (such as project developers, investors, farmers, or governments) to plan and estimate the productivity of bamboo accurately.

"Allometric equations are the most used tool to assess volume or biomass from forest inventory data (e.g., tree diameter and height) (FAO 2014)." Bamboos are hollow and have different growth characteristics than trees, but allometric equations represent an important tool for the estimation of bamboo biomass. Nevertheless, only very few tested allometric equations exist for bamboo. One example of an allometric equation for bamboo is presented below (*D-DBH, H-height, t-time* (in years)):

Allometric equations for Moso bamboo (*P. pubescens*)—Zhou et al. (2013):

$$D = 5.200 + 0.572\,t + 0.0452\,t^2 - 0.0056\,t^3 \quad R = 0.999$$
$$H = 0.5702 + 1.6426D - 0.0465D^2 \quad R = 0.727$$

The number of allometric equations for bamboo is limited and concentrated on a few species. This work needs to be extended to other species in order to allow stakeholders (such as governments or project developers) to quantify the impact of re-/afforestation measures with bamboo. The allometric equations should also be adjusted to external factors, such as elevation, climate, soil, or management. Generally, more developed allometric equations for bamboo would allow better and cost-effective impact predictions for managing bamboo stands.

)oos form extensive rhizome and root systems which can extend
Liese 1985). This rhizome system survives selective harvest of
The harvested biomass is usually replaced within a year (Lou
... 2010). This means that the bamboo ecosystem can be productive while
continuing to store and sequester carbon, as new culms will replace harvested
ones. These characteristics show that bamboos are different than trees with regard
to belowground carbon storage and its ability to be harvested. Bamboo is one of the
fastest-growing forest resources in the world: it has rapid growth rates, high annual
regrowth after harvesting, and high biomass production (Yen and Lee 2011).
Consequently—combined with its fast growth abilities—bamboo can be regarded
as highly renewable.

Generally, the share of belowground biomass of monopodial species is 43 % of
total biomass; for sympodial species that share is lower: 31 % of total biomass
(Kleinhenz and Midmore 2001). The study moreover states that "biomass can vary
substantially within individual species, even when cultivated at the same site." A
study in Northern Laos on a range of bamboo species estimated belowground
biomass with around 40 % of total biomass (Kiyono et al. 2007). Moreover,
biomass growth and composition of understory vegetation are significantly
impacted by culm density and stand composition (Lin 2002). In addition, it is
important to report the age and the management type of the bamboo stand, as
these factors can also influence its biomass storage capacity. The "root–shoot" ratio
describes the relation of aboveground biomass to belowground biomass. Table 4.11
shows the root–shoot ratio of a newly established *P. pubescens* plantation.

The data in Table 4.11 shows that the share of belowground biomass varies with
age of the individual culm. The presented data—however—does not indicate a
clear trend. Therefore, it is advised to further research the relationship of the age of
bamboo culms and the "root–shoot" ratio—also for other bamboo species. For
P. pubescens, other studies have estimated higher shares of belowground biomass
with 36.21 % (Jiang et al. 2011)—without a reference to culm age. The shoot–root
ratio of a bamboo stand, thus, also depends on the age composition of individual
bamboo stands. Figure 4.4 presents the findings of another study on *P. pubescens* in
which the fraction of carbon stored belowground is high: 66.6 % of total carbon is
stored in belowground carbon in the soil and root system of the Moso bamboo stand
(Isagi et al. 1997); the total system stores 179.9 t C/ha. The study shows, however, a
lower belowground biomass share: only 23.76 % (of total biomass) which converts
into a root–shoot ratio of 3.21.

Moreover, as Fig. 4.4 indicates, the complete carbon storage in the bamboo
ecosystem does not only include bamboo biomass, a considerable share of carbon is
also stored in the soil of the bamboo ecosystem. Another study estimated that even
69.95 % of total carbon stored within the bamboo ecosystem can be stored in the
soil (Zhou et al. 2010). A study in Vietnam estimated the soil organic carbon pool in
bamboo stands with 92 t/ha—"comparable to both forest and regenerated forest and

Table 4.11 Root–shoot ratio and share of belowground biomass of total biomass of *Phyllostachys pubescens* (Peng et al. 2002; Kuehl et al. 2013)

Age of culm	Root–shoot ratio	Belowground biomass of total (%)
1–2	3.196	23.83
3–4	2.693	27.08
5–6	2.402	29.39
7 (and older)	2.751	26.66

Fig. 4.4 Carbon stock and cycling in a *Phyllostachys pubescens* Stand (Isagi et al. 1997)

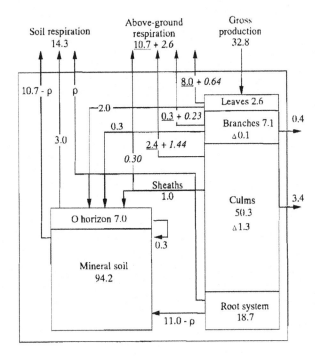

20 % higher than land cultivated with cassava or maize" (Ly et al. 2012). These findings indicate that soil carbon storage in bamboo is high. However, available data on carbon storage in the soil of bamboo ecosystems is still limited; therefore, it is recommended to extend related research.

4.3.2   Bamboo Biomass and Climate Change

This part introduces, reviews, and discusses bamboo's characteristics which are relevant to climate change mitigation. Photosynthesis represents an important factor when evaluating the carbon sequestration capacity of plants. Bamboo belongs to the class of C_3 plants (like trees), which indicates that the CO_2 fixation processes of bamboo are not quicker than fast-growing trees (Düking et al. 2011).

Kleinhenz and Midmore (2001) demonstrated that with increased age of culms and/or leaves, bamboo's photosynthetic rates decrease significantly. In contrast to Düking et al. (2011) who state that "the growth of the leafless culm does not originate from ongoing photosynthesis, but from allocation of organic material produced during the previous year and stored in rhizome system and older culms." Kleinhenz and Midmore (2001) observed that early growth stages of culms do not occur because of mobilization of energy reserves, but because of photosynthesis in leaves of older culms. These contradicting findings indicate a need for further research on the photosynthetic processes in bamboo stands. The leaf area index of mature bamboo stands is generally high; studies showed that bamboo's canopy can absorb up to 95 % of incident solar radiation (Scurlock et al. 2000). The cycling patterns of carbon within the bamboo ecosystem (see Fig. 4.5) are essential when analyzing the carbon sequestration potential of bamboo. However, scientific data in this field are still limited.

Studies (e.g., Lou et al. 2010) have shown that unmanaged bamboo plantations (i.e., without removal of biomass) reach a level in which no additional biomass is formed, as individual bamboo culms can already start deteriorating after around 10 years (depending on species). Figure 4.5 displays the patterns of new and deteriorating biomass in a newly established bamboo plantations—assuming that no biomass is removed through harvest.

As Fig. 4.5 shows, once a bamboo plantation reaches the steady state (deteriorating biomass is replaced by new biomass) and is not harvested, the total net biomass of a bamboo plantation does not change significantly (see also: Liese 2009); therefore, the annual increments are complicated to compare, as they largely depend on the age of the plantation. Therefore, it is suggested to consider total standing biomass at a given time and not the annual increment, when evaluating bamboo's potential in climate change mitigation.

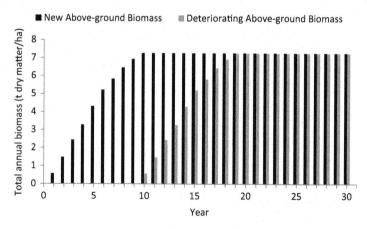

Fig. 4.5 New and deteriorating aboveground biomass of Moso bamboo—without harvest (Kuehl et al. 2013)

Unmanaged bamboo plantations do not sequester additional carbon—once they reach maturity. That is why studies conclude that "unmanaged bamboo stands do not store high levels of carbon, as their productivity is low and the accumulated carbon returns quickly to the atmosphere as the older culms decompose" (Kuehl et al. 2011). Figure 4.6 shows that a *P. pubescens* plantation does not store additional carbon after 19 years, as the deteriorating culms are not removed through harvest and not replaced by newly emerging culms.

Therefore, the presented and other studies (e.g., Lou et al. 2010) concluded that only managed plantations (i.e., through the regular removal of mature culms) can make use of bamboo's characteristics of fast growth and high renewability. In addition "managing bamboos increases the yield and quality of culms and increases farmers' incomes from their sale" (Kuehl et al. 2011). Figure 4.7 shows that in a managed plantation, *P. pubescens* can accumulate 1.68 times more carbon than Chinese fir—provided that harvested biomass (*P. pubescens* and Chinese fir) is used for durable products.

The data in Fig. 4.7 is in accordance with other studies: "compared with unmanaged stands, in managed stands, cultivation and harvesting practices enable much higher biomass production per unit area, at least doubling productivity" (Kuehl and Lou 2012). Also other studies presented similar results: Jiang et al. (2011) reported that *P. pubescens* ecosystem "fixed 1.69 and 1.63 times as much C as the Chinese fir and Masson pine forest ecosystems, respectively," and Yen and Lee (2011) conclude that *P. pubescens* "is a superior species for carbon sequestration when compared with Chinese fir", but emphasized the importance of harvest and regular removal of mature culms. Yen and Lee (2011) simulated a carbon sequestration of 8.13 t C/ha/year for *P. pubescens* compared to 3.35 t C/ha/year for Chinese fir. It is also important to note that "bamboos sequester more carbon in the early years of a plantation than comparable forest trees" (Kuehl and Lou 2012). This fact can be important when comparing land use options for climate

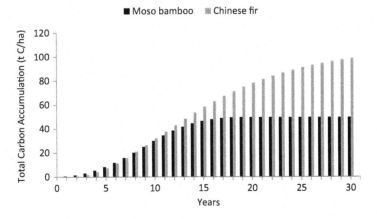

Fig. 4.6 Modeled aggregated carbon accumulation of newly established Moso bamboo and Chinese fir plantations—without regular harvest (Kuehl et al. 2013)

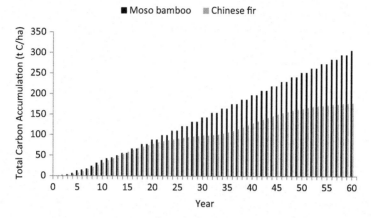

Fig. 4.7 Modeled aggregated carbon accumulation of newly established Moso bamboo—and Chinese fir plantations—with regular harvest (Kuehl et al. 2013)

change mitigation activities. Bamboo plantations can yield quick returns on investments and incomes (i.e., carbon credits) and can ensure quick additional benefits—such as erosion control, ground cover, or productivity.

The importance of harvest for the productivity and performance of bamboo stands for climate change mitigation purposes are linked to the usage and continued carbon storage of removed biomass. Bamboos potential can be maximized, if the harvested biomass is stored in durable products which have an expected lifespan of at least 20 years. The role of harvested wood products (HWP) in carbon accounting mechanisms is globally being discussed—recognizing their function as a carbon storage. For bamboo, it is important to note that "recent improvements in processing and the development of new types of products mean that many have lifespans of decades, meaning carbon can be stored for much longer, which increases the size of the bamboo carbon sink" (Kuehl et al. 2011).

The high share of carbon which is stored belowground (see previous chapter) also makes bamboo a secure carbon sink which is less susceptible to disasters such as fires and storms. The fast growth and high renewability of bamboo allow bamboo to quickly recover from disastrous events. Bamboo's extensive root and rhizome system implies that bamboos can be used to fight erosion, bamboos' extensive roots and rhizomes bind the soil, and as they can grow on poor soils, bamboos are most effective in areas prone to runoff such as steep slopes, river banks, or degraded lands. Bamboos are evergreen plants and the thick canopy and soil cover provided by dead leaves reduce direct and splash erosion and enhance infiltration" (Kuehl et al. 2011). Growing and utilizing bamboo can help people and ecosystems to adapt to changing climates.

Therefore, concluding, it can be summarized that bamboo can be a useful and specific tool for simultaneous efforts in climate change mitigation, adaptation, and development (bamboo's role in rural development has been demonstrated by a range of studies [e.g., Hogarth and Belcher 2013]).

References

AFCD (2006) Agriculture, fisheries and conservation department, Hong Kong. Nat Conserv Pract Note 2:1–6

Belcher B (1995) Bamboo and rattan production to consumption systems: a framework for assessing development options. INBAR working paper no. 4. INBAR, Beijing

Biswas S (1988) Studies on bamboo distribution in north-eastern region of India. Indian For 114:514–531

Buckingham K, Jepson P, Wu L, Rao R, Jiang S, Liese W, Lou Y, Fu M (2011) The potential of bamboo is constrained by outmoded policy frames. Ambio 40(5):544–548

Cai RQ, Wang KH (1985) Observations on the bamboo shoot growth of Phyllostachys fimbriligula. J Bamb Res 4:61–70 (in Chinese)

Cao Z, Zhou G, Wen G, Jiang P, Zhuang S, Qin H, Wong M (2011) Bamboo in subtropical China: efficiency of solar conversion into biomass and CO_2 sequestration. Bot Rev 77(3):190–196

Chen T (1993) Study on high-yield technique of cluster-growing bamboo for shoot and timber. J Bamb Res 12:30–34 (in Chinese)

Chen S, Wu B, Wu M, Zhang D, Cao Y, Yang Q (2004) A study of the inter-annual succession rule and influential factors of young stands structures of Phyllostachys pubescens. J Zhejiang For Coll 21(4):393–397 (in Chinese with English summary)

Cheng YL (1983) Potentialities of high yield and managerial techniques of annual-working bamboo groves of Phyllostachys pubescens. J Bamb Res 2:207–217 (in Chinese)

Chinese National Standard (GB/T 20391-2006) [Internet] (2006) High yield management techniques for Moso bamboo stands, issued on 13th June 2006 by the general administration of quality supervision, inspection and quarantine of the People's Republic of China and Standardization Administration of the People's Republic of China, effective from 1st December 2006 [cited 2012 June 15]. Available from: http://www.cn-standard.net/qtweb/debzfy/debzdetail/371/701C8D93.shtml

Düking R, Gielis J, Liese W (2011) Carbon flux and carbon stock in a bamboo stand and their relevance for mitigating climate change. J Am Bamb Soc 24:1–7

Fang XM, He JH, Ye L, Zheng YH, Liu HL, Huang YT (1997) Study on cultivation technique of mixed Phyllostachys heterocycla cv. pubescens in site III. J Bamb Res 16:23–29 (in Chinese)

FAO (2007) Food and agriculture organization of the United Nations. World Bamboo Resources. Non-Wood Forest Products 18. FAO, Rome

FAO (2010) Food and agriculture organization of the United Nations. Global Forest Resources Assessment 2010. Main Report. FAO Forestry Paper 163. FAO, Rome

FAO (2014) Food and Agriculture Organization of the United Nations. GlobAllomeTree: the international tree allometric equation platform [cited 2014 March 14]. Available from: http://www.fao.org/forestry/fma/83159/en/

Fu MY, Banik RL (1995) Bamboo production systems and their management. In: Rao IVR, Sastry CB, Widjaja E (eds) Bamboo, people and the environment. Propagation and management, vol 1. Proceedings of the Vth international bamboo workshop and the IVth international bamboo congress, Ubud, Bali, Indonesia, 19–22 June 1995. International Network for Bamboo and Rattan, New Delhi, India, pp 18–33

Fu MY, Xie JZ, Fang MY (1991) Fertilization studies in bamboo stands with different end uses II. High yield management method for bamboo shoot stands. For Res 4:238–245 (in Chinese)

Hogarth N, Belcher B (2013) The contribution of bamboo to household income and rural livelihoods in a poor and mountainous county in Guangxi, China. Int For Rev 15(1):71–81

Hu CZ, Pan XZ (1983) On the density of shoot producing Ph. pubescens forest. J Bamb Res 2:189–197 (in Chinese)

Hwang KK (1975). A study on the working-systems of Makino bamboo stand. Bulletin of Taiwan Forestry Research Institute No. 260 (in Chinese)

INBAR (2010) International Network for Bamboo and Rattan. China's bamboo. In: Yang Y, Hui C, Du F (eds) Culture/resources/cultivation/utilization. International Network for Bamboo and Rattan, Beijing, China

Isagi Y, Kawahara T, Kamo K, Ito H (1997) Net production and carbon cycling in bamboo Phyllostachys pubescens stand. Plant Ecol 130:41–52

Jiang P, Meng C, Zhou G, Xi Q (2011) Comparative study of carbon storage in different forest stands in subtropical China. Bot Rev 77(3):242–251

Kigomo BN, Kamiri JF (1985) Observations on the growth and yield of Oxytenanthera abyssinica (A. Rich) Munro in plantation. East Afr Agric For J 51:22–29

Kiyono Y, Ochiai Y, Chiba Y, Asia H, Saito K, Shiraiwa T, Horie T, Songnoukhai V, Navongxai V, Inoue Y (2007) Predicting chronosequential changes in carbon stocks of pachymorph bamboo communities in slash-and-burn agricultural fallow, northern Lao People's Democratic Republic. J For Res 12:371–383

Kleinhenz V, Midmore DJ (2001) Aspects of bamboo agronomy. Adv Agron 74:99–145

Kuehl Y, Lou Y (2012) Barbon off-setting with bamboo. INBAR working paper 71. INBAR, Beijing

Kuehl Y, Henley G, Lou Y (2011) The climate change challenge and bamboo: mitigation and adaptation. INBAR working paper 65. INBAR, Beijing

Kuehl Y, Li Y, Henley G (2013) Impacts of selective harvest on the carbon sequestration potential in Moso bamboo (Phyllostachys pubescens) plantations. Forests Trees Livelihoods 22(1):1–18

Liese W (1985) Bamboos – biology, silvics, properties, utilization. GTZ, Eschborn

Liese W (2009) Bamboo as carbon sink – fact or fiction? In: VIII world bamboo congress proceedings, Bangkok, Sept 2009

Lin QY (1995) Cultivation techniques for Dendrocalamopsis oldhamii. In: Rao IVR, Sastry CB, Widjaja E (eds) Bamboo, people and the environment. Proceedings of the Vth international bamboo workshop and the IVth international bamboo congress, Ubud, Bali, Indonesia, 19–22 June 1995. Propagation and management, vol 1. International Network for Bamboo and Rattan, New Delhi, India, pp 18–33

Lin H (2002) Study on dynamic change regulation for biomass of bamboo forest ecosystem. China For Sci Technol Suppl 16:26–27 (in Chinese)

Lobovikov M, Lou Y, Schoene D, Widenoya R (2009) The poor man's carbon sink – Bamboo in climate change and poverty alleviation. FAO working document no. 8. FAO, Rome

Lobovikov M, Schoene D, Lou Y (2012) Bamboo in climate change and rural livelihoods. Mitig Adapt Strat Glob Chang 17(3):261–276

Lou Y, Li Y, Buckingham K, Henley G, Zhou G (2010) Bamboo and climate change mitigation. INBAR technical report no. 32. INBAR, Beijing

Ly P, Pillot D, Lamballe P, de Neergaard A (2012) Evaluation of bamboo as an alternative cropping strategy in the northern central upland of Vietnam: above-ground carbon fixing capacity, accumulation of soil organic carbon, and socio-economic aspects. Agric Ecosyst Environ 149:80–90

Mohanan C (2002) Disease and disorders of bamboo in Asia. In: Kumar A, Rao IVR, Sastry C (eds) Bamboo for sustainable development. VSP, Utrecht

Nath AJ, Das G, Das AK (2009) Above ground standing biomass and carbon storage in village bamboos in North East India. Biomass Bioenergy 33:1188–1196

Oshima J (1931) The culture of Moso bamboo in Japan, Part II: methods for growing Moso bamboo shoots. J Am Bamb Soc 3:33–50

Patil VC, Patil SV, Hanamashetti SI (1994) Bamboo farming: an economic alternative on marginal lands. In: Bamboo in Asia and the Pacific. Proceedings 4th international bamboo workshop, 1991, Bangkok

Peng Z, Lin Y, Liu J, Zou X, Guo Z, Guo Q, Lin P (2002) Biomass structure and energy distribution of Phyllostachys heterocyla cv. pubescens population. J Xiamen Univ (Nat Sci) 41(5):579–583 (in Chinese with English abstract)

Riano NM, Londono X, Lopez Y, Gomez JH (2002) Plant growth and biomass distribution of Guadua angustifolia Kunth in relation to ageing in the Valle del Cauca – Colombia. J Am Bamboo Soc 16:43–51

Scurlock JMO, Dayton DC, Hames B (2000) Bamboo: an overlooked biomass resource? Biomass Bioenergy 19:229–244

Shanmughavel P, Francis K (1996) Above ground biomass production and nutrient distribution in growing bamboo (Bambusa bamboos (L.) Voss). Biomass Bioenergy 10(5/6):383–391

Shen CQ, Fan JM, Liu DN, Chen CG, Li CF (1993) A preliminary report of research on high production technique of Phyllostachys nidularia Munra. J Bamb Res 12:53–63 (in Chinese)

Tang JW, Wei DM (1984) Study on the cutting patterns of Bashania fargesii forests. J Bamb Res 3:102–111 (in Chinese)

Yen TM, Lee JS (2011) Comparing aboveground carbon sequestration between Moso bamboo (Phyllostachys heterocycla) and China fir (Cunninghamia lanceolata) forests based on the allometric model. For Ecol Manage 261(6):995–1002

Zhang X, Wang JP, Zhang XM (1997) Study on relation of rotation cycle density with stand productivity of Phyllostachys nidularia. J Bamb Res 16:68–78 (in Chinese)

Zheng MZ, Lin XS, Feng CY (1996) Comparison test on bamboo shoot and timber yield of on-off year and constant management. J Bamb Res 15:40–46 (in Chinese)

Zhou G, Jian R, Xu Q (2010) Advance in study of carbon fixing and transition in the ecosystem of bamboo stands. Science Press of China, Beijing (in Chinese with English abstracts)

Zhou G, Shi Y, Lou Y, Li J, Kuehl Y, Chen J, Ma G, He Y, Wang X, Yu T (2013) Methodology for carbon accounting and monitoring of bamboo afforestation projects in China. INBAR working paper 73. INBAR, Beijing

Chapter 5
Bamboo Silviculture

Ratan Lal Banik

Abstract This chapter describes different ecological factors influencing the distribution, occurrence and proper development of bamboo vegetation. The three-phase gregarious flowering behaviour of bamboos, reproductive biology, the existence of flowering diversities and importance of more than one flowering cohort in a species are discussed for frequent seed yield. Emphasis was given on the nursery techniques of planting material production, steps of raising different types of bamboo plantation and species suitability, and aftercare in the field along with the bamboo-based different agroforestry options. The scientific management procedure of bamboo stand is also outlined for higher productivity.

Keywords Bamboo • Distribution ecology • Three-phase gregarious flowering • Socio-economic and ecologic impact • Reproductive biology • Importance of flowering diversities • Aided natural regeneration • Planting materials • Nursery • Plantation raising • Bamboo stands management • Bamboo based agroforestry • Homestead bamboo management

5.1 Introduction

Grasses are the most useful of all plants, and no growing things on earth have so many and so varied uses as the tree grasses or bamboo. Bamboos are closely interwoven with the day-to-day life of the people, especially in the continent of Asia. Bamboos were, in the past, treated as less valued forest plants compared to timber species. In general it is felt that the existing practices of bamboo grove management are mostly traditional and have little scientific back up. With the present awareness of bamboo as a cash crop, people are showing interest in raising bamboo plantation, and therefore an updated knowledge of bamboo silviculture is an important basis of production and sustainable maintenance of this important resource of plant.

R.L. Banik (✉)
National Mission on Bamboo Applications (NMBA), New Delhi, India

INBAR, New Delhi, India

BFRI, Chittagong, Bangladesh
e-mail: bamboorlbanik@hotmail.com; rlbanik.bamboo@gmail.com

© Springer International Publishing Switzerland 2015
W. Liese, M. Köhl (eds.), *Bamboo*, Tropical Forestry 10,
DOI 10.1007/978-3-319-14133-6_5

5.2   Distribution of Bamboos and Related Ecological Factors

Bamboos are very unevenly distributed in the tropics, and mild temperate regions of the world, from sea level to the snow line. Some woody bamboos were recorded from latitudes as far north as 46° and as far south as 47° and occurring at elevation as high as 4,000 m, whereas the herbaceous bamboos are confined to the tropics and subtropics (Soderstrom and Calderon 1979). In the Western Hemisphere, the natural distribution of bamboos extends from 39° 25′ N in eastern United States to 45° 23′ 30″ S in Chile and even to 47° S in Argentina. However, the distribution of bamboo is rich and gregarious in the areas of Tropic of Cancer on the northern side and the Tropic of Capricorn on southern side, especially for clump-forming bamboos. In this zone, clump-forming bamboos exist in two categories (Uchimura 1987), namely, bamboo forest (1) in extensive areas and (2) in groups or colonies. The first category of bamboos is more or less confined within 15–25° of northern side, especially tropical monsoon Asia. On the other hand, only colonies of bamboo forests occur in other tropical American countries, tropical Africa and the rest of the above tropical countries including both sides of the equator between 30° N and S latitude. Worldwide, 22 million hectares of land is assumed to be covered by bamboos (ICBR 2004). Major species are found in Asia Pacific (China 626, India 102, Japan 84, Myanmar 75, Malaysia 50 and few others) and South America (Brazil 134, Venezuela 68, Colombia 56 and few others), while least (5) in Africa (Bystriakova et al. 2003) and none in Europe.

The controlling factor of its abundance, distribution of species, growth and development within this limit are mainly annual precipitation, relative humidity and nature of soil.

5.2.1   Topography and Soil

The culm yield and above-ground biomass of bamboos grown in the flatland constitute about four times as much as those in the hill side, while the litter fall is about twice in the former as compared to the latter site. Difference in longitude, latitude, elevation, slope direction and gradient will result in varied sun radiation and duration of sunshine, and there will be much differences in the condition of water, heat and light. There are more shoots in a bamboo stand at the south slope, but they are comparatively small in size and have short internodes in a culm. *Melocanna baccifera* clumps growing on the northern slope usually have large-sized culms with long internodes (Banik 2010a). Among the *steep slope, gentle terrain* and *level ground*, the land at gentle slope up to 30° is most favourable to the bamboo stand growth. The proportion of bamboo forest in China with a slope of 25° below is 90.39 %, while it is less than 3 % only with a slope of 35° above (Guan et al. 2012).

Most bamboos occur and exhibit normal growth in well-drained, sandy loam to clay loam soils, provided the drainage, rainfall and temperature conditions are favourable. Topography and the soil condition are the main factors affecting stand growth within the same climatic region. Loose and fertile soil with high content of organic matter, good water holding ability and water permeability facilitates rhizome growth. The growth of rhizome is retarded and is twisted with short and abnormal rhizomes in compact and heavy soil which has low water permeability and poor nutrient content. The geologically younger soils are more suitable for bamboo growth than the older soils (Lyall 1928). According to Uchimura (1980), soils high in N, P_2O_5, K_2O, CaO and SiO_2 promote best growth of bamboo culms. Soil N content is the most important factor affecting bamboo growth, but organic matter, texture, aeration, base exchange capacity and depth are also important (He and Ye 1987). Top soils that are suitable for bamboo vary in colour from yellow to brown–yellow, and heart soil colour are clear red, yellow–brown to blue–grey (Uchimura 1987). Soil pH ranges from 5.0 to 6.5 which is most suitable for bamboo; some species may grow even at pH 3.5. Bamboo can neutralise acidic soil. Bamboos are usually not found to grow naturally under saline coastal inundation in the tidal or mangrove forests. However, people in the coastal area of Bangladesh and India (Orissa, Konark) have been successfully cultivating *Bambusa vulgaris, B. balcooa* and *B. nutans* in offshore homestead/farmlands. *Bambusa atra* (*B. lineata*) is found near the tidal swamp forest of Andamans and therefore can be tried in coastal areas.

5.2.2 Temperature and Light

The majority of clump-forming bamboos grow at temperature ranging from 7 (sometimes 2–3 °C) to 40 °C. In general, high temperature accelerates the growth of bamboo and low temperature inhibits it. However, the maximum occurrence of tall and majestic clump-forming types is in the humid tropics, mainly in places where the average temperature is higher than 20 °C. There are a number of shrubby bamboo species such as *Arundinaria* and *Thamnocalamus* found in the cooler environment (snow line) of the Himalayas.

Bamboo, in general, prefers light for its healthy growth. A certain amount of overhead cover improves the quality of bamboos at the expense of quantity. The productive months of culm emergence have days of longer photoperiod (Banik 1993a). Thinning operations in the tree plantations facilitates entry of light to the forest floor, and as a result, bamboo appears as undergrowth.

5.2.3 Altitude

A number of bamboo plants, which are mostly short and represent the genera *Arundinaria, Chimonobambusa, Semiarundinaria, Sinobambusa* and *Thamnocalamus*, are present in the alpine belt to temperate forests of Nepal, Bhutan, North India (the Himalayan mountains zone) and Tamil Nadu in the south (Nilgiris). These zones have been classified into wet, moist and dry types and occur between 1,500 and 3,050 m. The region has extremely low temperatures in the winter, often below freezing. Dense mist prevails during monsoons and is the characteristic climatic feature of this region. The clump-forming type is observed to predominate in low and medium altitude from sea level to 700 m, while the nonclump-forming type occurs more abundantly at high elevation up to 3,300 m (Numata 1987). Altitude and temperature are closely related, and it is difficult to separate one from the other, for example, some species of *Phyllostachys* are cultivated at high elevations in India and Nepal but also occur at low elevations in countries of the temperate zone (Uchimura 1987). Thus, altitude also affects the distribution of bamboo with respect to form or type even in the tropical region.

5.2.4 Rainfall

Both bamboo and rice need sufficient amount of rainfall for survival and growth, and therefore most of the bamboo producing countries fall in rice-producing regions. High temperature and humidity requirements are common characteristics of bamboo plants originating in the monsoon areas of Southeast Asia. The distribution of bamboo in India is related to the rainfall (Gamble 1896). Bamboo hardly occurs in zones with less than 1,015 mm of rainfall. The upper limit is not known, but the species are (*M. baccifera, Melocalamus compactiflorus*) found to grow in zones with over 6,350 mm. The most common range is 1,200–4,000 mm per year. As the water requirement of bamboo is high, the amount of rainfall is also a limiting factor to distribution (Numata 1987). Rainfall promotes growth of the culm during speedy elongation period.

During cyclone and storm, isolated tall and thick leafy crowned bamboo clump often gets uprooted. In cyclone-prone southern area (Cox's Bazar) of Bangladesh, farmers have been cultivating a distinct biotype (locally known as 'Kanta Bizzya Bans') having short (7–15 m tall) and somewhat solid culms in comparison to that of common tall type (25–35 m) of *Bambusa vulgaris*, which can stand against the cyclone and tidal bore (Banik 1994a).

5.2.5   Influence of Flood

Generally a bamboo clump cannot tolerate water logging, especially in the growing season. The bamboos grown in the *flood plains* can tolerate short seasonal floods. During the devastating flood of 1998, water was flowing 1.0–2.0 m above the surface and stagnated the land up to 35–92 days in 52 districts out of total 64 districts of Bangladesh. It is apparent from a limited survey that about 30–50 % clumps of *Bambusa balcooa, B. vulgaris* and *B. tulda* were partly or fully died. About 40–60 % newly emerged culms of different bamboo clumps rot and died as they were sunk under water for more than a month. About 90 % clumps of *Melocanna baccifera* cultivated in the homesteads were damaged and died as underground rhizome system in the waterlogged condition suffered from oxygen deficiency (Banik 2000). Species such as *Ochlandra scriptoria, O. stridula* and *O. travancorica,* indigenous to southern India and northern Sri Lanka, are mostly found in marshy areas and riverbanks that get flooded in the monsoons. *Phyllostachys purpurata (Ph. heteroclada)* and *Ph. atrovaginata* are monopodial species that can grow in wet soils and waterlogged areas. Interestingly, the rhizomes of these monopodial species have air canals. It is reported from Vietnam that in extreme cases, some species such as *Bambusa stenostachya* can also tolerate submerged condition (flood) up to 1 month.

5.2.6   Influence of Fire

Bamboo can stand fire and grazing to a limited extent due to its well-developed underground rhizome system. Repeated firing over short cycles results in almost pure stands of bamboo over vast areas in the hills. The widespread distribution of *M. baccifera* throughout eastern India, Bangladesh, northern Myanmar and Thailand, species of *Thyrsostachys* in Thailand and species of *Schizostachyum* in Vietnam mainly occurs as secondary vegetation due to the destruction of tropical rainforest by fire, shifting cultivation, logging and warfare.

5.3   Flowering Periodicity and Diversities

5.3.1   Flowering Nature

Generally bamboos are not annually flowering plants, rather in most cases flowering is after long gap of vegetative phases of life, which is usually 15–60 years in tropical bamboos and 60–120 years in bamboos of temperate region. Evidence of regular flowering cycles of ca. 30 years was found for most of the neotropical woody bamboo species (*Chusquea, Guadua,* etc.) from northern

Mexico to southern Argentina and Chile (Guerreiro 2013). So seeds of a bamboo species are not available every year and thus seedling planting materials cannot be produced. Additionally the large-scale death of bamboos after flowering has been considered as an important limiting factor in sustainable production of bamboo resource and its silvicultural management. Therefore, the knowledge on various aspects of flowering nature and seed yield is discussed in the following sections.

5.3.1.1 Three-Phase Flowering and Flowering Wave

The gregarious flowering in major bamboo species of Asian tropics like *Bambusa bambos*, *B. polymorpha*, *Dendrocalamus asper*, *D. hamiltonii*, *D. strictus*, *Melocanna baccifera*, etc., usually takes place in three phases after a definite period of time. The three stages (Troup 1921) are (1) preliminary *initial sporadic* flowering, (2) *gregarious* flowering and (3) *final sporadic* flowering of the remaining clumps. The length of each phase may vary considerably, and the first and third phases may be so prolonged as to include the second, which may not be very marked. As, for example, in Myanmar, the flowering of *M. baccifera* was initially sporadic (*Si*-phase) distinctly for 4 years (1902–1905) which continued and gradually became gregarious (*Gr*-phase) during 1910–1913 and then again turned into final sporadic (*Sf*-phase) for a few years (1915–1916 years) before the mass-scale death of the clumps (Plate 5.1; Banik 1998, 2000). The flowering continued in a specific direction like a 'wave', with a period of 12, 14 and 17 years in Chittagong (Plate5.1a), Myanmar (Plate5.1b) and India (Plate5.1c), respectively, spreading over the whole forest area. During present time (1995–2012) in the above region, *Melocanna* forests also exhibited 3-phase flowering like earlier incidences (Banik 2010a; Plate 5.1). Like *M. baccifera* most of the populations of some major bamboo species in tropical Asia, like *Dendrocalamus asper, D. strictus, Bambusa bambos, B. polymorpha, Thyrsostachys oliveri,* etc., rarely flower sporadically; rather they were found to exhibit gregarious nature of flowering after a definite period of time. According to Holttum (1958), the gregarious flowering in bamboos, at interval of many years, does not normally occur in the equable climate of the Malayan region.

5.3.1.2 Flowering Population (Cohort) and Interseeding Period

A bamboo species may have more than one flowering cycle (interseeding period) differ and isolated from each other by reproductive time, and thus each seems to represent a 'flowering population' (Table 5.1). In the boundary areas, however, populations are likely to overlap with each other, and some of the clumps may flower after shorter, and others after longer, periods or in between (Banik 1994a, 2000, 2010a). It is not uncommon to find few clumps of *M. baccifera* remains with green leaves without any flowers growing in small patches (locally known as Mau-Hak in Mizoram) inside a vast majority of gregariously flowering population covering a large tract of land (Fig. 5.1). While studying flowering ecology of *Ph.*

a

Bangladesh (Chittagong south and Cox's Bazar; border to Myanmar): Four flowering incidences each one having 3-phase flowering wave

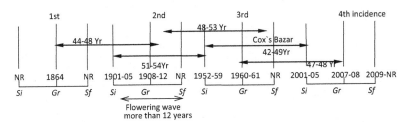

b

Myanmar (Arakan): Three flowering incidences each one having 3-phase flowering wave

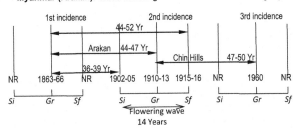

c

India (Assam, Tarai, Garo and Khasi hills, Mizo hills, Lushai hills, Manipur, Mizoram, Tripura): Five flowering incidences each one having 3-phase flowering wave

Plate 5.1 Estimation of flowering cycle in *Melocanna baccifera* from the illustration of a number of incidences of flowering in the different localities of India–Bangladesh–Myanmar region during the last 150 years (*note* 3-phase flowering nature: *Si* initial sporadic flowering, *Gr* gregarious flowering, *Sf* final sporadic flowering, *NR* flowering date not reported). (**a**) *Bangladesh (Chittagong south and Cox's Bazar, border to Myanmar)*: four flowering incidences each one having 3-phase flowering wave. (**b**) *Myanmar (Arakan):* three flowering incidences each one having 3-phase flowering wave. (**c**) *India (Assam, Tarai, Garo and Khasi hills, Mizo hills, Lushai hills, Manipur, Mizoram, Tripura):* five flowering incidences each one having 3-phase flowering wave

bambusoides during 1960 in the Chiba forest at Narita, Japan, Numata et al. (1974) mentioned that the stand contained a number of genetically different clones (cohort) and they did not flower synchronously in the stand. *Phyllostachys pubescens* also exhibits *sporadic* and *irregular flowering* in addition to periodic *gregarious flowering* (Watanabe et al. 1982). So a bamboo species may have more than one 'flowering genotype' (Banik 1997a).

Table 5.1 Diversities in flowering nature and interseeding period of some bamboos in Asia

Species	Flowering nature	Estimated interseeding period	References
Bambusa bambos	Gregarious, rarely sporadic and irregular	30 ± 5 (Coast, India) 30–32 (Brazil) 40 (Orissa, India) 45 ± 5 (North India)	Kurz (1876) Dutra (1938) Banik (1980, 2000) Blatter (1929)
B. balcooa	Rarely flowers, no seed	40 ± 5 (Assam, North India, Bangladesh)	Banik and Alam (1987)
B. longispiculata	Sporadic, part flowering	Period not known Sporadic (Chittagong)	Banik (2000)
B. nutans	Sporadic, part flowering, gregarious	Every year (Thailand) Sporadic, Prt- Flower (NEIndia, CHT) Long interval, sporadic (Dehra Dun) 35–37 (lower Himalayas, gregarious)	Anantachote (1987) Gamble (1896) Banik (2000) Gamble (1896) Bahadur (1980)
B. polymorpha	Gregarious, rarely sporadic	50 ± 5 (Myanmar) tsporadic, FRI Chittagong)	Brandis (1899) Banik (1987a)
B. tulda	Part flowering, sporadic, occasionally gregarious	Sporadic, isolated clump (NE India, W. Bengal, CHT, Myanmar) 20, 16, 19–30, all over subcontinent 40 ± 5 (Myanmar, Mizoram, Sylhet)	Gamble (1896) Banik (1991) Brandis (1899) Troup (1921), Banik (1987a)
B. vulgaris	Rarely flowers, no seed	80 ± 8 (Chittagong)	Banik (1979)
Dendrocalamus giganteus	Sporadic, iso. clump, gregarious	North India, Chittagong Kepong, Selangor 85 ± 5 (Dehra Dun)	Bahadur (1979) Banik (2000) Holttum (1958) Bahadur (1979)
Dendrocalamus hamiltonii	Spor./iso. clump Gregarious Gregarious Gregarious	Continuous Every year (Darjeeling) 30 (Assam) 23 ± 5 (Sylhet, Cachar hills, Mizoram) 45 ± 5 (Manipur, Sikkim, Arunachal)	Rogers (1900) Cavendish (1905) Banik (1999) Banik (2000)

D. longispathus	Often sporadic Occasionally gregarious	CHT, Myanmar, Tripura, Mizoram 26 (Tungo, Myanmar) 29 (Pegu, Myanmar) 30 ± 2 (CHT, Chittagong, Myanmar)	Gamble (1896) Banik (2000) Troup (1921) Brandis (1906) Banik (1991)
D. strictus	Gregarious and sporadic	25 (Mysore, South India) 40–45 (North, East, Central India) 45 ± 5 (Chittagong) 65 (West India)	Kadambi (1949) Gupta (1952) Banik (1981) Mathauda (1952)
Melocanna baccifera	Gregarious, rarely sporadic	30 ± 5 (Cachar hills, CHT, Chittagong) 45 ± 5 (Manipur, Mizoram, Sylhet Tripura, Arakan, Cox's Bazar) 60 (some part of Arakan)	Brandis (1899) Troup (1921), Banik (2010a) McClure (1966)
Schizostachyum dullooa	Sporadic and occasionally gregarious	47 (CHT) 15 ± 2 (Cachar, Assam) 37 (Chittagong)	Banik (2000) Gupta (1972) Banik (1999)
Ochlandra travancorica	Gregarious, sporadic	7 (?)	Varmah and Bahadur (1980)
Thyrsostachys siamensis	Gregarious	48 23–24	Varmah and Bahadur (1980) Banik (2000)
Ph. Pubescens	Gregarious, sporadic, irregular	60, 120	Watanabe et al. (1982)

CHT Chittagong Hill Tracts, N.E. India northeast India, Iso. Isolated

Fig. 5.1 Few clumps of
M. baccifera remain with
green leaves without any
flowers growing in small
patches (locally known as
Mau-Hak) inside a vast
majority of gregariously
flowering population
covering a large tract of
land in Mizoram

Kawamura (1927) and Janzen (1976) inferred that such variation in flowering periodicity is due to different clones (cohort) within the same bamboo species that are slightly 'out of phase' in flowering with each other. A bamboo 'flowering population', that is, a 'flowering genotype', may be described as a cohort, located at some geographical position, where individuals may have closely allied genetic makeup (Banik 2000) and may consist of a number of individuals (clumps) that are half sib in nature (only one parent is common), and due to this genetic proximity, the 'biological clock' runs almost synchronously in all the plants in the cohort (Pattanaik et al. 2002)

- *Importance of cohort diversities*: The existence of a number of cohorts with diverse duration of flowering cycles seems a buffering mechanism at a time death of all clumps of a bamboo species (Banik 2000). The genetic variability provides opportunities for the improvement of different bamboo species to enhance the productivity. It is not unlikely that a few individual clumps (cohort) may be discovered that flower but do not die after seeding or die partly and rejuvenate.

Cohorts of a bamboo species (seeds/seedlings/other propagules) having such diverse duration of interseeding periods need to be collected from different localities (including countries) and centralised in the plots, may be termed as 'seed stands' (Banik 1997a, 2000, 2008), and these centralised cohorts would offer possibilities of frequent availability of seeds in the next (future) flowering time as they have different interseeding periods.

5.3.1.3   Reproductive Biology

(a) *Part-flowering and complete-flowering nature of clumps:* Depending on the intensity of flowering among the culms along with the duration, a clump of a bamboo species may be either *complete flowering* or *part flowering* (Banik 1986). However, a bamboo species may have both types of flowering clumps. All culms of different age groups in a clump of majority of forest-grown natural bamboo species (*B. bambos, B. tulda, Cephalostachyum pergracile, D. asper, D. brandisii, D. hamiltonii, D. longispathus, D. strictus, M. baccifera, Schizostachyum dullooa, Thyrsostachys siamensis,* etc.) produce flowers simultaneously and die at a time within 1 year of blooming. Thus, the clumps in these species are generally *complete flowering* in nature.

In cultivated population of *B. tulda, B. nutans* and *D. longispathus*, a few culms (irrespective of age) of a clump may flower and die partly in the first year, while the remaining culms may complete flowering within next 2–3 or more years and then die exhibiting *part-flowering* behaviour (Banik 1986). Most of the clumps of *B. nutans* and *B. longispiculata* produce only a few partially (0.2–8.0 %) flowering branches mainly in lower mid-positions in the culms, and then gradually all branches produce flowers and the clump may die within 3–4 years or revive (*part flowering*). Rarely in some occasion, a clump of *B. nutans* may continue to flower every year for 17 years, exhibiting *continuous-flowering* nature and then stop flowering without any death. Similarly one clump of *B. longispiculata* flowered partly every year for 22 years, since 1978 till the year 2000, and did not die, exhibiting *continuous-flowering* nature (Banik 1997a). Irrespective of age, all culms in flowering clumps of *B. tulda* may produce both partially and fully flowered branches. Thus, some bamboo species may show both *part-flowering* and *continuous-flowering* nature. *Thyrsostachys siamensis* usually flower *completely* within 1 year, while a few clumps also showed *part-flowering* nature for 2½ years and then died (Banik 2000). *B. balcooa* and *B. vulgaris* are common homestead bamboo species rarely seen to flower without any seed yield, mostly in isolated clump complete flowering within 12–18 months (Banik 1979; Banik and Alam 1987). However, a few clumps of *B. vulgaris* also exhibited *part-flowering* nature for 3–5 years and stopped flowering and revived; such 'genotype' may be selected and centralised for cultivation (Banik 1979). *Bambusa vulgaris* do not produce any seeds as 70–92 % pollens are sterile (Banik 1997a). The bamboos in the homestead cultivation are thus selected earlier for their rarity and part-flowering nature.

(b) *Time/duration of floral shoot appearance to seeding:* The inflorescence in bamboo is an indeterminate compound panicle, usually large, in which spikelet-like branches developed, termed as pseudospikelets (McClure 1966). During flowering in many bamboo species, pseudospikelets are commonly developed on specialised branches, termed as *floral shoots* (Banik 1986, 1997a, 2000). These shoots are developed from the node (e.g. *Dendrocalamus* sp. (Figure 5.2a), *Schizostachyum dullooa, Thyrsostachys* sp. and in some *Bambusa* sp., etc.) or at the apex of vegetative branches in *M. baccifera*, and floral buds are arranged on one side of the floral shoot (Fig. 5.2b). Soon after flowering, the leaves below the floral shoot rapidly turn yellow and wither, and finally all branches on the culms become

Fig. 5.2 (**a**) *Dendrocalamus longispathus*, long floral shoots developed at the culm nodes; (**b**) *Melocanna baccifera*, floral shoots at the apex of vegetative branch and flora buds are arranged on one side of the floral shoot; (**c**) *Bambusa tulda*, floral shoots born directly on the leafy branches; (**d**) *Bambusa balcooa*, has florets with purple tips; (**e**) *Bambusa cacharensis*, anther filaments are long, free and hanging in nature; (**f**) *M. baccifera*, filaments are comparatively short

leafless and produce flowers. The inflorescences in *B. tulda, B. nutans, B. glaucescens* and *B. longispiculata* are a short panicle which usually do not develop on any floral shoots (or may have short floral shoots) and born directly on the leafy branches (Fig. 5.2c). The length and nature of floral shoots also vary with the bamboo species.

Initiation of flowers on the culm-node position within a clump also varies with the species. In some species (*B. glaucescens, B. tulda, D. longispathus, D. hamiltonii*), flowering starts from the top of the culm, and in others (*B. nutans, B. longispiculata*) it is from the bottom or mid-position of the culms. Conversely, random flowering throughout the crown (as in *M. baccifera, B. bambos*) stimulates greater movement of wind between inflorescence and different clumps.

The initiation of floral shoots and complete development, blooming and anthesis of flowers in different bamboo species usually take place within 3–8 weeks, mainly during the dry season of the year (September to March, autumn–winter, early part of spring) (Table 5.2; Banik 1986, 2000). Fruit setting starts within a week of pollination. However, fruit (seed) maturation is quicker in the early part of seeding season. Both flowering and fruiting in a clump are simultaneous.

(c) *Flowering and culm emergence in a clump:* Troup (1921) reported that checking of new culm production from a bamboo clump could be a reliable sign to predict flowering in the following years. Contrary to his statement, Banik (1986) observed that in *B. bambos* and *M. baccifera*, culms were also produced in the preceding year of flowering. However, *complete-flowering* clumps of these species did not produce any culms in the current year of flowering, whereas those of *B. cacharensis, B. tulda, B. vulgaris* and *D. longispathus* produced culms but fewer in numbers. All the *part-flowering* clumps of different species, except *B. glaucescens*, produce culms in the current year of flowering. Observation on the 1993–1994 flowering of *D. hamiltonii* in Arunachal Pradesh of India and 1997–1998 flowering at BFRI Bambusetum of Chittagong revealed that there was no new culm production in the flowering clumps of both the places (Banik 2000).

(d) *Blooming nature, anthesis and pollination related to seed yield*: Each pseudospikelet consists of a number of florets. The florets are green (*B. nutans, B. tulda,* etc.) and not showy (Fig. 5.2c) or bright, while *B. nepalensis* and *B. balcooa* (Fig. 5.2d) have purple tips. Anther filaments are long, free and hanging in nature (Fig. 5.2e) (e.g. *B. bambos, B. longispiculata, B. cacharensis, B. tulda, D. hamiltonii,* etc.) or may be comparatively short (e.g. *B. vulgaris, D. longispathus, M. baccifera* (Fig. 5.2f), etc.). The anthers are generally coloured (yellow, purple, reddish, etc.) or white. The florets of *M. baccifera* are diagnostically prominent, and it has an elongated purple-coloured stigma (Fig. 5.2f). Since bamboo is ane-mophilous, wind pollinated, it must have many flowers at anthesis at the same time for successful spread of the pollen. The cross-pollination is usually difficult for wind-pollinated plants in a low-wind forest understorey

Table 5.2 Floral shoot appearance, flowering and seeding period in some bamboo species at Chittagong (*Source* Banik 2000)

Species	Floral (Fl.) shoot		Flowering date	Fl. shoot to flowering (weeks)	Seed ripen (date)	Flowering to seed (weeks)
	Initiation date	Length (cm)				
Bambusa bambos	Dec 2nd week 1979	5–48	Feb 1st week 1980	6	Mar–Apr 1980	6–7
B. glaucescens	No fl. shoot, Mar 1st week 1978	–	Apr 1st week 1978	3	May–Jul 1978	6–7
B. longispiculata	No fl. shoot, Feb 2nd week 1978	–	Mar 2nd week 1978	3	Apr–Jul 1978	6–7
B. tulda	Dec 3rd week 1979	5–25	Feb 1st week 1980	5	Apr–Aug 1980	6–8
B. vulgaris	Jan 1st week 1979	10–45	Feb 4th week 1979	6	No seed	–
Dendrocalamus longispathus	Jan 2nd week 1977, 1978, 79	50–90	Mar 2nd week 1977, 1978, 1979	6	Apr–Jul 1977, 1978, 1979	6–7
Dendrocalamus hamiltonii	Oct–Dec 1st week 1996	50–250	Dec 4th week 1996, Jan 4th week 1997	6–8	Mar–May 1997	6–8
Melocanna baccifera	Sept 2nd week 1990	12–45	Nov 2nd week 1990	7	Apr–Sept 1990	weeks

environment. While collecting seed of *Dendrocalamus hamiltonii* across dense forest in East Nepal, clumps near to ridges had better seed production than those in furrows (Stapleton 1982). Dehiscence starts from the apex and moves longitudinally down of the anther. In *Bambusa bambos, B. tulda, B. glaucescens* and *Dendrocalamus longispathus*, opening of flower and pollen discharge take place only in the morning 0600–0900 h (Banik 1986) and in *D. strictus* between 0600 and 1300 h (Nadgauda et al. 1993). The flowers are open for about 2–3 h and then close, but close more quickly in dry weather. On touching or with gentle air movement, the anthers of *B. tulda* and *D. longispathus* discharge pollen grains in the form of a pollen cloud. The filaments continue to elongate and expose to facilitate pollination, and the anthers start drooping followed by dehiscence. The pollination in bamboos is not by insects but mainly by wind movement (Faegri and Vander Pijil 1979; Banik 1986; Koshy and Harikumar 2001). In *D. strictus* the gynoecium matures 3–4 days before the androecium (protogynous), effectively preventing self-pollination (Nadgauda et al. 1993). *Phyllostachys heteroclada, Ph. nidularia* and *Ph. nuda* are also protogynous. Stigmas were receptive and had pollen deposited on them 2 days before anther dehiscence (Huang et al. 2002). In *M. baccifera*, anthers come out in the morning and burst in the afternoon between 1500 and 1700 h, as the filaments are short and anthers do not emerge out much, whereas the style is much elongated and the major portion is exposed, so that more pollen from the wind can be trapped to favour cross-pollination and more viable fruits are produced (Banik 1998, 2010a). Heavy rainfall during anthesis would wash away pollen grains and hamper the success of pollination and final seed yield. Ueda (1960) observed pollen fertility varied from 91.3 to 97.0 % in some Indian bamboo species and 90.3–98.6 % in Japanese bamboo species. The pollen cannot germinate during exposure to sunshine or dry conditions for about half an hour.

Majority of bamboo species flowering gregariously over a large tract of land facilitates cross-pollination, resulting in higher production of viable seeds with wide genetic base having lots of variation in progenies. The seed setting is also much higher in the complete-flowering clumps than part-flowering clumps (Banik 2000).

Blooming in a bamboo clump is not continuous; rather it occurs in three distinct successive flushes (*flush period*) with two non-blooming periods (*rest period*) in between (Banik 1986; Table 5.3). In the case of *part-flowering* clumps, where flowering continues for 2–3 years, the *flush* and *rest* periods are distinctly identifiable generally in the first year of flowering. The first two *flush periods* of *Bambusa glaucescens* and *D. longispathus* with viable seed yield are longer, whereas the final (third) *flush* period is short and seeds (caryopses) are mostly empty (Banik 1986). Generally seed collection should be done in the first two flushes to obtain maximum amount of viable seeds (Table 5.3). The *rest periods* in between the flush periods probably are required for resuming the physiological potentiality by repositoring the meristem at the base of pseudospikelet. In *B. bambos* duration of all the three flush periods is more or less equal and short, and due to this, the species produced viable seeds throughout the whole flowering period. In Orissa, Nicholson (1945) also observed that *B. bambos* flowered *successively 3 times* in the flowering year and produced plenty of seeds. During 1990–1991 flowering in *D. strictus* clumps did *number of flushes* interspersed with *short rest* periods (Nadgauda et al. 1993).

(e) *Seeding season, adaptation and seed dispersal:* Generally in tropical and subtropical bamboo species, 'seed' production is optimum during spring to early part of rainy season (mid-February to June) and poor in autumn to early part of winter (from later part of September to November). Bamboo seeds are short lived and loss viability within 1–2 months of collection. Immediately after ripening, seeds fall on the ground during monsoon (later part of May–August) and start germinating within a week. If seeds would mature much earlier to the rainy season, there would be no or poor germination due to their

Table 5.3 Successive nature of flowering with seed yield and germination in five bamboo species

Species	Flowering period (dates) Blooming	Non-blooming	Seed per clump (g)	Average germination (%)
B. bambos	FL—1 (3 Feb–14 Feb) 12 days	RT—1 (15 Feb–20 Feb) 60 days	30–80	49. 5
	FL—2 (12 Feb–7 Mar) 15 days	RT—2 (8 Mar–21 Mar) 14 days	10–25	45. 0
	FL—3 (21 Mar–3 Apr) 13 days		4–6	62.4
B. glaucescens	FL—1 (5 Apr–24 May) 50 days	RT—1 (25 May–29 May) 5 days	15	25.2
	FL—2 (30 May–18 June) 20 days	RT—2 (19 June–30 June) 12 days	6	55.2
	FL—3 (1 Jul–9 Jul) 9 days		Empty seed	0
B. vulgaris	FL—1 (25 Feb–10 May) 75 days	RT—1 (11 May–28 May) 18 days	Seed not produced	0
	FL—2 (29 May–4 Aug) 68 days		Seed not produced	0
	FL—3 (17 Aug–29 Sept) 44 days	RT—2 (5 Aug–16 Aug) 12 days	Seed not produced	0
B. balcooa	FL—1 (08 Dec–18 Feb) 70 days	RT—1 (19 Feb–10 Mar) 20 days	Seed not produced	0
	FL—2 (11 Mar–09 May) 58 days	RT—2 (10 May–23 May) 13 days	Seed not produced	0
	FL—3 (24 May–12 Aug) 42 days		Seed not produced	0

(continued)

Table 5.3 (continued)

Species	Flowering period (dates)		Seed per clump (g)	Average germination (%)
	Blooming	Non-blooming		
D. longispathus	FL—1 (10 Mar–3 May) 55 days	RT—1 (4 May–22 May) 19 days	15	25.2
	FL—2 (23 May–17 Jun) 26 days	RT—2 (18 Jun–27 Jun) 10 days	25–40	48.8
	FL—3 (28 June–6 Jul) 9 days		Empty seed	0

FL flush period, RT rest period

short-lived nature. Thus, ungerminated seeds remain exposed to the predator animals like birds, bisons, wild bores, porcupines, deer, rodents, etc. Again if the distribution of seeding time, seed maturation and seed fall is prolonged towards October to January (autumn–winter), seeds would fail to germinate due to dry and cool climate that would result in failure in natural regeneration of bamboos. Therefore, the seeding season of tropical bamboos, like other tropical trees (*Dipterocarpus, Artocarpus* spp.), seems to be an adaptation to nature for protecting and maintaining their races in the total ecosystem.

The bamboo seeds, usually, are eaten heavily by rats, birds, wild boars, porcupines, deer and other animals and also by the local hill tribes and usually carry these far away from the seeding mother, thus assisting in dispersal. Mature seeds (fruits) of *M. baccifera* may also disperse far away from mother clumps by rolling over the hill slopes due to their heavy weight, smooth and somewhat round shape. Bamboo seeds are also moved along the flow of rainwater through hilly streams (Banik 1998) and dispersed widely covering large areas of land.

5.3.2 Flowering and Monocarpic Death

Bamboos are in general monocarpic, meaning plants (clumps) flower and fruit once in their life time and then die. Death is attributed to reproductive exhaustion caused by the utilisation of food reserves from the vegetative parts.

5.3.2.1 Impact of Monocarpic Death

The possible impact of flowering and large-scale death of different bamboo species is discussed in the following segments.

5.3.2.1.1 Socioeconomic Effect

(a) *Shortage of bamboo and loss of livelihood:* Next to agriculture, bamboo-based economic activities generate a large amount of employment which spells ready cash for the population inhabiting in the hilly and rural areas of the Indian subcontinent, Myanmar, Thailand, Vietnam, Cambodia, China and many other countries of Asia and South America. The entire standing stocks of bamboo crops are eliminated within few years of gregarious flowering, creating *scarcity of bamboo resources and raw material* for housing, construction, handicraft, cottage and processing industries and pulp and paper mills (Banik 2004). This creates *loss of employment and earnings* which result in socioeconomic sufferings of the local people and political unrest as happened during the 1960s in Mizoram, India.

(b) *Shortage of edible shoot production:* Local tribal communities have been consuming the tender shoots as one of the major food items during rainy season. Recent flowering (1995–2010) and large-scale death of *D. asper* and *T. siamensis* in Thailand and *M. baccifera* and *D. hamiltonii* in the northeast India, Chittagong Hill Tract and northwest Myanmar including sporadic death of *B. tulda, S. dullooa* and *D. longispathus* have resulted in an alarming *shortage* of *edible bamboo shoots* and thus influence the nutritional level and health condition of local hill people.

(c) *Localised famines and diseases:* Mass flowering and death of bamboos in Mizo hills during the 1960s led to an explosion in the population of rats. These devoured the grain in the field and in village warehouses, causing scarcity of food for humans which created a famine in the area. The explosion of rat population have a devastating effect on the jhum (slash and burn) cultivation on which a majority of the indigenous people depend for growing food, thus affecting the already precarious food security of the rural hill people. The farmers may be inspired to grow plant crops that rats do not like to eat, such as ginger, turmeric, cotton and costly medicinal plants, so as to earn cash to buy food. In the past, such adverse socioeconomic effect prompted many people to move from their native hilly village to settle somewhere far away.

5.3.2.1.2 Effect on the Forest Ecosystem

The entire standing stocks of bamboo vegetation are eliminated within a few years of time due to gregarious flowering. The bamboo vegetation starts shading leaves and within 2–3 months becomes leafless and culms turn pale green to yellow and simultaneously produce seeds. In most of the cases, no new culm emergence and no new photosynthetic reserves are created due to the absence of green foliage. Thus, the C_2 cycle in the existing continuous bamboo vegetation is disrupted, and the stored carbohydrates in the rhizomes and culms are utilised for flower and fruit production. All flowered bamboo plants in the vegetation die within 18–24 months, the culms dried and become brittle for gradual rotting in to humus. Sudden opening of 20–50-year-old continuous dense green overhead cover of bamboo vegetation allow easy penetration of sunlight through open canopy, influence the gregarious growth of weeds detrimental for the survival of regenerating bamboo seedlings and affect the ground moisture content, microflora and forest composition by influencing the regeneration of local tree communities and invasion of other species. During 4–6 months, draught season increases the chances of forest fire due to large presence of dead and dry culms. Exposed hills increase the amount of soil erosions, landslide, etc., and also adversely affect the water catchment areas of many perennial streams and rivers. Many animals like monkey, porcupines, buffalo, wild boar, elephant, etc., suffer from scarcity of food and prompted them to migrate in nonflowering areas.

5.3.2.1.3 Possible Opportunities from Bamboo Flowering

(a) *Collection of seeds by local people and selling as commodities in the market:* In the past, least amount of bamboo seeds was collected for raising bamboo plantations, as the nursery technology was unknown for most of the bamboo species. In those days, seed collection was also very difficult due to lack of road communication and wilderness of the forests. As a result, in the past, huge amount of seeds used to remain in the forest floor and attracted the wild animal. But in recent years of bamboo seeding in Asian countries, the situation has changed; emphasis has been given in raising large-scale bamboo plantation. So people are collecting bamboo seeds and quickly dispatched to the local markets, buyers procuring these with cash, and thus bamboo seed has become a commodity for earning money. In Betcherra area of north Tripura (India), about 530 days of employment was generated during 2004 through seed collection of *M. baccifera*.

- *Formation of community nursery:* The people residing in the flowered area can be trained to form *community nursery as local enterprises* for producing quality bamboo planting materials for selling and *raising large-scale plantation* and to *protect the regenerating bamboo seedlings*.
- *Employment for indigenous community:* The local indigenous people should be employed in harvesting, storing and value-added utilisation of dead bamboos and restocking of bamboo resource.

5.4 Aiding and Maintenance of Natural Regeneration

Monsoon starts the profuse regeneration of bamboo seedlings from the fallen seeds on the ground. The seedlings of *B. bambos, D. hamiltonii, D. strictus* and *M. baccifera* take 4–5, 4–7, 4–6 and 3–4 years, respectively, to establish, and the new bamboo crop is ready in 8–12 years for systematic and scientific harvest for the next life cycle of 30–50 years. To ensure the success in establishing the regenerating seedlings, proper aiding and management are necessary.

5.4.1 Related Factors and Activities in Aiding the Natural Regeneration Process

At the initial stage (from year 1 to 3), the wild seedlings of different bamboo species (from year 1 to 5) need proper care and nursing for survival in the process of obtaining the successful natural regeneration (Banik 1988). The following aiding operations (Banik 1988; Fu Maoyi and Banik 1996) have been found essential for hastening the success of natural regeneration (Aided Natural Regeneration) into a bamboo forest again.

i. *Survey, identification and demarcation of the area* should be done to get acquainted with the topography and extent of land under seeding, and a map may be drawn for monitoring the regenerating process.
ii. W*eeding, vine cutting and maintenance of partial shade:* Mature bamboo seeds fall randomly on the ground from the newly seeded mother clumps, germinate profusely on the onset of rains and start growing. Wild seedlings look like rice or wheat seedlings and are often seen as a thick mat on the ground just below the flowering mother clumps. Due to the death of seeded mother clumps, the population of weeds and vines increases on the forest floor that suppress and may even kill the regenerating bamboo seedlings. Only 1 month of weed suppression was found to kill most of the seedlings of *B. tulda, D. longispathus, B. bambos, D. strictus, D. hamiltonii, S. dullooa* and *G. andamanica*, while *M. baccifera* seedlings can stand suppression up to 6 months as seedlings of this species are fast growing and aggressive in nature (Banik 1988). In 4–5 years, *M. baccifera* seedlings develop into merchantable clumps, whereas it takes 10–12 years in the natural condition if left unattended. In Burma, the regenerating seedling population became thin in unattended sites, and bamboo areas become squeezed in the next generation (Kurz 1877). *First weeding* and vine cutting have to be done at the beginning of rainy season to minimise the weed suppression, and again at the end of rainy season, a *second weeding* may be done. This may be followed in the *second year*, and at the *third year*, during rainy season, one thorough weeding and vine cutting have been found very useful.

The bamboo seedlings grow *better under partial shade,* so to *provide partial shade* to the regenerating bamboo seedlings, harvesting of dead mother clump may be delayed at least 3–4 months in *M. baccifera* and 6–9 months in other bamboo species (Banik 1988). The regenerating seedlings under *complete overhead shade* gradually degenerated, but sprung up readily in the crown gaps (partial shade).

iii. *Fencing the area so that cattle cannot enter and graze:* The tender leaves and rhizomes of the bamboo seedlings are very delicious fodder to the *animals* and thus remain vulnerable to the *grazing and trampling*. So in initial years, fence or trench may be made at *the vulnerable entry side* to prevent the entry of animals. Bamboos cultivated in the villages, during seeding, need special attention for protecting the seeds against the domestic fowls, birds and the regenerating seedlings from grazing.

iv. *Filling up the gap area by substitute seed sowing and seedling planting*: The seeds are not evenly disbursed throughout the regenerating area, and as a result some patches of land in between remain very thinly populated or without any bamboo seedlings. *Gap filling* is usually done by seedling planting at even spacing or by substitute direct sowing of seeds (e.g. in case of *Melocanna baccifera* seeds 2 per pit) in planting pits during rainy season to obtain adequate evenly distributed plant density of regenerating stock. Depending on the situation, this operation may again be carried out in the second and third year of regeneration process.

v. *Protection from fire and monitoring:* During winter and dry season, there is always a chance of fire in the forest floor due to the presence of dead mother culms that may kill the regenerating bamboo seedlings. So a *fire line* (5 m wide) may be made by scraping the ground with spade and cutting jungle on both sides in the beginning of dry season to control fire and local watcher may also be engaged. This may also serve as *inspection path*. Within 12 months time, the seedlings develop dependable rhizome systems which help them to revive even after death of above-ground shoots due to fire. Awareness campaign is needed among the local hill/forest people against the possible *fire hazards, animal predation* and *grazing*.

Gradually in 2–3 years time, clusters begin to form, and eventually in 5 years or more, the area carries a healthy homogeneous crop of more or less equally spaced young clumps of bamboo. The naturally selected seedlings are superior and give rise to a vegetation composed of diversified genetic materials.

5.5   Artificial Regeneration

5.5.1   *Production of Planting Materials and Nursery Management*

5.5.1.1   Nursery for Production of Planting Materials

A full-fledged nursery facility is essential to obtain success in propagation activities in bamboos. These are summarised below.

5.5.1.1.1   Permanent and Temporary Nursery

A nursery area should be at a gentle slope of about 5° to ensure water run-off, may be from 1,000 to 5,000 m² depending on the amount and extend of production of planting materials. The nursery site should have the road accessibility and needs electric connection to install facilities of intermittent mist, seed storage and other installations. A *permanent nursery* is needed for producing large number of planting stocks year after year, not less than 7–10 years. The nursery site should have year-round water facility. *Propagation beds* are to be prepared on properly levelled (no water stagnation) ground or cemented platform (Fig. 5.3a). The ideal size of each propagation bed is 1.2 m wide, 6.0 m long and 21 cm deep. About 0.75 m wide foot path is provided between 2 adjoining beds. There should be workable distance (1.0 m) between two platforms; a drain line has to be made surrounding each of the platforms to drain out excess water. More number of such platforms may be constructed to prepare the desired number of beds on them.

A *seedbed* is usually smaller than propagation bed, where seeds are sown for germination and seedling production (Fig. 5.3b). The common size for a seedbed is 5 m long and 1.2 m wide. The bed is usually raised 10–15 cm to ensure good drainage. The germination medium in the seedbed should contain soil, sand and cow dung or FYM or vermicompost at the ratio of 2:2:1 for proper germination and healthy growth of the seedlings. The seedbeds may be drenched with 0.01 % Aldrex and 0.05 % Bavistin to prevent termite and fungal attack, respectively.

Overhead Agrinet has to be placed on a strong frame at 2.5 m height in the centre and 1.5 m at the sides on the cemented platform for maintaining 60 % light on the beds below, also controls wind speed and can be used as hardening chamber. One or two cemented platforms may be kept for using as *hardening nursery*, on which 5–10 beds may be prepared. The freshly transferred rooted cuttings from propagation beds to *polyethylene bags* (poly bags) need to be placed for a week or two on such beds to harden. The nursery tools including trolley for carrying the plants and other materials have to be ensured.

A number of nursery beds need to be prepared on the cemented platform without any overhead net shed cover (*open nursery*) for placing the poly bags having well-rooted cuttings or seedlings under open sky.

Fig. 5.3 Nursery beds: (**a**) *propagation beds* prepared on properly levelled (no water stagnation) ground or cemented platform; (**b**) *seedbed* where seeds are sown for germination and seedling production

A ***misting facility*** has to be installed in the permanent nursery.

A ***temporary nursery*** is set up mainly to produce limited number of planting stocks near the planting sites, for 1–2 years of activities. It is constructed by using very low-cost materials. Watering may be done manually. *Propagation beds* and *seedbeds* may be placed directly on the levelled ground covered with thick black polyethylene sheet. Overhead shade may be provided using palm or bamboo roofing. A low-cost shed structure can also be made using bamboo poles and strips, with overhead transparent polyethylene film cover.

In both cases, occasionally wash the area with a 2 % formaldehyde solution or sodium hypochlorite solution (1-part to 9-part water).

5.5.1.1.2 Sexual Method for Producing Bamboo Planting Materials

(a) ***Seeds and seedlings***: Depending on the availability, bamboo seeds/seedlings are ideal planting stocks particularly for large-scale plantation, especially, in forests. Clumps developed from seeds have maximum duration of vegetative state and thus continue to produce culms for longer period. Importantly it also maintains the genetic variability.

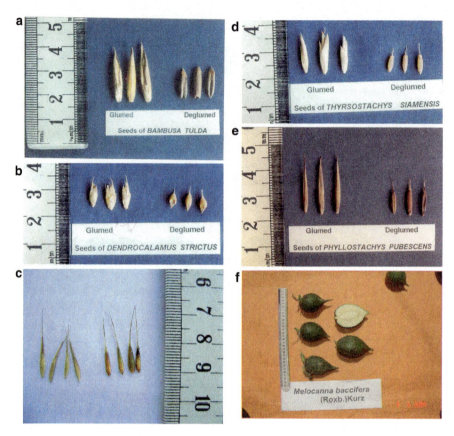

Fig. 5.4 Seed character of different bamboo species: (**a**) *B. tulda,* like wheat grain; (**b**) *D. strictus,* small like coriander seed, shiny grain-like somewhat ovoid to subglobose; (**c**) *Schizostachyum dullooa,* narrow, elongated grain-like, somewhat broader and flat base with cylindrical top terminating in a long beak formed by the persistent base of the style, deglumed seeds are *blackish brown*; (**d**) *Thyrsostachys siamensis,* small like paddy grain; (**e**) *Phyllostachys pubescens,* narrow elongated grain-like; (**f**) *M. baccifera,* very large and obliquely ovoid, thick, *fleshy green* and smooth

Bamboo produces one-seeded fruits with thin pericarp adnate to the seed coat, known as *caryopsis* (hereafter 'seed'), and covered with a number of persistent glumes. These are *small, light in weight and dry grain-type seeds* and can tolerate some desiccation (*orthodox*). There are few bamboo species (genera *Dinochloa, Melocalamus, Melocanna, Ochlandra,* etc.) which produce *large fruits,* usually green or brown, smooth or wrinkled and sensitive to desiccation (*recalcitrant*). The fruit of *M. compactiflorus* appears as 'a small wood apple'; of *Melocanna baccifera* as large, fleshy and 'berrylike'; and of *Dinochloa scandens* as 'spherical'. In general the number of seeds per kilogram varies from species to species. The seed character, shape and weight of some major bamboo species are presented below (Fig. 5.4a–f; Table 5.4).

Table 5.4 Seed characters of some bamboo species of Asia

Species	Seed shape and weight
Bambusa balcooa	Do not produce seed
Bambusa bambos	Small like wheat grain, 990 seeds per 10 g
B. bambos var. *spinosa*	Small like wheat grain, 1,325 seeds per 10 g
B. glaucescens	Like wheat grain, 151 seeds per 10 g
B. longispiculata	Like wheat grain, 145 seeds per 10 g
B. nutans	Like wheat grain, 100 seeds per 3 g
B. polymorpha	Small like wheat grain, 1,250 seeds per 10 g
Bambusa tulda	Like wheat grain, 150 seeds per 10 g
B. vulgaris	Do not produce seed
Dendrocalamus giganteus	Ovoid to oblong, hairy at the upper end, 200 seeds per 10 g
D. hamiltonii	Small, broadly ovoid, rounded at the base 264 seeds per 10 g
D. longispathus	Small like coriander seed, 1,350 seeds per 10 g
D. membranaceus	Ovoid, rounded at the base 474 seeds per 10 g
D. strictus	Small like coriander seed, shiny grain-like somewhat ovoid to subglobose, 515 seeds per 10 g (Chittagong, Bangladesh), 258 seeds/10 g (Chhindwara, Madhya Pradesh (MP), India), 223 seeds/10 g (Betul, MP, India), 244 seeds/10 g (Hoshangabad of MP, India), 265 seeds/10 g (Bilaspur of MP, India)
Gigantochloa andamanica	Small like wheat grain, but narrow 265 seeds per 10 g
Melocalamus compactiflorus	Like chestnut or betel nut. The weight of a seed varies from 2 to 20 g
Melocanna baccifera	Very large and obliquely ovoid, thick, fleshy onion-shaped and the apex terminating in a curved beak, green with smooth surface and not covered with glumes. The weight of a seed varies from 7.0 to 150 g, length from 35 to 110 mm and diameter from 22 to 60 mm, 45 to 70 seeds per kg
Ochlandra travancorica	Large, 5-cm long, 2–3-cm broad, brown, oval–oblong, wrinkled, with 4–5-cm long stiff beak, pericarp fleshy enclosing seed, the whole surrounded by the persistent glumes and palea. About 40 deeds per kg
O. travancorica var. *hirsuta*	Description more or less similar to *O. travancorica*, the weight of a seed 24–32 g
O. scriptoria,	640 fruits per kg
Schizostachyum dullooa	Narrow, elongated grain-like, somewhat broader and flat base, with cylindrical top terminating in a long beak formed by the persistent base of the style, deglumed seeds are blackish brown, 393–410 seeds per 10 g
Thyrsostachys siamensis	Small like paddy grain, 248–490 seeds per 10 g
Phyllostachys pubescens	Narrow elongated grain-like, length 21–27 mm, diameter 1.2–1.8 mm; 4,840 seeds weighing about 73 g (Watanabe et al. 1982)

Seed Collection and Processing: Bamboo seeds collected from the ground generally give low germination percentage, while mature and healthy seeds from the crown collected by shaking give higher germination percentage. In case of *M. baccifera*, seeds have to be plucked directly from the culms or the clumps have to be shaken (Banik 1998). For tall culms like *B. bambos* and *D. strictus*, a long pole with a hook is needed to pull and shake the culms. For small culms like *T. siamensis* and *S. dullooa*, manual shaking is a very easy method. The mature seeds should be protected from predators.

Empty seeds can be separated by floating in water. The seeds should be kept and stored in gunny or hessian cloths bags and kept open in shade in ventilated atmosphere to prevent heat from building up inside which affects the viability of seeds. After this, tie the bag with proper tagging with name of species (both local and scientific), locality, date of collection and collector's name for subsequent identification of raised seedlings.

Seed Longevity and Storage: Dormancy in bamboo seed is not known, and seeds are short-lived and difficult to store (Banik 1987b). The seeds of *D. strictus* could be stored over silica gel or anhydrous calcium chloride in a desiccator or at 3–5 °C ambient temperature after reduction of its moisture content to 8 % (Varmah and Bahadur 1980). Fresh seed lots with 67 % germination capacity were stored under the above three conditions and exhibited 51, 54 and 59 % germination, respectively, even after 34 months. The soaking–drying treatment with low concentration of disodium hydrogen phosphate (10^{-4} M) proved to be better than water in maintaining the vigour and viability of seeds of *D. strictus* (Sur et al. 1988). Seed drying may be done in well-ventilated condition rather than directly in the sun. The seed longevity period of *B. tulda* could be increased up to 18 months by storing over silica gel in a desiccator (Banik 1987b). The fleshy seeds of *M. baccifera* when stored in an air-conditioned room retained viability up to 45 days and 60 days when stored with dry sand in jute bags, while it was only 35 days at normal room condition (Banik 1994b). Dried seed of *D. hamiltonii* stored in a sealed container with silica gel and dried seed stored in a refrigerator without silica gel gave 25 and 22 % germination, respectively, after 62 weeks. Seeds of *T. siamensis* stored at normal room temperature lost viability within 21 months, but when stored under cold room temperature (2–4 °C) and deep freezer (−5 °C) germinated 89.2–92.5 % even after 27 months (Ramyarangsi 1988). The seed viability of *Phyllostachys* sp. was preserved by storing the seed over calcium chloride at room temperature.

Seedbed and Germination Process: Preparation of a *seedbed* has been discussed under Sect. 5.5.1.1.1. Big and healthy bamboo seeds should be sown in *seedbeds* or polythene bags as soon as collected. Dry grain like seeds may be soaked in water overnight and sown in a line, with a seed-to-seed gap of 10 mm and between the lines 5 cm, in horizontal/laying down position rather than scattering or broadcasting. After sowing, a thin layer

Fig. 5.5 Abnormalities in bamboo seedlings in *M. baccifera:* (**a**) abnormal germination, (**b**) albino seedling

(10–15 mm) of the medium mixture should be applied on the sown seeds across the bed so that seeds never get exposed. The medium should be moist and well drained, but not waterlogged. Regular watering (light misting) around the sown seeds in the seedbed is needed. Seeds start germination within 3–7 days of sowing. The suitable temperature is 25–35 °C and bamboo seeds also need overhead partial shade (shed net) as they are *negatively photoblastic* in nature (Banik 1991, 1994c).

Different *types of abnormalities* (Fig. 5.5a) such as rootless plumules, stunted radicles and radicles growing upward are not uncommon (Banik 1994b) in the seedlings produced from light weight (4–10 g/seed) and sometimes too big and heavy (more than 150 g) seeds of *M. baccifera*. So, medium to big seeds should be selectively collected for raising plantation of the species.

In the caespitose types of bamboos with grain-like seeds, when germination takes place, the plumule emerges in the form of a pointed conical bud with sheathing scale-like leaves, which rapidly develops into a thin, wiry stem bearing single foliage leaves arising alternately at the nodes, the bases of the leaves sheathing the stem. Meanwhile fibrous roots develop from the base of the young shoot. Seed remains attached to the seedling 2–3 weeks of age. A rhizome system starts to develop in the seedling after 1–2 months of germination, and at a young stage, the rhizome movement is strongly geotropic. Adequate watering and cleaning of weeds from the beds and bags are to be practised regularly.

Transfer of Seedlings in Bags and Nursing: The seedlings of 4–6-leaved stage are to be pricked up from the seedbed and transferred to poly bags—of sizes 10×15 cm or 15×23 cm depending on the seedling size. The bags should be filled with soil and farmyard manure (FYM) at the ratio of 3:1. Regular weeding, watering and occasional adding of soil mixture to the bags should be practised. Initially up to 3–5 months, seedlings do best in partial shade compared to direct sunlight (Banik 1997b). The increase in total biomass of bamboo seedling in N treatments was found to be much greater (88 %) than in phosphorus treatments (24 %). A two-split application of $N_{100}P_{50}$ at 4 and 8 weeks after germination was superior.

Frequent shifting of seedling bags from one bed to other is an essential practice, otherwise the roots and rhizomes of one bag penetrate the adjacent bags, get intermingled with each other and thus cannot be transplanted in the field plantation without damage and injuries.

Seedling Character: In most of the bamboo species, the tufted form of the young plant commences to show itself at an early stage in the bags due to the production of short rhizomes that curve upwards and form aerial shoots. The buds and rhizomes, and the shoots arising from them, become successively larger and larger. The earlier shoots are thin, but subsequently a time comes when woody culms are produced. In *M. baccifera*, the germinating plumules are thick (4–6 mm diameter); within 1–4 weeks plumules elongate rapidly into stems, which are soft and succulent with vigorous growth, bearing single leaves arising alternately; and at 3 months of age, they become most elongated (175 cm) with thick (0.8 cm, diameter) stem.

Variation in Seedling Form: Different types of growth form like grassy, grassy erect, erect and very erect have been seen among the seedling population of *B. glaucescens, B. tulda* and *D. strictus* (Banik 1980) and bushy seedlings in *B. bambos* (Kondas et al. 1973). The erect and very erect type grew faster with elongated nodes and wider culm diameter, while the somewhat bushy forms are with many small culms. The erect and very erect type should be selected for raising plantation due to their growth potential. Sometimes a few albino seedlings (Fig. 5.5b) may be found in a population collected from isolated flowering clumps of *B. tulda, B. bambos, D. hamiltonii, D. strictus, M. baccifera*, etc.

In many cases, seedling populations of bamboos show genetic diversity (McClure 1966) after each gregarious flowering and, thus, provide opportunity for selecting tall and vigorously growing superior seedlings with desired combination of characters (*juvenile selection*) having a high yield potential (Banik 1997a, 2008). These seedlings can be multiplied further through macroproliferation technique (discussed under 5.5.1.1.4d), and high-yielding bamboo plantation may be raised through *juvenile selection*.

(b) *Utilisation of wild seedlings:* The densely populated bamboo seedlings com-
pete strongly in the wilderness for survival and should be thinned out to
minimise competition (Banik 1988). Two- to four-leaved stage of wild seedlings of different bamboos are sometimes collected and transplanted in the
polythene bags and used as planting stocks. Immediately after collection, wild
seedlings should be brought to the nursery and transplanted to polythene bags
containing soil (sandy loam) mixed with FYM/cow dung (3:1). At the beginning, seedlings have to be properly watered or misted keeping under partial
shade for 3–5 days for hardening. Sixty to seventy per cent light may be
provided by placing overhead shed net for healthy growth compared to direct
sunlight. Seedlings need regular weeding and daily watering at the nursery
stage.

5.5.1.1.3 Asexual Method (*Macro-Vegetative* and *Micropropagation*)

Due to the scarcity of seeds, bamboo has been generally propagated by various
vegetative methods. Desired special characteristics are often not reproduced when
grown from seed, so it is important to try to conserve them vegetatively (clones).
Many of the bamboos in cultivation, especially in homestead, were introduced as
clones of that species. A bamboo propagule must develop all the three morphologic
structures—the leafy axis, rhizome and root; its inability in the development of any
of these phases leads to complete failure of a bamboo propagule (Banik 1980).

5.5.1.1.4 Basic Environmental Conditions Required for Macro-Vegetative
Propagation

High Air Humidity: Vegetative pieces of bamboos, e.g., branch and culm cuttings,
lose water rapidly especially through cut ends. Such pieces have no roots to take up
water to replace the water loss. The cuttings must not be allowed to show wilting for
any length of time due to drop of the humidity. Death due to desiccation before
rooting is the major cause of lack of success in propagating bamboos. Intermittent
misting provides showering and bathing effect that cools the body temperature of
the plant piece (cuttings) and thus slows down the rate of respiration, so the stored
food is less utilised (Banik 2010b). More stored food in the plant piece means better
chances and good amount of rooting. Continuous misting usually causes rotting of
the cuttings. Waterlogging hinders root initiation by limiting the oxygen supply in
the bed and also promotes moulds and rotting of cuttings. So the propagation bed
should always be well drained. Use water free from pathogens as much as possible
for misting.

Air and Rooting Medium Temperature: Temperature above 38–40 °C in the
rooting medium and air even for short time is likely to result in the death of the
cuttings. The optimum temperature for both air and rooting media should be within
25–35 °C; it helps rooting by initiating cellular activity. High air temperature (more

than 35 °C) accelerates the bud break and sprouting, and thus stored food in the cuttings is utilised rapidly for shooting less amount remains for root production (Banik 2010b).

Propagation Bed and Rooting Media: Preparation of a *propagation bed* has been discussed under Sect. 5.5.1.1.1. *Rooting medium* is the material or mixture of materials in which unrooted plant parts (nodes, branches, rhizomes) of bamboo are stuck to produce roots. The suitable rooting medium should fulfil the following objectives (Banik 1983):

- Support plant pieces so that they remain in particular position.
- Should provide full moist condition and be porous to maintain well-aerated bed condition and maintain uniform medium temperature.
- Medium material should be chemically inert; neutral pH condition is better for rooting.
- Should not harbour any insect/harmful microorganism which would encourage diseases in cuttings and be free from toxic substances that hinder the rooting process.
- Easy to dug up and shift rooted cuttings with no/less root damage.
- Readily available, cheap, long lasting and can be reused.

A number of field studies and practical experiences in the bamboo propagation nurseries of the region show that sand has been found to fulfil almost all the above-mentioned criteria of an ideal type of rooting medium (Banik 1984, 1995) and, therefore, nowadays used as popular rooting medium in many bamboo nurseries of the region. Besides soil, other rooting media like sawdust, wood shavings, peat moss, bark dusts, coconut fibres, etc., were also tested as rooting media, but most of these provided limited rooting success and are not popular in the bamboo nurseries. Sand has to be cleaned and washed by hot water before using in beds.

Different vegetative propagation methods have been practised (Banik 1995) to produce mass-scale planting materials of desired bamboo species. Among all the known methods, (a) offsets, (b) part-clump, (c) culm-segment cutting and branch cutting, (d) macroproliferation of seedlings/cuttings and (e) micropropagation (tissue culture) are popular and being practised in the production of bamboo propagules. These are discussed below.

(a) **Offsets** (Fig. 5.6a): An offset propagule is composed of basal parts of a culm along with an underground rhizome portion bearing viable buds. The desired length of the culm part of an offset is 1.0–1.5 m with three to four nodes bearing alive and healthy branch buds. The offset should be 1–2 years old and collected from a healthy mother clump during spring (mid-February–April). If collected during June–August, when buds on the rhizomes are sprouting and emerging, become liable to damage. The rhizome bearing buds at the basal portion of an offset should not be injured at the time of excavation; these are essential for field survival. The viable bud is straw coloured and will feel tough when pressed with the fingertip. The collected offsets should be maintained in the temporary nursery (*transit nursery*) beds with sand rooting media) under

Fig. 5.6 Different types of bamboo planting materials. (**a**) *Offset* is made by detaching from mother rhizome, (**b**) *part-clump*, (**c**) *culm cutting*, (**d**) *branch cutting*, (**e**) *macroproliferation*: (**i**) a seedling with exposed rhizome system; (**ii**) many individuals are created through rhizome separation

high air humidity through watering for better protection against desiccation before field planting. After 3–4 months in the transit nursery beds, these are taken out and planted in the field during rainy season and usually have 80–90 % survival depending on the species.

Offset planting is the most conventional method of cultivating thick-walled bamboos in homesteads and small farms as it is bulky, heavy and expensive. However, *Thyrsostachys oliveri*, an important commercial bamboo species in Tripura, India, has been extensively cultivated by offset method as the offsets of the species are light in weight (1.2–3.0 kg per offset), small and not costly.

(b) **Part-clump (rhizome assembly)**: This is a rhizome assembly having 2–3 offsets connected with each other to be collected during spring at a time as a propagule (Fig. 5.6b). The individual rhizome in a part-clump propagule should not be separated or damaged during collection from the soil and transportation. Like offset the part-clump plant materials should also be maintained for 3–4 months in *transit nursery* before outplanting during rainy season. As the propagule contains more than one rhizome, the planting pit should be wider in size. This method is suitable *mostly for thin-walled bamboo* species, like *M. baccifera, Schizostachyum dullooa, S. lima, S. lumampao,* etc., with 40–55 % survival.

(c) **Culm cutting and branch cutting** (Fig. 5.6c, d): A *culm cutting* is a culm segment which may have 1 or usually 2–3 nodes bearing healthy buds or branch bases, while a *branch cutting* is only a detached stout branch from the culm. Always *select the healthy and productive mother clump* of the desired bamboo species for collecting these types of cuttings.

   • **Selection, collection and preparation of cuttings:** The upper narrow part of the selected bamboo culm is discarded. The remaining wider portion of the culm has to be cut at the middle of internodes (with the sharp saw to avoid splitting) into 1 node (for long internode species) and 2–3 node segments (for short internode species) having healthy buds on the nodes. The branches along with leaves on each culm segment are pruned by secateurs to a length of 2–3 cm without damaging the buds at the culm nodes. These segments are used as *culm cuttings* for rooting.

   Thick and stout branches of many tropical bamboo species (*Bambusa, Dendrocalamus, Gigantochloa, Guadua,* etc.) have swollen rhizomatous bases, often with in situ roots. These branches with aerial roots sometimes also produce in situ rhizomes at the base and also on the injured and broken culms. These branches are ideal for producing planting materials by inducing active root system in propagation beds, usually termed as *prerooted and prerhizomed branch cutting* (Banik 1980, 1987a). The species belonging to the genera like *Cephalostachyum, Melocanna* and *Schizostachyum* have numerous small, subequal *thin branches* on the culms and do not have swollen bases and aerial roots and so *cannot be used as branch cuttings.* Selected branches for making cuttings should possess alive and well-developed buds on its nodes and root primordial structures and buds at rhizomatous swelling base and are collected by separating at the point of conjecture of the branch and the culm using a saw to avoid splitting and injury of branch base.

The cut ends of both type cuttings need to be waxed or wrapped with moist gunny bags or placed in moist sawdust or coconut husk or straw to minimise water loss during transportation to the nursery before placing into propagation beds.

- **Optimum age and collection time:** Both culm and branch cuttings should be selected from 1½- to 2½-year-old culms. Spring to early part of rainy season is the best time for culm collection for making the culm cuttings. However, collection and propagation work may be continued up to mid-November (before dry winter) but with less rooting success.

  After a few pre-monsoon showers, more new aerial roots develop at the branch base on the culm nodes, and these are good branch cuttings for producing roots. The collection may be continued till the later part of rainy season.

- **Cutting materials on the suitable position of culms**: The lower to mid-culm zone is the suitable position from where culm segments are to be collected. Branch cutting can be taken from any nodal position of the culm provided it contains healthy living buds and swollen rhizomatous base with fresh root primordia. However, in *B. vulgaris* and *B. vulgaris* var. *striata*, most of the parts of a culm and even branch nodes can root successfully.

- **Sizing and condition of cutting:** Both the terminal intermodal cut end portions of culm segments are pruned to a length of 5 cm, and thus size of a cutting gets little smaller. While sizing branch cutting, the apical narrow part and side branches with leaves are to be removed. The retained thick lower portion of branch should have 2–3 nodes (for long internode species) or 5–7 nodes (for short internode species), and the exposed portions may be waxed. Then the culm or branch cuttings are to be washed thoroughly with clean water and treated with fungicide (0.1 % Bavistin solution for 15 min) just before putting into the rooting medium already placed in propagation beds.

- **Treating cuttings with rooting hormones:** Hormonal treatment, sometimes found useful only in a few bamboo species (*B. blumeana, B. polymorpha, B. pallida, B. tulda, D. giganteus, T. oliveri, T. siamensis,* etc.), especially in thin-walled species (like *S. dullooa, Ochlandra* sp.), is difficult to root. Rooting hormone solution, 100–600 PPM of IBA (indole-3-butyric acid) or combination of IBA and NAA (naphthalene acetic acid), or the IBA in talc (root-promoting powder) is applied to the cuttings before placing into the propagation bed. The concentration and duration (from few seconds to 24 h) of hormone treatment depend on the rooting ability (difficult or less difficult) of a bamboo species. Cuttings treated with rooting hormones may also be mixed with a little amount of Bavistin powder (0.1 %) for protecting from microbial infection.

- **Placing of cuttings in the propagation bed**: The exposed lumen portions of the culm cutting segments are filled with sand and then *placed* *horizontally* at 10–15 cm gap in a row inside the propagation bed and 1–2 cm below the surface of sand rooting medium thoroughly soaked with water. Branch cuttings are *placed erect* at 7–10 cm spacing, putting only the rhizomatous base in 6–8 cm deep below the surface of sand rooting medium.

- **Rooting and management of cuttings in the beds**: The cuttings start producing branches and leaves within 7–10 days and profuse roots in the propagation bed depending on the season and species within 4–8 weeks (for cuttings taken during spring–summer–rainy season) or 8–10 weeks (for cuttings taken during early autumn).

  During the first 4–5 weeks, maintaining the high air humidity surrounding the cuttings is needed to reduce water loss from the leaves by watering manually or through misting. The dead cuttings should be removed from the propagation beds to maintain sanitation around the cuttings.

  Medium to good rooting success and survival (40–98 %) of both *culm* and *branch cutting* have been obtained in *B. balcooa, B. bambos, B. blumeana, B. cacharensis, B. multiplex, B. nutans, B. polymorpha, B. tulda, B. vulgaris, Dendrocalamus asper, D. giganteus, D. hamiltonii, D. hookeri, D. longispathus, D. membranaceus, D. strictus, Gigantochloa levis, T. siamensis*, etc., in Indian subcontinent, other countries of tropical Asia and Africa; so both the methods attained popularity in the nurseries (Banik 1994c).

- **Handling and shifting of cuttings from bed to polythene bags/container and maintenance in the nursery till outplanting**: All the cuttings in the propagation bed never root at a time, so the rooted cuttings are removed batch by batch almost in every day. The rooted cuttings should not be kept in the sand medium for longer time as sand is not growth medium. Immediately after taking out from the propagation bed, the rooted cuttings should be placed/kept in water to prevent desiccation.

  The suggested size of poly bag or container is 15 × 23 cm and 40 × 50 cm for branch cutting and culm cutting, respectively. The bags should be filled with growth mixture of sandy loam and FYM at the ratio of 3:1 and kept ready before removing the rooted cuttings from propagation bed, so that transplanting is not delayed. Transplanted cuttings in polythene bags need placement under shed net roof (for providing 50–60 % partial shade) for hardening. Watering in the form of fogging the leaves and stems of the cuttings has to be done frequently to prevent wilting.

  Within a week or so, gradually the bags are shifted in the *nursery beds under open sky*. Regular watering and weeding have to be maintained till the field planting.

(d) **Macroproliferation of bamboo seedling/cuttings**: *Macroproliferation* is a method for multiplication of bamboo seedling/cuttings through rhizome separation (Fig. 5.6e; Banik 1987a; Kumar 1991). One 5–9-month-old seedling of *B. bambos, B. tulda, B. cacharensis, D. hamiltonii, D. strictus, Gigantochloa ligulata, S. dullooa and T. siamensis* can be multiplied 3–5 times in number through this technique. Gently wash the soil from root and rhizome system, being careful not to cause damage. The culms are then separated by cutting the rhizome with secateurs so that each individual includes roots, old and young rhizome and rhizome buds. Thus, each separated individual is transplanted into a separate poly bag and kept under shade for 3–5 days and then in full sun. Although this method can be used through the year, results are best during spring–summer–rainy season. Macroproliferated seedlings will again develop multiple culms within a year. So, every year the seedling can be multiplied at the same rate and a big portion of them may be planted while keeping a stock for future *macroproliferation*. The survival rate is 90–100 %. Seedlings raised in large-sized poly bags (15 × 23 cm) produced a higher number of multiplied seedlings (5–7 times), whereas seedlings in a small-sized bag (10 × 15 cm) could produce only 2–3 multiplied seedlings. Large-size poly bag provides adequate room space for proper development of rhizomes.

(e) **Micropropagation (tissue culture)**: Research on the tissue culture of bamboos got momentum during 1980s in different countries of the region, especially in India, Thailand and China. Through micropropagation technique, a large number of plants can be produced within a short period of time round the year inside the laboratory conditions without much seasonal constraints, although their field transfer is season specific and delicate and needs special care. The details of cultural requirements (explant type, culture media, hormone, environment condition, etc.) for each of the major bamboo species were also summarised in many published documents (Rao et al. 1989; Zamora 1994; Banik 2000). A brief on the methodology is presented below.

The Explant Selection: Various explant tissues have been successfully used for regenerating the plantlets of different species of bamboo in in vitro conditions. For direct development and proliferation of axillary shoots (through micropropagation) in different species of bamboos, newly germinated seedling stem segment, dormant culm buds, stem bud and 10–15 mm one-node segment from fresh lateral branches have been used.

Tissue Sterilisation: Surface sterilisation of explants is usually done by soaking in 70 % ethanol; diluting sodium or calcium hypochlorite, mercuric chloride and antibiotics like streptomycin; and finally repeated washing in distilled water. The soaking and washing vary from a few minutes to hours depending on the type of explant and species nature. The effectiveness of most sterilising chemicals has been enhanced by adding a small amount of detergent, such as Tween20 (polyoxyethylene sorbitan monolaurate), into the sterilising solution.

Nutrient Medium, Growth Regulator and Procedures: The artificial nutrient medium used for bamboo tissue culture has mostly been the MS medium, and only the embryo cultures were induced on other media such as B_5 (Banik 1987a; Zamora 1994). The sterilised bamboo explants are inoculated usually on MS medium supplemented with growth regulators such as cytokinins—commonly BAP (benzylaminopurine)/BA (benzyl adenine 0.2–7 mg/l), Kn (kinetin 0.2–4 mg/l) and coconut milk (10–15 %) for direct proliferation (Banik 1983). Growth regulators may be used individually or in combination with auxins—IAA, IBA, NAA, etc. Cluster of shoots develop after 3–4 weeks. These clusters are again divided into 2–4 smaller clusters (each containing at least three shoots), called subcultures, and the process is repeated to multiply the shoots exponentially every 4–6 weeks. It is better to avoid old culture. Successive transfer of culture has been found to be one of the important procedural steps for obtaining optimum proliferation in bamboo culture. Browning of explants and media (due to phenolic compound exuded from the explants) has been found to be common problem, especially culturing the vegetative explants taken from the adult bamboo plants. Activated charcoal (AC) or polyvinylpyrrolidone (PVP) is commonly supplemented with the media for overcoming such a problem. The culture medium for rooting generally contains high auxin, low or no cytokinin, low sugar and low salts. Various auxins such as IAA, IBA or NAA can be tried either individually or in combination at various concentrations (ranging from 0.1 to 10 ppm). The rooting process generally takes 2–3 weeks.

For inducing callus in the culture 2,4-dichlorophenoxyacetic acid (0.5–25 mg/l) has been commonly used in a number of bamboo species. Plantlets could be regenerated from callus in a few species (*B. bambos, D. hamiltonii, D. strictus*, etc.) by subculturing the same on media containing BAP and NAA.

Cultural Condition: Commonly the culture is maintained under a 14–16 h daylight regime (1,500 lx from cool white daylight fluorescent tubes) at 27 ± 2 °C for better bud proliferation.

Micropropagation protocols have been developed for different bamboo species (*B. bambos, B. balcooa, B. glaucescens, B. nutans, B. tulda, Dendrocalamus asper, D. giganteus, D. strictus, D. hamiltonii, D. membranaceus, D. longispathus, T. siamensis*, etc.), and a large number of planting materials are produced in many Asian countries. The multiplication rates differed between species. In East Africa (Kenya), micropropagation technique developed for *Dendrocalamus yunnanensis* where 3,500 plantlets are produced within 8 months (Hunja 2009).

- *Handling and field plantations of tissue culture (TC) plants*: For field planting, TC plants should be fully hardened and healthy. These should be of:

  - Minimum 60–70 cm in height
  - Minimum 3–4 shoots with 4–8 leaves per shoot
  - Well-developed root and rhizome system with adequate number of secondary, tertiary roots and root hairs to enable growth and survival under field planting conditions

5.5.1.1.5 Management of Planting Materials (Seedlings, Cuttings, TC Plants, etc.) in the Nursery Beds

Nine- to twelve-month-old seedlings/cuttings show higher survival rate (about 80–90 %) when planted in the field. During this period, regular watering/misting to the plant material is essential, but excess causes damping. Desired temperature (25–30 °C), light and well aeration/ventilation have to be maintained in the nursery area, especially in the beds having planting stocks. Some of the important management activities are mentioned below:

- Regular weeding and daily watering of planting materials at the nursery stage.
- Every one month interval loose the soil in the poly bags by light hoeing from the top by sharp tools, then add soil mixed with FYM (30–45 g) and water adequately. Planting materials may need application of both foliar and ground fertilisers through water. The increase in total biomass was found to be much greater (88 %) due to N treatments than phosphorus treatments (24 %). The highest total biomass (7.31 g/plant) of *B. tulda* seedling was obtained in $N_{100}P_{50}$ as compared to 2.94 g/plant in control. A two-split application at 4 and 8 weeks after germination or rooting in cuttings was superior to single application.
- Protecting planting materials from grazing animals, monkeys, etc.
- Controlling pest and diseases—cleaning and sanitation measures are essential.
- In the nursery when bamboo planting materials (BPM) in poly bags are left for 9–12 months, they produce strong rhizome system and roots which penetrate the ground below and the neighbouring polythene bags. During lifting/transportation of bags, rhizomes and roots of BPM are damaged. Providing cemented platform below the nursery beds is a better option as there is no or less damage of cuttings due to rhizome and root penetration in the ground. Shifting of bags may be done from one bed to other at 2–3 months interval.
- Keeping record on survival and health of seedlings/branches or culm cutting, species name, date of raising and source of material and also marking each batch of planting material in the bed.

Clumps raised through macropropagation methods (offsets, part clump, cuttings, etc.) are liable to flower simultaneously with the flowering of parent clumps, and thus their life may be short or uncertain in the plantation.

5.5.1.1.6 Advantages and Disadvantages of Different Types of Planting Materials

Method	Advantages	Disadvantages
Offset	Survival good for many species; appropriate for small scale and homestead plantation	Labour intensive, expensive, bulky and heavy, limited material available, low survival for some species
Part clump	Survival good for species with narrow, long internode and thin-walled culms	Time consuming, expensive, bulky and heavy, limited material available, low survival or inappropriate for some species

(continued)

Method	Advantages	Disadvantages
Culm cutting	Survival good, appropriate for many species	Not appropriate for long internode species, low rooting for some species. Many culms required for making the culm segments
Branch cutting	Survival good, only for species having stout branches. Inexpensive, producing many planting stocks, low transportation costs and damage	Not applicable for species having thin and narrow branches
Ground layering	Only good for multiplying isolated clumps	Rare and labour intensive, requires flat open area around mother clump, produces only few plants and low survival
Seedling layering	Produces a lot of materials for several years	Rare and successful in only a few species; takes 9–12 months to root
Macroproliferation	Appropriate for seedlings/TC plants and only cuttings having more than 3 shoots, survival good, produces a lot of materials for several years, low transportation costs	Depends on the availability of seedlings, same material/stock should not be multiplied for more than 10 years. Not applicable for *Melocanna baccifera* seedlings

5.5.2 Plantation Raising and Aftercare

- Extent, Purposes and Types of Plantation: The raising of bamboo plantation and maintenance cost is very low in comparison to any other agricultural or horticultural crop. Unlike timber plantation, a bamboo plantation starts production very early from the 5th year of plantation, and after each harvest replanting of crop is not required. Harvesting can be continued every year till the death of the plant due to flowering (35–50 years for tropical species or 120 years for temperate species like *Phyllostachys pubescens*).

Specifically the planter of a commercial bamboo plantation needs to have some market study to determine the demand of species and also identify clients/buyers. On the basis of such information, he/she can select bamboo species on the basis of climatic suitability, marketing surety and commercial profit and size of area for raising plantation.

The grower must consider the following parameters in raising a bamboo plantation:

- Marketing opportunity of the produced crop from the proposed plantation
- Availability of required amount of suitable land planting materials of desired bamboo species with number
- Easy availability of labours
- Investing capital in the right time

Planting Season: The rainy months are the best season for outplanting of all types of bamboo planting materials.

After completion of planting, earth mounting is to be provided at a radius of 0.50 m and height of 0.10–0.20 m around each of the propagules.

(a) *Forest plantation*: The bamboo plantations are usually raised in several hundred to thousand hectares of land except for a few cases owned by the government. Except for the production of culm timber, pulp wood and edible shoots, most of the bamboo plantations in the forests are raised for the ecological reasons, such as creation of eco-parks, river catchment area plantation, soil erosion and landslide control, quick regreening and improvement of deforested and degraded hilly lands, e.g. rehabilitation of degraded forest ecosystems in the lower Mekong countries (Gilmour et al. 2000).

In Nepal bamboo plantation was highly preferred by local community to reduce the impacts of landslide (Paudel and Kafle 2012). Species like *B. bambos, B. balcooa, B. vulgaris, B. tulda, D. hamiltonii, D. membranaceus, D. strictus, D. longispathus* and *M. baccifera* are usually planted at closer space on the lands susceptible to floods and along riverbanks with higher clump and culm densities for embankment protection. *Schizostachyum dullooa* is an ideal species for raising plantation in gullies of the hills along the side of the streams at close spacing to prevent soil erosion and improve the water catchment area. In villages people have been cultivating bamboo grooves at close spacing along the river, stream and canal banks to reduce or prevent soil cutting from riverbanks, resulting in reduced flooding during peak rainy season. The clumping bamboos often tend to 'mound', leading to an increase in the height of banks as well, and the strong rhizome systems can control soil erosion and water flow. On the banks of the Luochuan River, in central Yunnan, thousands of *Neosinocalamus affinis* clumps are grown to protect the paddy fields from flood damage. In Ecuador, farmers commonly plant *Guadua* in the slope above farms since it increases water availability.

Raising Mosaic Plantation Utilising Seeds/Seedlings of Different Cohorts: While raising a plantation of a bamboo species, it is better to use the seeds from different available genetic sources having diverse duration (short flowering cycles, 30–35 years, long flowering cycles 45–50 years and others) of interseeding periods (cohorts) and others. Avoid planting the seeds of the same cohort (having the same duration of interseeding period) continuously covering large tract of land; rather plantation should be raised with seeds/seedlings of dissimilar cohort (flowering population) side by side in patches or blocks. Mosaic plantations may also be raised by mixing with more than one local bamboo species, as, for example, a major species like *M. baccifera* planted on the hill tops, slopes, valleys and riverbanks and seedlings of *S. dullooa as a minor partner* species in gullies and along the shady stream bank. Sometimes *mosaic plantations may be raised* with seedlings and cuttings of different bamboo species. In the next flowering time, all these populations raised in a *mosaic plantation* are likely to flower after different intervals of time, and so all the clumps of the bamboos will not flower and die at a time.

(b) *Commercial plantation:* There is growing interest to raise bamboo plantations for commercial profit. The area of a privately owned *commercial plantation* generally extends from 1 to 25 ha of land. The forest department, with a specific objective, may also raise commercial plantation covering 50–100 ha of land. Many commercially important bamboo species have been cultivated in different parts of the world in a wide variety of soil and climatic conditions.

(c) *Homestead plantation*: Generally the homestead bamboo cultivation is always in a small scale from 1 to 50 clumps grown together or in scattered manner and owned by the families. On the basis of objective and need, homestead plantations may be of different types. These are *backyard planting, lawn planting* and *hedge planting*. The *homestead plantation* is raised usually at the *backyard* for small-scale production of bamboo resource to meet the family and construction needs in most of the South Asian and Far East Asian countries; the common bamboo species cultivated in villages are clump forming, usually congested in nature, with large, tall, branchy culms. In China, there are millions of farmers who grow bamboo as a component in integrated farming systems. The *lawn planting* has been practised for landscaping and ornamentation of the homestead area. For greening and beautifying, the environment bamboos are often selected as excellent planting materials and usually planted individually or in small lots (a few clumps). Bamboos have high ornamental and aesthetic values. In many cities of India, China, Thailand, Japan, the USA and Australia, people have been concentrating on planting bamboos in the gardens, lawns and hotels, in pots or as screens, and tall timber bamboos are planted by those who have the room. The species with colour stripes or spots, square culm, nodes or internodes of peculiar shape, etc., are also chosen for ornamental use and commonly grown for landscaping. Maintenance mainly involves irrigation in dry season, pruning and pest control. Their growth has been maintained by planting them on a mound, encircled by a shallow trench, about 25 cm wide and 20 cm deep, where the rhizomes can be cut easily when they travel. The *hedge planting* is usually done at the periphery borderline of the homestead for maintaining the demarcation of the homestead area. It also serves as border fence and privacy curtain. Generally species having straight, close culms of attractive stem colour with narrow leaves (*Thyrsostachys* sp., *B. polymorpha*, etc.) or sometimes normal bigger leaves (*B. pallida, B. nutans, B. tulda*, etc.) are selected. The clumps of *B. glaucescens* possess short (1.0–2.5 m high), thin and closely grown culms and thus are cultivated as hedge bamboo in the homestead lawns and in the boundaries of flower gardens. *Protective fence planting* is done on the boundary and borderline with dense thorny species *B. bambos*.

5.5.2.1 Species Selection

The selection and cultivation of a bamboo species is mainly dependent on farmer's choice, its utility for construction and agricultural implements, market value and obviously the climatic suitability for growing the species in the locality. Some common tropical bamboo species suitable to different types of homestead plantations are shown below:

Backyard plantation (for timber): *Bambusa balcooa, B. blumeana, B. tulda, B. vulgaris, B. nutans, B. bambos, B. cacharensis, B. pallida, B. polymorpha, D. asper, D. brandisii, D. strictus, D. hamiltonii, D. membranaceus, Gigantochloa* sp., *Pseudoxytenanthera stocksii, T. oliveri, T. siamensis,* etc.

Lawn plantation: *B. vulgaris* var. *striata, B. ventricosa, B. vulgaris* cv. *wamin, Lingnania* (*Bambusa*) *chungii, T. oliveri, T. siamensis, D. giganteus,* etc.

Hedge plantation: *B. glaucescens, B. jaintiana, T. oliveri, T. siamensis, B. bambos,* etc.

Most of these species can also be raised in **commercial plantation**. However, some common bamboo species for shoot production (**for food**) are *Dendrocalamus asper, D. hamiltonii, D. brandisii, D. longispathus, D. latiflorus, Bambusa polymorpha, B. blumeana, M. baccifera, T. siamensis, T. oliveri* and *Gigantochloa albociliata,* and some bamboo species such as *B. polymorpha, B. cacharensis, B. blumeana, B. glaucescens, T. siamensis, T. oliveri, G. hasskarliana, Schizostachyum dullooa, S. humilis, S. lima, S. lumampao, Cephalostachyum virgatum, M. baccifera,* etc., are well known for making baskets and handicraft. In raising **forest plantation,** emphasis should be given on local bamboo species, and some of the above species may be planted depending on the site suitability and market demand.

5.5.2.2 Outsourcing and Procurement of Quality Planting Materials (QPM)

Large-scale planting requires sufficient number of QPMs (healthy, productive and no disease) (Banik 1993b) of desired bamboo species from one place to minimise the transportation cost. Forest departments are, most of the time, well equipped to raise huge numbers of bamboo planting stocks from their own nurseries. Private planters are, nowadays, more concerned about the quality of the planting stocks of desired species, which are often not available in any nurseries and market. Therefore, they prefer to establish nursery to raise QPM of desired species for their own plantation and some stocks are also sold as side business. Reputed and experienced nurseries can only take such responsibility of advertising the availability of bamboo PM (species, quality, quantity available, unit price, specification, etc.). On time and safe delivery systems should be there (Banik 2008).

5.5.2.3 Site Selection and Land Preparation (Weeding and Vine Cutting)

The following factors need to be considered in selecting the sites for bamboo plantation:

- Bamboos do not survive under deep shade. Underplanting may be done in the well-thinned or widely spaced deciduous to semi-deciduous forest plantations. Sites having full sun exposure are very suitable.
- Should have moist environment, be well drained and not be in waterlog.
- Gentle terrain and lower slopes of the hills, fertile land and sandy loam are suitable, avoiding hill tops and upper slopes.
- Dry with rocky or too sticky soil is not suitable.
- Do not survive in low-lying marshy areas, saline habitats or costal land subjected to daily inundation.
- New canal banks are also good sites for raising bamboo plantation.
- Northwest part of the homestead is the preferred site for bamboo plantation because the raised clumps act as windbreak during storm.

While raising plantation, the land needs to be cleared by cutting the jungles, weeds and vines in dry months so that twigs and debris can be burnt. The overall loosening of soil up to 40 cm depth by clearing rocks, tree roots and stumps has to be ensured to improve the land environment for better aeration and complete mixing of the organic and supplied inorganic fertiliser with the soil for supporting the plant growth.

The strip land preparation is generally adopted on hill slope planting. To prevent the soil erosion and water loss, strip land preparation can be made parallel with the contours. The strip width and distance between strips is about 3.0 m for big- or medium-sized bamboo species (*Bambusa* spp., *Dendrocalamus* spp., *Gigantochloa* spp., etc.) and 2.0 m for small-sized bamboo species (*B. multiplex*, *Thyrsostachys* sp. and other similar species).

5.5.2.4 Planting Space, Pit (Hole) Making and Soil Preparation

Digging planting holes according to desired spacing can be adopted. The planting pits of 60 cm cube are dug at 5 m × 5 m or 6 m × 6 m spacing for cultivating large-sized caespitose bamboo species—*Bambusa balcooa, B. bambos, B. blumeana, B. nutans, B. tulda, B. polymorpha, B. vulgaris, Dendrocalamus asper, D. longispathus, D. brandisii, D. giganteus, D. latiflorus, D. hamiltonii, D. membranaceus, D. strictus, Pseudoxytenanthera stocksii* and *Gigantochloa* sp. So in a hectare, either 400 or 280 propagules are necessary. However, in the Philippines, bamboo planting is usually done at 8–10 m spacing (Lantican et al. 1987).

In the steep hill slopes, planting pits are dug on the contour lines at desired spacing. The spot can be cleared of weeds with 2.0 m radius around the planting pit. In raising shelterbelt, river or canal bank, border fence and erosion control planta-tion, very close spacing (1.5 × 1.5 m, 2 × 2 m, 3 × 3 m) is adopted. The culms are thinner in size in the high-density planting because of the severe competition among the plants for food, water and light. Higher-density planting (closer spacing) results in lower number of culms per clump, but the total number of culms per hectare is comparatively high. In the barren hills and scrubby forests, close planting of *M. baccifera* seeds is suggested, and the planting pits are to be dug at 3 m × 3 m spacing, so 1,111 pits per ha. The seed planting pit size may be 30 cm × 30 cm × 30 cm; 2 seeds may be sown in each pit on the contour lines in hills (Banik 2010a).

When a large number of tall and straight culms with relatively smaller diameter are desired, closer spacing is practised. In Tripura farmers have been maintaining the closer spacing either 1.25 or 1.5 m in raising *Thyrsostachys oliveri* plantation in their farm to produce straight, less branchy and more number of culms per hectare. About 2,960 offset propagules are planted per hectare, and after 3 years of planta-tion age by selective felling, nearly 8,000 culms are harvested.

Pits should be dug and prepared 10–15 days before planting. At the same time, as pit preparation, all herbaceous and woody weeds within 1 m of the pit should be removed. When digging a pit (planting hole), the underneath and surface soil should be separated and put at the two sides of the hole. The dug earth must be used to fill 2/3 height of the pit, and during planting the surface soil must be added near the root zone of the plants. The earth in the hole must be broken small, thoroughly mixed with farmyard manure (FYM).

5.5.2.5 Application of Insecticide and Fertiliser

Following the soil preparation, a treatment against ants and termites may be carried out by spraying the soil with Decis (deltamethrin) at 15 g/m^2 or with Diazinon (240 g/l). Fertiliser should be applied at least 2–3 weeks before planting. The pits are to be filled with cow dung/FYM (10 kg), urea (20 g), triple superphosphate (20 g) and muriate of potash (10.0 g) and soil.

5.5.2.6 Planting of Plants in the Pits

Planting too deeply or too shallowly is unfavourable for its growth. Then fill the hole with soil in layers around the root to keep the root in tight contact with the soil and water it. The newly planted plants should not be shaken. Care should be taken to ensure that planted materials are protected from damage by people and animals.

5.5.2.7 Aftercare and Tending Operations

(a) *Watering and drainage:* Irrigation is needed in the early stage of life especially in the drier months. However, in the hilly forest plantations, irrigation may not be possible; therefore, planting should be done in the rainy days. In areas of low rainfall, bamboos are planted in sunken pits specially designed for moisture conservation. Drip irrigation in dry areas has been found effective, while in other areas, furrow irrigation at 40 CPE (cumulative pan evaporation) + mulch has been found effective for proper growth of bamboo plants. During continuous and heavy rain, plots may get flooded, so quick drainage of stagnant water is needed.

(b) *Protection from grazing and fire:* During the first 3 years of planting, all the bamboo plants have to be protected from grazing and fire. Porcupine, wild boar, monkey and deer are usually found to graze and eat young tender bamboo shoots. Such injury and damages make the underground rhizome system vulnerable to further infections from fungi and insects. Therefore, it is necessary to protect the plantation/farm by constructing fence in the boundaries depending on the availability of fund. Locally available *Prosopis* sp. could be successfully used as the live fence of the bamboo plantation garden against the grazing. Bushy clumps of thorny bamboo like *B. bambos* may also be planted in 2–3 lines at periphery of a bamboo plantation area to repulse the entry of animal and thieves.

   Hunting dogs as farm watchers have been found effective in some oil palm garden in Malaysia. Mere barking of dogs repulses the entrance and movement of porcupine, wild boar, monkey and deer in the garden. A 5-m wide *fire line* may be laid at the beginning of dry season around the vulnerable side of the plantation area (discussed detail under Sect. 5.4.1 v.).

(c) *Weeding:* After planting, three to two times weeding and vine cuttings are done in July–October during first year depending on the intensity and occurrence of weeds. Then weeding has to be done two times in the second year. Afterwards weeding around the bamboo plant has to be done only once in the third and fourth year depending on the species nature and weed occurrence. A 3-m radius ring weeding is preferred than complete weeding of land.

(d) *Vacancy filling:* With good care, about 80–90 % plant may survive in the field, so 10–20 vacant pits have to be filled up during the early part of the rainy season with the propagules of the same species and age, preferably from the same stock which was used for raising plantation.

(e) *Loosening of the soil and mounding:* During the second year in the month of April to May, *mounding* or heaping fresh, lose soil around and over the base of the plant in 10 and 15 cm height has to be carried out so that rainwater does not retain. Thus, underground rhizome remains covered and emerging shoots are protected. So mounding may be practised only in marshy or waterlogged sites to accelerate drainage and create a favourable condition of O_2/CO_2 exchange for underground rhizome system.

(f) *Application of fertiliser:* Fertiliser dose in combination of $N_{50}P_{50}K_{50}$, $N_{100}P_{25}K_{50}$ and $N_{50}P_{25}K_0$ as per the fertility of site may be adapted in different zones for proper growth of bamboo species. The N can be supplied in the form of urea, P in the form of single super phosphate (SSP) and K in the form of muriate of potash (MP). The pits are to be filled with 10–15 kg cow dung or FYM, and urea, SSP and MP are to be thoroughly mixed with soil in the pit (per plant).

In the first year of plantation, nitrogen should be applied in 2 split doses, viz., 50 % in spring and 50 % in rainy season, and phosphorus and potassium should be applied as basal dose while filling the pit. In second and third year, the N application may be also continued in 2 split doses, while P and K should be applied in the month of spring only. Green and barnyard manure can be applied just before the rainy season. However, about 150 kg N per hectare as effective component of chemical fertiliser is applied in the spring or early rainy season for sympodial bamboos and in rainy season (July or August) for monopodial bamboos during the first 1–2 years after planting. 350 kg N in chemical fertiliser per hectare is needed in the third year, and 15–30 t of organic fertiliser is applied in winter.

(g) *Mulching:* As a common practice of mulching, organic materials like fallen tree leaves, barks, bamboo leaves, rice or wheat stalk and chaff or hay, cut grass, etc., alone or mixed, are placed around the stems (clumps) of the bamboo plant covering the exposed ground of the pit to reduce the loss of moisture due to evaporation from the planting holes. After each ring weeding and at the end of the rainy season, proper soil work is done around each of the planted seedling/cuttings. Then mulch materials are, usually, placed at 7.0–10.0-cm-thick layer and spread at 1.0–1.5-m radius on the ground around the bamboo plant. Mulching conserves moisture in the pit and underground temperature in drought period and also checks the weed growth around the bamboo plant. Mulches increase soil temperature to varying degrees (2–5 °C), notably during winter (December–January), and boost the physiological activity of underground rhizomes of bamboo plants and stimulate the early sprouting of shoots and also prolong the shooting period. Mulch also serves as a green manure.

(h) *Intercropping:* The practice of intercropping is found to be beneficial to bamboo plants, especially in the early part of plantation before the full-grown canopy formation of bamboo clumps. Intercropping in newly established bamboo stands not only suppresses the weed growth but also increases productivity and soil fertility and economic returns. Crops should not be planted too close to the bamboo plants, as this could negatively affect their growth. Some legume crops, like pea (*Pisum sativum*), mung (*Phaseolus aureus*), lentil (*Lens esculenta*), soybean (*Glycine max*), arhar (*Cajanus cajan*), mustard (*Brassica* spp.), potatoes, etc., are grown in between the planting rows. In comparatively wetland, rice (*Oryza sativa*), *Sesbania* sp., wheat (*Triticum aestivum*) and maize (*Zea mays*) are also cultivated in widely spaced (9 m × 9 m, 10 m × 10 m) bamboo plantation at the early stage of establishment. However, crops should not be planted too close to the bamboo plants; usually 1.0 m space is left from the base of a bamboo plant. After 3–4 years of plantation, i.e., when plants become tall and crown comes closer, some shade-tolerant cops like ginger (*Zingiber officinale*), turmeric (*Curcuma longa*) and medicinal plants can be grown successfully.

Intercropping with trees is discussed in Sect. 5.6.2 (**Some Models** Group I (2)).

- *Ecologic and Socioeconomic Impact of Bamboo-Based Land Use:* Over the long term, a bamboo-based cropping system compared favourably to several other land use alternatives in the degraded mountains of northern Vietnam (Ly et al. 2012). Compared to cassava, rice and maize, bamboo provides 49–89 % higher average return to labour. Carbon content in above-ground biomass of standing bamboo is 17 t ha^{-1}, 18 % of that of forest. The soil organic carbon pool under bamboo amounts to 92 t ha^{-1} to 70 cm depth, comparable to both forest and regenerated forest, and is 20 % higher than land cultivated with cassava or maize. The study reveals that a shift in land use from annual crops to bamboo provides an annual net gain of soil organic carbon of approximately 0.44 t ha^{-1}. Such a shift is constrained however by income insecurity in the early stages of plantation.

During last two decades, homestead bamboo cultivation in Tripura, India, has been replaced by rubber plantation as the farmers are getting financial incentives from the private rubber planters. As a result, soil erosion increased, the soil moisture and organic carbon started depleting, and biodiversity in homestead gardens and farms is decreased.

5.6 Bamboos in Agroforestry

Bamboo-based agroforestry plantations are raised for socioeconomic and ecological consideration. Bamboos in agroforestry models mainly have light crowns such as *Dendrocalamus, Phyllostachys* and *Thyrsostachys* species. Fu and Banik (1996) described bamboo agroforestry models (mentioned in sect. 5.6.2) are divided into three groups depending on their functions and products mostly on the basis of experiences from South China, more or less similar to the environment and bamboo types growing in South Asian countries.

5.6.1 Basic Principles

Basic principles for model establishment should have clear management objective, suitable management strategy and positive relationship between the model and environment and maintain appropriate control of the compatibility and competition relationship among the species and full use of resources.

5.6.2 Some Models

Group I: Bamboo Agriculture (Forestry) Model

(1) *Bamboo + tea plant:* In South China bamboo is often planted at a spacing of
    6×4 m and tea plants at 2×0.5 m. Intercropping of seasonal agricultural
    crops, such as soybean and vegetables, can be done for 1–3 years after planting.
    But intercropping should leave enough space for the unhindered growth of
    bamboo and tea plants and ensure adequate nutrition supply to them.

(2) *Bamboo + conifer and broadleaf timber trees:* This model can be established by
    either converting seminaturally mixed stands or planting new ones. The ratio of
    bamboo to trees is important, and in seminaturally mixed stands, this may be
    7:3 or 8:2 for bamboo and broadleaf trees. The planting time for bamboo and
    trees should be determined based on the growth rate of the tree species
    involved. The tree species best adapted to bamboo crops are *Albizia* sp.,
    *Gmelina arborea, Tectona grandis, Lagerstroemia parviflora, Anogeissus
    latifolia, Phyllanthus emblica, Zizyphus xylocarpa, Bombax ceiba,
    Stereospermum suaveolens, Melia azadirachta, Aegle marmelos, Albizia
    procera, Lannea grandis, Spondias pinnata, Erythrina indica, Eucalypts, Pop-
    lar, Dalbergia sissoo,* etc., owing to their peculiar deciduous light crowns.
    Areca catechu with its narrow crown is also found to grow satisfactorily
    together with bamboo clumps. All species with deep umbrageous crowns like
    Adina cordifolia, Ficus spp., *Mangifera indica, Artocarpus heterophyllus,
    Litchi chinensis,* etc., should be avoided (Banik 2000).

(3) *Bamboo + agricultural crops:* Large-sized bamboo species are planted at a
    spacing of 5×5 m or 6×6 m and 3×2 m for small ones. Intercropping, with
    crops employed in bamboo + trees model, should show a bamboo plant to have
    1 m^2 in area to ensure nutrition supply and can be done for a maximum of 4 years
    after bamboo planting. In planting trees or bamboos, full soil preparation may be
    employed on plain land. On sloping land, strip preparation—leaving alternate
    unprepared strips to prevent water and soil erosion—is recommended. It is
    necessary to place adequate fertiliser in the planting pits before planting is
    done. The seedlings of *B. tulda* produced elongated culms, five to seven times
    more in height, when grown under the partial shade of pigeon pea (*Cajanus
    cajan*) and dhaincha plant (*Sesbania* sp.) during the first 2 years of outdoor
    planting (Banik 2000). Crops such as watermelon, soybean, sugar cane, potatoes,
    sweet potato, mustard and other vegetables can be intercropped within 3 years
    after planting. Raising of pineapple, ginger, turmeric, shade-tolerant variety of
    sweet potato, etc., within a stand of adult bamboo clumps is technically feasible
    and economically viable (Banik 1997d). In India (in Jabalpur) seedlings of
    D. strictus, B. bambos and *B. nutans* were successfully intercropped with either
    maize or soybean. In Thailand, the bamboo species are also intercropped with
    maize and peanut. Through wider spacing of bamboo clumps and judicious
    manipulation of the bamboo canopy, it appears that the period of intercropping
    could be extended further.

The 'talun-kebun' system in Indonesia consists of a 6–7-year management cycle in which a 4–5-year fallow period of perennial clump bamboo is alternated with 2 years of food crop production. Clear cutting, raking the forest floor and slashing into piles for burning and hoeing the soil to a depth of 25 cm reduce the vigour of the bamboo to the point at which it poses no competitive threat to the first year of planted food crops. These crops are (typically) cucumber, bitter solanum and hyacinth (pole) beans. Ash from the burned slash piles plus some animal manure and NPK fertiliser are used to increase the production of these vegetables. The field is planted with cassava (a less nutrient-demanding root crop) in the second year. After 2–3 years of cultivation, the field is abandoned and permitted to revert to an unmanaged stand of bamboo for 4–5 years (Christanty et al. 1997). The bamboo recovers much of the nutrients leached deeper into the soil profile during the 2 years of cropping and deposits them at or near the soil surface as above-ground litter and dead fine roots. The biogeochemical role of bamboo in sustaining the productivity of this agroforestry system reflects the rural farmer's saying: 'without bamboo, the land dies'.

Group II: Bamboo–Fishery Model

(1) *Bamboo + fish pond + crop:* This model is usually made on the plain or lower and wetter lands where fish ponds are built. One to three rows of shoot-producing sympodial bamboos may be planted on the banks of the pond, and crops such as soybean and rye intercropped between bamboo clumps to form a complete food chain. Crops can be harvested as food of fish feed. Bottom mud from the ponds may be dug out in winter and used as fertiliser for bamboo clumps. The clumps of major cultivating bamboo species (*B. balcooa, B. bambos, B. nutans, B. multiplex, B. vulgaris, D. hamiltonii, D. strictus, Thyrsostachys, Gigantochloa* sp., etc.) of South and Southeast Asian countries may be replanted at 20–25 years of age for improving the productivity through rejuvenating the growth.

Group III: Special Purpose Model

(1) *Bamboo + edible fungi:* There are a large number of edible fungi (*Dictyophora tomentosa, Pleurotus ostreatus* and *Auricularia auricula-judae*) regarded as natural food rich in vegetable proteins. In China these can be cultured in bamboo stands, which satisfy the fungi's need for humidity, shade and a fertile bed. A bed of decayed bamboo litter and cotton shells, placed evenly on the ground up to a height of 10 cm, is used as the substrate. Inoculation of *Dictyophora* sp. is done in September for varieties that grow in normal temperature and in May–June for those that require a higher temperature. Harvest is after 4–8 months, depending on the fungus variety. *P. ostreatus* is inoculated in March and harvested 2 months later. *A. auricula-judae* needs to be cultured in bags filled with the growth medium and hung on the bamboo.

(2) *Bamboo + medicinal plants*: This model is suitable for hilly areas in the sub-tropical, monsoon climatic zone that has a mild climate and adequate rainfall.

The medicinal plants should be chosen to suit the topography of the site. Plants belonging to Aracae and Zingiberaceae and having shade-tolerant habit may be selected for intercropping.

5.7   Bamboo Stand Management

A '*bamboo stand*' composed of the standing bamboos at different ages, in which culm component of various ages is called age composition. The difference in vigour and capacity of accumulating organic matter of bamboos in a stand will result in difference in capacity of production and biomass formation. Under tropical rainforest conditions, bamboo possesses luxuriant growth with a vigorous underground rhizome system and grows either in pure stands or mixed with trees as understorey and/or intercrop. The indicator species is the dominant species, and accordingly the stands are divided into pure stand and mingled (mixed) stand. In the natural forest, bamboos are intermixed with tree, shrub and herbaceous layers of vegetation. These are habitats for numerous species of insects at the soil layer and spiders, butterflies, birds and other higher life forms at the tree layer. The mix of plant species is important for maintaining high levels of nutrients in the soil and a high degree of resilience of the ecosystem to weather events and disease and insect infestation. A pure bamboo stand usually can have one dominant species (like *Phyllostachys pubescence*, *M. baccifera*, *B. bambos*, *D. hamiltonii*, *D. membranaceus*, *D. strictus*, etc.) but sometimes may contain not more than 10–20 % of 2–3 other bamboo species; and several species cover vast tract of land like a 'bamboo sea'.

5.7.1   Factor to Indicate the Stand Condition

(a) *Clump health:* The growth of the stand is characterised by the organ increase in volume, number and weight, including emergence of shoot, growth of new culm, spreading of branch and leaf and the development of rhizome system including buds and roots. Higher economic output can be achieved from a bamboo stand having the evenness of culm distribution and no congestion, higher leaf area index (LAI), younger culms (1–2 years of age), size of the culms and rhizome composition. Thus, the maintenance of the suitable age composition of bamboos is a basic factor to maintain high yield. *The 1–2-year-old culms are soft and not economically useful but important in the stand for the continuation of production of new bamboos.* Studies in the clumps of *D. longispathus* (Banik 2000; Banik and Islam 2005) and *M. baccifera* (Ueda 1960) showed that a number of leaves and leaf biomass are highest in 2-year-old culm and then gradually decrease as the culm gets older. So, where there are more young and strong culms in a stand, there is high yield. The size of culms is an important factor to indicate the stand condition, usually expressed by mean breast-height diameter.

The rhizomes of 2–3-year-old culms are rich in reserve nutrients as well as starch and nitrogen, and these culms, because of their plentiful leaves, play a functional role in the plant's metabolism and maintain the healthy composition of a bamboo stand. If the young and strong rhizomes are dominant in a system, it will have the greater capacity to absorb mineral elements and store higher amount of carbohydrate, which influence more shoot production. In general a tall and big bamboo can produce and accumulate more nutrients for the stand because it occupies more space, shares more sunshine and absorbs mineral elements.

(b) *Nutrient cycling and ground condition:* The uninterrupted adequate cycling of nutrient is one of the important indicators for scientific management of a stable bamboo stand. This is a very essential consideration to maintain the productivity. During commercial production, harvest and management conditions result in soil disturbance, which stimulates a chain reaction that reduces total soil content of nutrients and increases nutrients in the labile fraction of soil biomass and finally in the mineralised form (Silgram and Shepherd 1999). Exclusive application of inorganic fertilisers has the same effect, and therefore, adding organic fertilisers to bamboo is required. Part of this organic fertiliser is self-provided in the form of bamboo litter: on average, about 8 t/ha of biomass, which is approximately 7 % of the total biomass of plants including 63, 6 and 42 kg/ha of N, P and K, is annually recycled in bamboo stands. The insects, bacteria, other microbes and animals also affect the growth and performance of a stand directly or indirectly. Among them, some are favourable to the stands, such as these microbes, and bacteria can decompose the litters and help to loosen the soil, while some others are not beneficial to the stands. The presence of vesicular arbuscular mycorrihza (VAM), *Glomus albidum, G. fasciculatum, G. mosseae, G. reticulum, G. intraradices, G. magnicaulis* and *Gigaspora* sp., was found from healthy clumps of *B. bambos* samples collected from Kerala (Appasamy and Ganapathy 1992).

5.7.2 Purpose of Stand Management

Although bamboos mature within a few years of planting, they should be managed continuously to obtain the maximum yield. In order to achieve the expected fast growth, high quality and maximum yield, certain species should be selected that have adaptability or can adapt quickly and bring economic benefits and are suitable for particular environmental conditions, soils and topography. In order to reach the targets, it is necessary to know the requirements of species with reference to climate, topography, soil and others. Most sympodial bamboos have lower cold resistance, while monopodial bamboos are generally cold resistant and can be planted even in warm temperate zones. According to the currently prevalent intensive production management practices, bamboo stands can be categorised into five end-use types (Fu and Banik 1996): timber stand (including timber-and-shoot stands), shoot stand, pulp stand, ornamental stand and water/soil conservation stand. The species involved and the management system will vary with the intended end use.

5.7.2.1 For Timber

The end product of sympodial species, such as *Bambusa balcooa, B. bambos, B. polymorpha, B. tulda, B. vulgaris, D. giganteus, D. hamiltonii and D. strictus,* is 3–4-year-old culms. The recommended stand density is 700 clumps per hectare, each clump containing 10–20 culms at 1–3 years of age. The annual yield of culm timber generally ranges from 3 to 10 t, sometimes reaching 15–33 t/ha. Some species can be grown for both culm timber and edible shoots, thus leading to the term 'timber–shoot stand' or 'multipurpose stand'. In the hills of many Asian countries, the tribal people manage the bamboo groves of different species (*B. bambos, B. polymorpha, D. asper, D. hamiltonii, D. longispathus, M. baccifera,* etc.) for culm timber and edible shoots. Each clump of sympodial species is grown at 4 × 5 m or 5 × 5 m spacing and can have 6–8 culms of 1–2 years of age. A few 3-year-old culms may be retained to provide mechanical support to the clump for future production of timber culms. Cultural measures for multipurpose stand differ somewhat with those employed for single-purpose stand.

5.7.2.2 For Pulp Production

To obtain high-quality paper, the raw materials for pulping must have long fibres and low silica and lignin content. The species of *Bambusa* and *Phyllostachys* show better pulping properties than *Dendrocalamus* and small-sized bamboos like *Schizostachyum* sp. *Melocanna baccifera* stand is the main raw material supplier for the best pulp and rayon production in the factories of Assam (India), Chittagong, Sylhet (Bangladesh) and Myanmar. Since ancient times, people in China, Japan and Korea had been using 1-year-old culms for handmade paper. These practices had a harmful effect on the total annual bamboo culm production by reducing the number of younger culms in the stand.

5.7.2.3 For Edible Shoot

Bamboo Shoots: Over 500 bamboo species can produce edible shoots, including some monopodial bamboo species such as *Phyllostachys* spp. and sympodial species such as *B. polymorpha, Dendrocalamus asper, D. latiflorus, D. longispathus, D. hamiltonii,* etc. *Gigantochloa apus and G. atter* stands in

Indonesia produce more edible shoots than that of *D. asper*, but these stands differ in quality. Generally a shoot stand consumes more mineral nutrients from the soil than a timber stand, and hence, the application of organic or chemical fertilisers is important. In Asian regions, where the potassium content in the soil is generally high, the NPK ratio could be 4:3:1 or 5:2.1. Up to 1.5 t/ha of NPK fertiliser and 3 t of FYM/Dung are needed annually (application 1–3 times). Irrigation might be required for intensively managed shoot stands when there is no rain for more than 10 days, especially during shooting period. Shoot must be covered with soil before the middle of May (summer) when the temperature goes up by 34–38 °C or above in Indian subcontinent. Covering the shoots with wet soil will prevent damage from excessive heat and sunlight and improve their quality and grade. It was observed that the shoot production both by number and weight per clump increased in the first, second and third year after which it declines, if continuously harvested for several years. Therefore, continuous harvesting of all shoots for more than 3 years from a clump should be discouraged (Banik 1997c, 2000). After 3–4 years of harvest, move to a new clump. Bamboo stands managed in this way can produce 10–20 t of edible shoots per hectare annually for monopodial species or 10–30 t for sympodial ones.

Sometimes farmers and indigenous people harvest earlier sprouting and vigorous shoots and leave later sprouting, weak or unhealthy shoots as mother bamboos. Such selective harvesting is detrimental in the longer term to healthy and productive bamboo forest. Breaking shoots off by hand instead of unearthing them with hoes or other tools also degrades the bamboo forest.

5.7.2.4 For Environmental Purposes

(a) *Soil Water Conservation:* Naturally mixed or pure bamboo stands are usually distributed and also developed in the remote high mountainous regions near the banks of river, hilly streams, lakes and even in seashore. These stands not only conserve soil and water but also provide a fine living environment for wildlife. Moreover, uncontrolled felling has been adversely affecting the growth and development of bamboo. In addition, periodic gregarious flowering after 30–50-year interval kills most of the bamboo clumps in the forest, creating denudation of the hill slopes. As a result, the rate of soil erosion has increased. Bamboos are not deep-rooted plants; rather roots are mostly distributed within 30 cm of soil surface and rhizomes lay in soils 20–50 cm deep, thus helping in binding the surface soil and thereby controlling soil erosion. The dense natural stand of *M. baccifera* in the hills extended over the vast area of northeast India, Chittagong Hill Tracts of Bangladesh and Arakan ranges of Myanmar has been controlling soil erosion and landslide through the underground extensive vastly spread, branched, long-necked interconnected ramifying rhizome networks (Banik 1997e). The stem flow rate and canopy intercept is about 25 % which means the compact and continuous green crown cover of this bamboo greatly reduces the rainwater run-off and conserves more water on the ground and for subterranean flows. The thick litter absorbs rainwater and retained moisture gradually gets released into the ramifying rhizome nets and stored inside the microtunnels of dead roots (root diameter varies from 0.04 to 0.48 cm) below the ground and recharges the springs and streams originating in the water catchment areas in the region (Banik 1989, 2010a). This type of stand has both ecologic and economic value by producing fertile soil, continuous water flow in the rivers, healthy culms and edible shoots. The bamboo plantations for flood prevention and strong wind protection on Jiulongjiang river bank of Fujian Province and on Dayingjiang river bank of Yunnan Province are famous examples (Fu and Banik 1996).

(b) *Shelterbelt and Windbreak:* Bamboos, particularly the clumping type (sympodial), are an effective shield against the onslaught of wind. The flexibility of the culms (for green culms, the modulus of elasticity is about 9,000–10,000 N/mm^2 and the modulus of rupture 84–120 N/mm^2) helps them to bend without breaking even in relatively strong winds. Bending over of the bamboo until it touches the ground is a common observance in very strong winds, cyclones and typhoons. Because of this, bamboos are commonly used as wind barriers along boundaries of farms and to protect agricultural land from wind erosion during fallow periods. Bamboo is also now being planted as an inner line plantation behind coastal mangrove and casuarinas to shield the interior from the effects of strong winds and cyclones (Banik 2000).

(c) *Protective Fencing:* The thickly branched and very thorny bamboo (*Bambusa bambos*) stand is created along the periphery of homestead in 3–5 rows at close spacing (1½ × 1½ m or 3 × 3 m) as a protective fence in the northern district of Bangladesh and in some rural farm lands of UP, Central to South India (Kerala, Andhra Pradesh, Orissa, Madhya Pradesh). The densely interlacing thorny branches and branchlets make it a close, almost impenetrable hedge. Gamble (1896) wrote 'against such a hedge nothing but explosives would be much effect'. The clump of *B. glaucescens* possesses short, thin and closely grown culms and thus is cultivated as hedge bamboo in the homestead lawns and in the boundaries of flower gardens.

(d) *Aesthetic and Landscaping:* Bamboo stands for ornamental purpose are often characterised by beautiful appearance and colour combination of clump, culm and foliage. There are some species (mentioned under Sect. 5.5.2.1) commonly grown for landscaping. A number of small block plantations (from 0.25 to 2.0 ha) of suitable bamboo species may be raised scatteredly in the highly populated metro cities like Bangkok, Chennai, Colombo, Delhi, Dhaka, Kolkata, Manila, Mumbai and others. The bamboo crown generates oxygen, provides low light intensity and protects against ultraviolet rays and is an atmosphere and soil purifier. Bamboos, being semi- to evergreen plants, provide shade to the people during hot summer, and such small green permanent plantations also absorb substantial amount of atmospheric carbon from the cities and provide a park-like place for rest and tranquillity with least amount of air and noise pollution. Bamboos are also grown individually in the homestead lawn, and maintenance mainly involves irrigation in dry season, pruning and pest control. The existence of sacred groves of different bamboos in the indigenous communities also plays the similar role in addition to preservation of traditional cultural diversity and conservation of important bamboo species.

(e) *Energy Plantation:* Because of a high growth rate (typically matures within 5–7 years) plus a number of important fuel characteristics such as low ash content, alkali index or heating value, bamboo is a promising energy crop (Scurlock et al. 2000). To increase the biomass production, closer spacing is used in plantation. A field trial on *D. strictus* in Karnataka revealed that planting at closer spacing (1 m × 1 m), 9.9 culms were produced per clump as compared to 6.5 and 5.2 in the medium (2 × 2 m) and wider (3 m × 3 m) spacing, respectively, at 572 days after planting. The closer spacing resulted in a higher leaf area (477.4 dm^2/clump), leaf area index (4.77), leaf area duration (779 days), rate of dry matter production (10.72 g/clump/day) and crop growth rate (108.4 mg/dm^2/day). The dry matter increased from 4.0 t/ha in control to 12.5 t/ha with an application of $100 + 50 + 50$ NPK/ha/year (Patil and Patil 1988). A local clone selected in India from *Bambusa balcooa*, locally known as *Beema bamboo*, has high calorific value of 4,000 k cal/Kg, low ash content between 0.4 and 1.0 % and biomass yield 100 to 125 t/ha/annum. The bamboo is multiplied in large numbers through micropropagation for commercial plantation in India and abroad.

5.7.3   Tending Operations

5.7.3.1   Adjusting the Structure and Density of Bamboo Stands

Occasionally *sanitary felling/thinning* is done by taking out (felling) the broken, damaged, malformed, weaker, rotten and dead culms from the clumps to improve the health condition and productivity of the stand for facilitating proper growth and development of new shoots. These operations reduce competition and infestation of pests and diseases and encourage new sprouting (culms).

Digging out of old stumps may be practised only in a small bamboo stand on a more or less flat land as the risk of landslide is less. Not more than 60 % of old stumps are dug out by leaving the remaining stumps scatteredly in the clump to maintain the mechanical strength of clump anchorage. Occasionally *vacant* or *gap spots* are created inside the bamboo stand due to localised forest fire, illicit felling, damage of clumps by elephants and clump death. Filling up of such vacant spots is beneficial for maintaining the culm density of the stand by planting offsets and/or part clump of the same bamboo species during pre-monsoon time. In areas where bamboo is the predominant species, isolated trees overtopping the bamboos are either girdled or carefully felled, and the forest is converted into a pure bamboo type.

5.7.3.2 Clump Congestion

The culms are packed tightly together with many coppice shoots and often twisted in congested condition, frequently seen especially in village bamboo groves due to browsing by cattle and uncontrolled cutting of the young culms around the clump periphery. Continuous removal of the young shoots for food, digging up culms with rhizomes attached for making walking sticks, frequent fire and dry or hardened and poor soil create clump congestion in the bamboo forest. Such constant injury to the periphery of the clump develops a dense mass of dead rhizomes that prevents the growth of rhizomes from spreading outwards and the new culm production gets reduced and clump gets congested. Regular thinning of bamboo is a prerequisite for sustaining the vigour of rhizomes. It is important to thin the culms yearly, especially for congested clump of *D. strictus, B. bambos, B. balcooa, B. polymorpha, T. oliveri,* etc., because this helps to increase the number and quality of culm production for the next two or more years. It also lessens fire and insect danger.

5.7.3.3 Pruning

The branches of *B. bambos* bear thorns in all the nodes. As a result, it becomes difficult to harvest culms from the clumps and also to take care and manage the clumps by coming closer to the plants. The 6–8-month-old mature branches of this bamboo are cut/pruned from lower mid to base of the standing culms during winter (Fig. 5.7). The pruned-out thorny branches provide supplementary income for the farmers in Kerala and drier parts in India as these branches are utilised to make fence at the boundary of homestead, farms and crop land, to support small climbing vegetable annuals and to make firewood. The thorny branches are also used as reinforced materials at the centre of mud wall for the construction of houses. In *B. balcooa, B. tulda* and *B. vulgaris,* stout long branches growing on the lower

Fig. 5.7 *B. bambos,* pruning of thorny branches from mid to lower part of culm

nodes of more than 2-year-old culms are also often pruned and utilised. In the Philippines, the removal of spines from culms close to the ground increases the shoot production by reducing the shoot mortality of *B. blumeana* (Tesoro and Espiloy 1988).

5.7.3.4 Weeding and Cleaning

In a well-developed bamboo stand, sometimes vines and climbers may infest the clumps, especially in the gap spots and boundary areas. These vines and weeds are to be cut between the spring and pre-monsoon time to minimise the fire hazard.

5.7.3.5 Making and Maintenance of Fire Lines

Forest fires are a potential risk to highland bamboo natural stands, especially during the dry season. Dry leaves, branches and twigs should be collected and used as mulching material around the clump which aids in preventing the spread of fires. Corridors 10–15 m wide are usually sufficient to stop fire from spreading throughout the plantation.

5.7.3.6 Soil Works and Fertilising

Loosening the soil in early summer to a depth of 6–12 cm and removing weeds around bamboo clumps are to be done to expose bamboo buds to higher temperatures and sunlight. This stimulates shoot sprouting, prevents bamboo roots from getting entangled and increases nutrient supply for shoot production. About 10–20 kg FYM and 0.5–0.75 kg urea or ammonium sulphate may be added in the ditch dug around a clump and then cover the ditch with earth. The NPKSi compound commercial fertiliser (375 Kg per hectare) may be applied in early spring to increase the yield of *Ph. pubescens* stand. Adding of new soil (mounding) to the clumps is done for maintaining the health of bamboo groves. Recent study at Xianju, Zhejiang Province of China (Shi and Peng 2010), showed that injecting fertiliser into bamboo cavity could effectively enhance yearly average number of bamboo shoots per bamboo and output of bamboo shoots per unit area. The net income was increased by 37.7 and 24.1 %, respectively, as compared with that of no fertilisation and conventional fertilisation, and the labour cost of fertilisation was greatly saved.

5.7.3.7 Watering

It always influences shoot production and elongation. However, in the hilly areas, watering in bamboo stands is not feasible. In draught areas, especially in the small farms and homesteads, watering should be done in the drier months. There was some indication that irrigation has reduced the strength properties of bamboo culms of *B. blumeana* (Midmore 2009).

5.7.3.8 Mulching

In a well-developed bamboo stand, litters act as mulch on the ground. It helps to maintain soil organic matter and manipulate soil temperature and soil moisture. It was observed that the frequency of irrigation was reduced in both '2 cm mulch' and '4 cm mulch' and reduced the total amount of the irrigated water 20 % less than no mulching. However, the '2 cm mulch' was enough to prevent the decrease of the soil water content.

5.8 Bamboos in Homestead

Home gardens are located close to the houses and form a part of the intensively managed household management system. Bamboos have a long history of use and play an integral part in the daily lives of rural people of Asian countries. Furthermore, bamboos are mainly utilised for construction works and making agricultural implements by the rural people. About 1.0 billion people worldwide live in bamboo houses. Due to large size and strong nature annually about 606.2 millions poles of these cultivated bamboos are used for constructing rural houses, temporary bridges, scaffolding, and pandals for religious and social gathering, cultural meetings; and about 4.0 million for making rickshaw hood, boat roof, punt pole and bullock cart in Bangladesh only (Banik 2000). It is assumed that these figures are much more high for such utilization in India. Thus homestead bamboos in south Asia may be treated as 'Plants of mass people.' Therefore, people of tropical Asia have been cultivating bamboo, mostly clump forming (species mentioned under Sect. 5.5.2.1), as one of the homestead crops from the unrecorded past. Bamboos in the village homesteads usually occupy the backyard and the periphery of the holdings. There are millions of people who depend on bamboo for part or all of their income. For example, in India, about two million traditional bamboo artisans depend almost entirely on the harvesting, processing and selling of bamboo and bamboo products such as baskets, mats and handicrafts (Belcher 1996). In China, there are millions of farmers who grow bamboo as a component in integrated farming systems.

References

Anantachote A (1987) Flowering and seed characteristics of bamboo in Thailand. In: Rao AN, Dhanaranjan G, Sastry CB (eds) Recent research on bamboos. Proceedings of the international bamboo workshop, 6–14 Oct 1985. Hangzhou, China, IDRC, Canada, pp 136–145

Appasamy T, Ganapathy A (1992) Preliminary survey of vesicular arbuscular mycorrhizal (VAM) association with bamboos in Western Ghats. BIC-India Bull 2(2):13–16

Bahadur KN (1979) Taxonomy of bamboos. Indian J For 2(3):222–241

Bahadur KN (1980) A note on the flowering of *Bambusa nutans*. Indian For 106(4):314–316

Banik RL (1979) Flowering in biajja bansh (*Bambusa vulgaris*). Letter to the editor, Bano Biggyan Patrika 8(1&2):90–91

Banik RL (1980) Propagation of bamboos by clonal methods and by seeds. In: Lessard G, Chouinard A (eds) Bamboo research in Asia. Proceedings of a bamboo workshop, Singapore, 28–30 May 1980. IDRC, Ottawa; IUFRO, Vienna, pp 139–150

Banik RL (1981) A short note on the flowering of *Dendrocalamus strictus* Nees (Lathi bansh) in Chittagong - Short communications. Bano Biggyan Patrika 10(1&2):94–96

Banik RL (1983) Macropropagation of bambusoid grass (*Phragmites communis* (cav.) Trin and Stend.) and micro-propagation of bamboo (*Bambusa glaucescens* Siebold). M.S. thesis. Department of Horticulture, University of Saskatchewan, Saskatoon, pp 1–155

Banik RL (1984) Macropropagation of bamboos by prerooted and prerhizomed branch cuttings. Bano Biggyan Patrika 13(1&2):67–73

Banik R L (1986) Observations on special features of flowering in some bamboo species of Bangladesh. In: Higuchi T (ed) Bamboo production and utilization. Proceedings of the project group P5.04. XVIII IUFRO World Congress; Ljubljana, Yugoslavia, pp 56–60

Banik RL (1987a) Techniques of bamboo propagation with special reference to prerooted and prerhizomed branch cuttings and tissue culture. In: Rao AN, Dhanarajan G, Sastry CB (eds) Recent research on bamboos. Proceedings of the international bamboo workshop, 6–14 Oct 1985, Hangzhou, China. The Chinese Academy of Forest; IDRC, Canada, pp 160–169

Banik RL (1987b) Seed germination of some bamboo species. Indian For 113(8):578–586

Banik RL (1988) Management of wild bamboo seedlings for natural regeneration and reforestation. In: Rao IVR, Gnanaharan R, Sastry CB (eds) Bamboos -- current research. Proceedings of the international bamboo workshop, 14–18 Nov 1988. KFRI, Cochin, India, pp 92–95

Banik RL (1989) Recent flowering of muli bamboo (*Melocanna baccifera*) in Bangladesh: an alarming situation for bamboo resource. Bano Biggyan Patrika 18(1&2):65–68

Banik RL (1991) Biology and propagation of bamboos of Bangladesh. A Ph.D. thesis, University of Dhaka, pp 1–321

Banik RL (1993a) Periodicity of culm emergence in different bamboo species of Bangladesh. Ann For 1(1):13–17

Banik RL (1993b) Selection and multiplication of bamboos for rural and industrial planting programmes. In: Vivekanandan K et al (eds) Proceedings of a workshop on the production of genetically improved planting materials for afforestation programmes, June 1993. ICFRE/FAO/UNDP-FORTIP, Coimbatore, India, pp 76–97

Banik RL (1994a) Diversities, reproductive biology and strategies for germplasm conservation of bamboos. In: Ramanatha RV, Rao AN (eds) Bamboo and rattan genetic resources and use. Proceedings of the first INBAR biodiversity, genetic resources and conservation working group, 7–9 Nov 1994. IPGRI-APO, Serdang, Malaysia, pp 1–22

Banik RL (1994b) Studies on seed germination, seedling growth and nursery management of *Melocanna baccifera* (Roxb.) Kurz. In: Proceedings of the 4th international bamboo workshop on bamboo in Asia and the Pacific, 27–30 Nov 1991. FORSPA Publication No 6 IDRC FAO-UNDP, Chiangmai, Thailand, pp 113–119

Banik RL (1994c) Review of conventional propagation in bamboos and future strategy. In: Constraints to production of bamboos and rattan. Report of a consultation held 9–13 May 1994, Bangalore India, INBAR technical report no 5, New Delhi, pp 115–142

Banik RL (1995) A manual of vegetative propagation of bamboos. INBAR technical report no 6, New Delhi, India, pp 1–66

Banik RL (1997a) Domestication and improvement of bamboos. INBAR/UNDP/FORTIP, New Delhi; Guangzhou; Eindhovan, pp 1–53

Banik RL (1997b) Growth response of bamboo seedlings under different light conditions at nursery stage. Bang J For Sci 26(2):13–18

Banik RL (1997c) The edibility of shoots of Bangladesh bamboos and their continuous harvesting effect on productivity. Bang J For Sci 26(1):1–10

Banik RL (1997d) Bamboo resource of Bangladesh. In: Alam MK, Ahmed FU, Ruhul Amin SM (eds) Agroforestry: Bangladesh perspective. APAN/NAWG/BARC, Dhaka, pp 183–207

Banik RL (1997e) *Melocanna baccifera* (Roxb.) Kurz – a priority bamboo resource for denuded hills of high rainfall zones in South Asia. In: Karki M, Rao AN, Rao VR et al (eds) The role of bamboo, rattan and medicinal plants in mountain development. Proceedings of a workshop, Institute of Forestry, 13–17 May 1996, Pokhara, Nepal. INBAR technical report no. 15 INBAR/IPGRI/ICIMOD/IDRC, pp 79–86

Banik RL (1998) Reproductive biology and flowering populations with diversities in muli bamboo, *Melocanna baccifera* (Roxb). Kurz Bang J For Sci 27(1):1–15

Banik RL (1999) Flowering in *Dendrocalamus hamiltonii* Nees & Arn. ex Munro and *Schizostachyum dullooa* (Gamble & Majumdar) in Chittagong, Bangladesh. Bang J For Sci 28(2):69–74

Banik RL (2000) Silviculture and field-guide to priority bamboos of Bangladesh and South Asia. BFRI, Chittagong, pp 1–187

Banik RL (2004) Fatal flowers. In: World bamboo Congress & CIBART/INBAR Communiqué (News Letter), vol 1(1), New Delhi, pp 6–7, Feb 2004

Banik RL (2008) Issues on production of bamboo planting materials—Lessons and Strategies. Indian For (Bamb Iss) 134(3):291–304

Banik RL (2010a) Biology and silviculture of muli (*Melocanna baccifera*) bamboo. Department of Science & Technology, National Mission on Bamboo Applications (NMBA), TIFAC, New Delhi, pp 1–237

Banik RL (2010b) Physiology and practices in propagation of bamboos with special reference to rooting in cuttings. In: Nath S, Singh S, Sinha A, Das R, Krishnamurthy R (eds) Conservation and management of bamboo resources. Institute of Forest Productivity (ICFRE), Lalgutwa, pp 111–121

Banik RL, Alam MK (1987) A note on the flowering of *Bambusa balcooa* Roxb. Bano Biggyan Patrika 16(1&2):25–29

Banik RL, Islam SAMN (2005) Leaf dynamics and above ground biomass growth in *Dendrocalamus longispathus* Kurz. J Bamb Rattan 4(2):143–150

Belcher B (1996) The role of bamboo in development. In: Belcher B, Karki M, Williams T (eds) Bamboo, people and the environment. Socio-economics and culture, vol 4. Proceedings of the 5th international bamboo workshop and the 4th international bamboo congress, 19–22 June 1995. Ubud Bali, Indonesia. INBAR/EBF/IPGRI/IDRC, INBAR, New Delhi, pp 1–9

Blatter E (1929) Indian bamboos brought up to date. Indian For 55(541–562):586–612

Brandis D (1899) Biological notes on Indian bamboos. Indian For 25(1):1–25

Brandis D (1906) Indian trees. Periodical experts. Book Agency, Delhi, p 767

Bystriakova N, Kapos V, Lysenko I, Stapleton C (2003) Distribution and conservation status of forest bamboo biodiversity in the Asia-Pacific region. Biodivers Conserv 12:1833–1841

Cavendish FH (1905) A flowering of Dendrocalamus hamiltonii in Assam. Indian For 31:479

Christanty L, Kimmins JP, Mailly D (1997) Without bamboo, the land dies': a conceptual model of the biogeochemical role of bamboo in an Indonesian agroforestry system. For Ecol Manage 91 (1):83–91

Dutra J (1938) Bambusees de Rio Grande du sud Revista. Sudamerica de Botanica 5:45–152

Faegri K, Vander Pijil L (1979) The principles of pollination ecology. Pergamon, Oxford

Fu Maoyi and Banik RL (1996). Bamboo production system and their management. Bamboo production system and their management. In: Rao IVR, Widjaja E (eds). Bamboo, people and the environment. Propagation and management, vol I. Proceedings of the 5th international bamboo workshop and the 4th international bamboo congress, 19–22 June1995, Ubud Bali Indonesia. INBAR/EBF/IPGRI/IDRC, INBAR, New Delhi, pp 18–33

Gamble JS (1896) The Bambuseae of British India. In: Annals of the royal botanic garden, Calcutta, vol 7. Printed at the Bengal Secretariat Press Calcutta, London, pp 1–133

Gilmour DA, Van San N, Tsechalicha X (2000) Conservation issues in Asia: rehabilitation of degraded forest ecosystems in Cambodia, Lao PDR, Thailand and Vietnam- an overview. IUCN/WWF/GTZ, Thailand, pp 1–45

Guan F, Fan S, Liu J et al (2012) Study on topography differentiation characteristics of spatial distribution and dynamic change of bamboo forest. In: Luo J (ed) Soft computing in information communication technology. AISC, vol 161. Springer, New York, pp 523–530. doi:10. 1007/978-3-642-29452-5_74

Guerreiro C (2013) Flowering cycles of woody bamboos native to southern South America. J Plant Res (Published online at 27 Oct 2013). doi:10.1007/s10265.013.0593.z. http://link.springer. com/journal/10265

Gupta MLS (1952) Gregarious flowering of *Dendrocalamus*. Indian For 78:547–550

Gupta KK (1972) Flowering of different species of bamboo in Cachar district of Assam in recent times. Indian For 98:83–85

He LM, Ye ZJ (1987) Application of multiple statistical analysis to the study of bamboo (*Ph. pubescens*) soils. J Bamb Res 6(4):28–40

Holttum RE (1958) The bamboos of the Malay Peninsula. Garden Bull 16:1–135, Singapore

Huang S-Q, Yang C-F, Lu B et al (2002) Honeybee-assisted wind pollination in bamboo *Phyllostachys nidularia* (Bambusoideae: Poaceae)? Bot J Linn Soc 138:1–7

Hunja M (2009) Mass propagation of bamboo, and its adaptability to waste water gardens. A Ph.D. thesis. Department of Horticulture, Jomo Kenyatta University of Agriculture and Technology

International Center for Bamboo and Rattan (ICBR) (2004) Bamboo and rattan industry. www. icbr.ac.cn/english/industry/2004-08/28/ics_316.htm

Janzen DH (1976) Why bamboos wait so long to flower. Ann Rev Ecol Syst 7:347–391

Kadambi K (1949) On the ecology and silviculture of *Dendrocalamus strictus* in the bamboo forests of Bhadravati Division, Mysore state, and comparative notes on the species *Bambusa arundinacea, Ochlandra travancorica, Oxytenanthera monostigmata* and *O. stocksii*. Indian For 75:289–299, 334–349, 398–426

Kawamura S (1927) On the periodical flowering of the bamboos. Jpn J Bot 3:335–349

Kondas S, Sree Rangasamy SR, Jambulingam R (1973) Performance of *B. arundinacea* Retz. seedling in nursery. Madras Agric J 60(9–12):1719–1726, Tamil Nadu Agriculture University, Coimbatore, India

Koshy KC, Harikumar D (2001) Reproductive biology of *Ochlandra scriptoria*, an endemic reed bamboo of Western Ghat, India. Bamb Sci Cult J Am Bamb Soc 15(1):1–7

Kumar A (1991) Mass production of field planting stock of Dendrocalamus strictus through macroproliferation–A technology. Indian For 117(12):1046–1052

Kurz S (1876) Bamboo and its use. Indian For 1(3):219–269, and 1(4):355–362

Kurz S (1877) Forest flora of British Burma. In: Introductory. Bamboo jungles and savannahs, vol 1. Reprinted in 1974 by Singh BSMP. Dehra Dun& Periodical Expert, New Delhi, pp 29–30

Lantican CB, Palijon AM, Saludo CG (1987) Bamboo research in Philippines. In: Rao AN, Dhanarajan G, Sastry CB (eds) Recent research on bamboos. Proceedings of the international bamboo workshop, 6–14 Oct 1985, Hangzhou, China. The Chinese Academy of Forest; IDRC, Canada, pp 50–60

Ly P, Pillot D, Lamballe P et al (2012) Evaluation of bamboo as alternative cropping strategy in the northern Central Upland of Vietnam: above ground Carbon fixing capacity, accumulation of soil organic carbon and socio-economic aspect. Agr Ecosyst Environ 149:80–90

Lyall JH (1928) The distribution of sal and bamboos in South Palamau Division, Bihar and Orissa. Indian For 54(9):486–490

Mathauda GS (1952) Flowering habits of bamboo-*Dendrocalamus strictus*. Indian For 78:86–88

McClure FA (1966) The bamboos: a fresh perspective. Harvard University Press, Cambridge, MA, pp 1–347

Midmore DJ (2009) Overview of the ACIAR bamboo project outcomes. In: Midmore DJ (ed) Silvicultural management of bamboo in the Philippines and Australia for shoots and timber. Proceedings of a workshop Los Baños, Philippines, 22–23 Nov 2006. ACIAR Proceeding no. 129, pp 7–12

Nadgauda RS, John CK, Mascarenhas AF (1993) Floral biology and breeding behaviour in the bamboo *Dendrocalamus strictus* nees. Tree Physiol 13(4):401–408

Nicholson JW (1945) Flowering of Bambusa arundinacea in Orissa—a letter to the editor. Indian For 71(12):435–436

Numata M (1987) The ecology of bamboo forests—particularly on temperate bamboo forests. Bamb J 4:118–131

Numata M, Ikusima I, Ohga N (1974) Ecological aspects of bamboo flowering. Ecological studies of bamboo forests in Japan, XIII. Bot Mag Tokyo 87:271–284

Patil VC, Patil SV (1988) Performance of bamboo under varying spacing and fertility. In: Rao IVR, Gnanaharan R, Sastry CB (eds) Bamboos--current research. Proceedings of the international bamboo workshop, 14–18 Nov 1988. KFRI, Cochin, India, pp 107–111

Pattanaik S, Pathak KC, Trivedi S et al (2002) Cohort mapping in bamboo resource management. In: Pattanaik S, Sing AN, Kundu M et al (eds) Proceedings of expert consultation on strategies for sustainable utilization of bamboo resources subsequent to gregarious flowering in the northeast, 24–25 Apr 2002. RFRI Jorhat, book no 12-2002. UNIDO, India, pp 11–17

Paudel P, Kafle G (2012) Assessment and prioritization of community soil and water conservation measures for adaptation to climatic stresses in Makwanpur district of Nepal. J Wetl Ecol 6:44–51

Ramyarangsi S (1988) Techniques for seed storage of Thyrsostachys siamensis. In: Rao IVR, Gnanaharan R, Sastry CB (eds) Bamboos—current research. Proceedings of the international bamboo workshop, 14–18 Nov 1988. KFRI, Cochin, India, pp 133–135

Rao IVR, Yusoff AM, Rao AN et al (1989) Propagation of bamboo and rattan through tissue culture. IDRC Bamboo and Rattan Research Network, Singapore

Rogers CG (1900) Flowering of bamboos in the Darjeeling District. Indian For 26(7):331–332

Scurlock JMO, Dayton DC, Hames B (2000) Bamboo: an overlooked biomass resource? Biomass Bioenergy 19:229–244

Shi X-h, Peng J-l (2010) Preliminary experimental study on imported technology of injecting fertilizer into bamboo cavity. Acta Agric Jiangxi 2010-05. doi:CNKI:SUN:JXNY.0.2010-05-025

Silgram M, Shepherd MA (1999) The effects of cultivation on soil nitrogen mineralization. Adv Agron 64:267–311

Soderstrom TR, Calderon CE (1979) A commentary on the Bamboos (*Poaceae: Bambusoideae*). Biotropica 11(3):161–172

Stapleton CMA (1982) Bamboo in East Nepal: preliminary findings. Forest research & information centre report, Department of Forests Kathmandu

Sur K, Lahiri AK, Basu RN (1988) Hydration dehydration treatments for improved seed storability of bamboo (*Dendrocalamus strictus* L.). Indian For 114(9):560–563

Tesoro FO, Espiloy ZB (1988) Bamboo research in Philippines. In: Rao IVR, Gnanaharan R, Sastry CB (eds) Bamboos--current research. Proceedings of the international bamboo workshop, 14–18 Nov 1988. KFRI, Cochin, India, pp 15–21

Troup RS (1921) The silviculture of Indian trees, vol III, Gramineae. The Clarendon Press, Oxford, pp 978–1013

Uchimura E (1980) Bamboo cultivation. In: Lessard G, Chouinard A (eds) Bamboo research in Asia. Proceedings of a bamboo workshop, Singapore. IDRC, Ottawa, Canada, pp 151–160

Uchimura E (1987) Growth environmental and characteristics of some tropical bamboos. Bamb J 4:51–60

Ueda K (1960) Study on the physiology of bamboo with reference to practical application. Prime Minister's office, Resources Bureau, Science and Techniques Agency, Tokyo, Japan, pp 1–167

Varmah JC, Bahadur KN (1980) India - a country report. In: Lessard G, Chouinard A (eds) Bamboo research in Asia. Proceedings of a bamboo workshop, Singapore. IDRC, Ottawa, Canada, pp 19–46

Watanabe M, Ueda K, Manabe I, Akai T. (1982) Flowering, seeding, germination and periodicity of *Phyllostachys pubescens*. J Jpn For Soc 64(3):107–111

Zamora AB (1994) Review on micropropagation research on bamboos. In: Constraints to production of bamboos and rattan. Report of a consultation held in Bangalore, 9–13 May 1994. INBAR technical report no. 5, New Delhi, pp 45–100

Chapter 6
Pests and Diseases of Bamboos

Jinping Shu and Haojie Wang

Abstract This chapter assesses the diversity and characteristics of bamboo insect pests and diseases and their control. Based on available data, the number of insects that feed on bamboos is estimated to be more than 1,200 and that of fungi and saprophytes to be more than 400, while there are less than 100 insect pests and ten diseases that cause heavy damage to bamboos. In addition, this chapter describes the characteristics and status of main groups of bamboo insects including bamboo shoot and culm borers, defoliators, branch and culm pests, bamboo seed pests, and postharvest pests. Finally, this chapter discusses the control methods including cultural, physical, biological, and chemical control against bamboo insect pests and diseases, and it is necessary to develop the IPM programs for bamboo pests in the future.

Keywords Bamboo pest insects • Bamboo diseases • IPM

6.1 Introduction

Bamboo is a kind of fast-growing woody grass that grows in the tropics and subtropics and has received increasing attention in many countries because of its availability, rapid growth, easy handling, and desirable properties. Bamboos have a wide range of societal and industrial uses as foods, building materials, crafts, and high-quality paper, as well as for landscaping and soil conservation, and the bamboo industry is closely related to people's daily lives and plays an important role in national economies (Wang et al. 1998).

During their whole lifetime, bamboos like other plants are subject to damage by various kinds of herbivorous insect pests and diseases. Insect pests and diseases are believed to be the main reasons for causing considerable bamboo losses in natural stands and plantations. The number of insects that feed on bamboos is estimated to be more than 1,200 (Chang 1986; Chen 1989; Xu and Wang 2004) and that of fungi and saprophytes to be more than 400 (Balakrishnan et al. 1988; Mohanan 1997). Though bamboos grow in many countries and regions, only few countries and

J. Shu (✉) • H. Wang
Research Institute of Subtropical Forestry, Fuyang, Zhejiang, People's Republic of China
e-mail: shu_jinping001@163.com

© Springer International Publishing Switzerland 2015
W. Liese, M. Köhl (eds.), *Bamboo*, Tropical Forestry 10,
DOI 10.1007/978-3-319-14133-6_6

regions have researched bamboo insects and diseases in details (Dayan 1990; Xu and Wang 2004). In natural bamboo forests with reasonable biodiversity and stable population, the damage caused by insects and diseases is usually not serious because various and abundant natural enemies play an effective role in the control of the bamboo pests. However, because of the excessive human intervention of bamboo plantation and the mass application of broad-spectrum pesticide, some pests, especially bamboo locusts and the shoot borers, often break out in large areas (Yu et al. 2011). For example, bamboo shoot weevils, the yellow-spined bamboo locusts, and shoot-boring noctuids occur in about 80,000 ha of bamboo plantations in China and cause large quantities of loss of bamboo culms and edible shoots every year (Xu and Wang 2004). On an average, bamboo plantations lose about 10 % of their potential turnover because of insect pest infestations. In addition, some pests, such as bamboo shoot wireworms, regarded as minor pests in bamboo forest before, have caused economically severe losses of bamboo production in Zhejiang, China, in recent years (Shu et al. 2012). In order to reduce the damage caused by bamboo pests and diseases, it is essential to clear the current pest status of bamboo, as well as characteristics and control strategies of the important species.

Therefore, in the first part of this chapter, we introduce the biodiversity of bamboo pests and the current status of the important species. Secondly, we assess the damage of major bamboo diseases. Finally, control strategies of bamboo insect pests and diseases are discussed.

6.2   Bamboo Insect Pests

6.2.1   Diversity of Bamboo Insect Pests

More than 800 insect species have been recorded to feed on bamboos in the whole world, and most of them are in Asia (Wang et al. 1998). In China, at least 683 species, belonging to ten orders, 75 families, and 363 genera, were reported to attack bamboos (Table 6.1). About 180 insect pests associated with bamboo are discovered, and 80 pests are recorded on bamboos in Japan. As well, not more than ten species that injure bamboos are noted in South Korea, Thailand, and Nepal (Stapleton 1985; Kim and Lee 1986; Choldumrongkul 1994). The high number of bamboo insect pests in Asia may be attributed to the high diversity of bamboo. Forty-four genera (60 % of the world's total number) of bamboo occur throughout tropical, subtropical, and temperate Asia. This enormous diversity of bamboo species in an area is likely to support an equally diverse Insecta. The lower number of bamboo insects described from non-Asian regions may also be attributed to limited surveying.

So far, a detailed survey of bamboo insect pests has not been carried out yet, and more attention is focused on the insects that injure the economic and important bamboos. Therefore, the number of insect species that damage bamboos is

Table 6.1 Taxonomic composition of bamboo insect pests in China (Xu et al. 1993)

Order	Orthoptera	Phasmatodea	Isoptera	Homoptera	Hemiptera	Thysanoptera	Coleoptera	Diptera	Lepidoptera	Hymenoptera	Total
Family	3	3	2	17	7	2	12	6	18	5	75
Genus	17	3	5	101	54	4	70	16	85	8	363
Species	32	3	8	250	78	5	117	22	156	12	683

estimated to be more than 1,200,[1] and the homopterans are the most abundant in all presented bamboo insect species, followed by lepidopterans and beetles (Coleoptera). And most dangerous and disastrous species belong to Lymantriidae, Noctuidae, Pyralidae, Curculionidae, Bostrichidae, Cerambycidae, Pentatomidae, and Locustidae.

6.2.2 Characteristics and Status of Main Groups of Bamboo Insects

Bamboos are subject to injury by various kinds of herbivorous insects. The damages of most insects are not serious and have little effect on the growth of bamboo, though they are feeding on bamboo. About 100 of them can have a serious negative impact on the growth of bamboo. And no more than 100 of them can cause considerable economic losses and even cause quantities of bamboo stands to die when they break out in large areas. Bamboo insect pests can be differentiated as bamboo shoot and culm borers, defoliators, branch and culm pests, bamboo seed pests, and postharvest pests according to the injury period and damage parts of bamboo.

6.2.3 Bamboo Shoot and Culm Borers

More than 100 insects, including bamboo shoot weevils (Coleoptera: Curculionidae), bamboo shoot wireworms (Coleoptera: Elateridae), shoot-boring noctuids (Lepidoptera: Noctuidae), and shoot flies (Diptera: Anthomyiidae and Tephritidae), are presented to damage bamboo shoot and live culms. Recently, wireworms, shoot weevils, and noctuids occur extensively and have caused economically severe losses of bamboo production in Asia.

(1) Bamboo Shoot Wireworms
Wireworms, the common name for soil-dwelling larvae of click beetles (Coleoptera: Elateridae), occur extensively and have caused economically severe losses of bamboo production in South China in recent years. Larvae injure underground buds and root systems of bamboos, resulting in germination failure, ratooning failure, and losses in stand (Xu and Wang 2004; Zhou et al. 2008; Shu et al. 2012; Fig. 6.1). Figure 6.2 shows that the damage area and density of bamboo shoot wireworms increase rapidly in 8 years and the area where bamboo shoot wireworms are found in 2012 is over ten times than that in 2005. Moreover, the density of wireworm per shoot increases six times.

---

[1] Shu et al. (Data unpublished)

Fig. 6.1 Damage symptom of wireworm (*left*: *Phyllostachys praecox* shoot bored by wireworms; *right*: dead stands caused by wireworms)

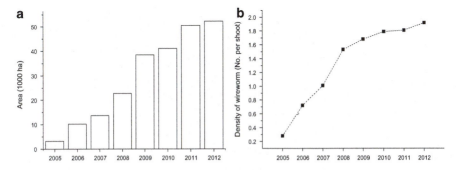

Fig. 6.2 Damage area and density of bamboo shoot wireworms in China during 2005–2012. (**a**) Damage area; (**b**) no. of wireworms captured in per shoot in sample site

About ten species of wireworms are estimated to feed on bamboo shoot, and only reported in China. *Melanotus cribricollis* is the predominant species in South China, and more than 60 bamboo species including *Phyllostachys* spp., *Dendrocalamus* spp., *Bambusa* spp., *Sinobambusa* spp., and *Pseudosasa* spp. are found to be infested by the wireworms (Xu and Wang 2004; Deng et al. 2010), and the *Phyllostachys* spp. are the most susceptible. *Melanotus cribricollis* completes a life cycle in 4 years and overwinters with larvae and adults in soil. Adults emerge from the soil in spring (mid-April). Larvae generally have two intense activity periods that may result in significant bamboo shoot damage over 1 calendar year, according to suitable temperature and moisture conditions in the different soil layers. They occur from March to May and from September to October (Zhou et al. 2008). The depth at which *Melanotus cribricollis* larvae can be found depends mostly on the temperatures and on soil moisture. It can be as deep as 60 cm in the winter, while in spring and early summer, most of the larvae are in the upper 20 cm of the soil (Song 2009).

Fig. 6.3 Bamboo shoot-boring noctuids. (**a**) *Apamea kumaso*; (**b**) *Oligia vulgaris*; (**c**) *Apamea apameoidis*; (**d**) *Apamea repetita conjuncta*

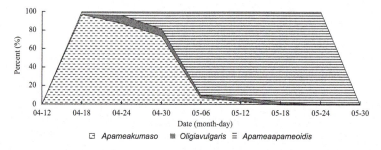

Fig. 6.4 The percent proportion of three bamboo-boring noctuids (Huang et al. 2009)

(2) Shoot-Boring Noctuids

Eight bamboo shoot-boring noctuids (Lepidoptera: Noctuidae) distributed in China, Japan, and India have been reported to feed on various *Phyllostachys* species (Xu and Wang 2004; Yoshimatsu et al. 2005; Choudhury and Ahktar 2007). Among them, *Kumasia kumaso*, *Apamea apameoides*, *Oligia vulgaris*, and *Apamea repetita conjuncta* are the most important, which can cause up to 90 % death of new shoots in China (Fig. 6.3; Huang et al. 2009). It is common to find several of these species occurring together in bamboo shoots. The damage is caused by larvae, which bore inside new shoots and cause the death or damage of shoots in most cases. Thus, damaged shoots and culms will have several feeding holes and tunnels.

All the shoot borers have a very similar life history and damaging habits (Fig. 6.4). There is one generation per year, and the pest overwinters as eggs. The overwintering eggs hatch in late March, and the larvae bore into the bamboo shoot for feeding in early April. Larvae develop fully and fall to the ground to pupate 20–30 days later. Adults emerge and lay eggs in June. And the damage period of larvae depends on the spring temperature.

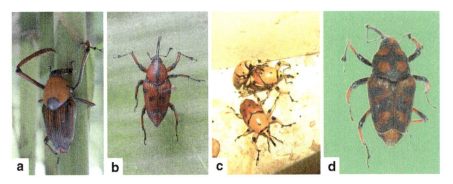

Fig. 6.5 Bamboo shoot weevils. (**a**) *Cyrtotrachelus longimanus*; (**b**) *Otidognathus davidis*; (**c**) *Otidognathus* sp.; (**d**) *Otidognathus rubriceps*

Fig. 6.6 Damage symptom of bamboo shoot weevils

(3) Bamboo Shoot Weevils

There are about 18 weevil species (Coleoptera: Curculionidae) distributed in China, Bangladesh, Japan, India, Myanmar, Brazil, and Sri Lanka attacking bamboo shoots (Wang et al. 1998; Choudhury and Ahktar 2007). Five species, *Cyrtotrachelus buqueti*, *Cyrtotrachelus longimanus*, *Otidognathus davidis*, *Otidognathus* sp., and *Otidognathus rubriceps*, are reported to cause serious damage to bamboos (Fig. 6.5). Both adults and larvae of these weevils feed on shoots, although larvae, which bore holes in bamboo shoots, are responsible for most of the damage. The damage caused by weevils usually results in the death of young shoots or deform and stunt growth of new culms with very closely formed nodes at the feeding site (Fig. 6.6). The larger species, such as *Cyrtotrachelus buqueti* and *C. longimanus*, are the most common and destructive on sympodial bamboos, and the small species, for example, *Otidognathus davidis*, *Otidognathus.* sp., and *O. rubriceps*, are found on monopodial bamboos.

Bamboo shoot weevils finish one generation per year and overwinter as adult in cocoons in the soil. The adults that damage monopodial bamboos begin to swarm from the soil in June when bamboo shoots are available and end in early October. Those weevils that injured monopodial bamboos appear from late April to mid-June.

6.2.4 Bamboo Leaf Defoliators

More than 400 insects are found to eat bamboo leaves, and about 39 species including locusts, leaf rollers, puss moths, tussock moths, and sawflies are reported to cause a huge economic loss in China (Zhang et al. 2002; Xu and Wang 2004).

(1) Bamboo Locusts

The locust is one of the most important pest insect groups that affect bamboo. About 40 bamboo locusts have been reported to feed on bamboo leaves in Asian countries. They are classified into a number of genera, of which *Hieroglyphus* are the most common. Both adults and nymphs feed on bamboo leaves and outbreaks usually cause complete defoliation of bamboo stands (Fig. 6.7). Heavy and repeated defoliation will result in the death of bamboo plants (Anonymous 1960, 1979). In recent years, the damage area by the yellow-spined bamboo locusts, *Ceracris kiangsu*, increases quickly.

Ceracris kiangsu has one generation per year and overwinters as eggs in capsules below ground. The nymphs hatch in April and take 40–44 days to develop fully, passing through five instars, to become adults in July. Adults start to lay eggs in August and end in October. Bamboo locusts have the habit of migration, and the population is affected significantly by temperature and rainfall.

(2) Leaf Rollers

Leaf rollers (Lepidoptera: Pyralidae) belong to the most important groups of leaf feeders on bamboo. More than ten species of bamboo leaf rollers have been reported as attacking various bamboos in Asia (Wang et al. 1998). Among them, four species, *Algedonia coclesalis*, *Crocidophora evenoralis*, *Demoboty pervulgalis*, and *Circobotys aurealis*, are the most important. Several species often occur together. The damage is caused by larvae, which tie leaves together as leaf cases and feed on the upper tissues of the leaves (Fig. 6.8). Outer leaves of the rolled leaf cases often wither and eventually fall off. Outbreaks are often reported in China, India, Japan, and Korea causing serious defoliation that results in reduced vigor and even the death of culms. Damage is found to be more severe in plantations than in natural stands and individual plantings.

Moreover, environment of forest, especially temperature, rainfall, and soil moisture are the key factors that impact the population of bamboo leaf rollers.

Fig. 6.7 Damage symptom of bamboo locusts. (**a**) Leaves of mao bamboo are eaten by locusts; (**b**) Adults of the yellow-spined bamboo locusts

Fig. 6.8 Damage symptom of leaf rollers. (**a**) Leaf roller; (**b**) larvae in the roller

6.2.5   *Bamboo Branch and Culm Pests*

More than 300 species distributed in China, India, and Japan are reported to feed on bamboo branches and culms, and most of them are sapsucking insect pests (Fang and Wang 2000; Xu and Wang 2004). The most important species include stink bugs (Hemiptera: Pentatomidae), coreid bugs (Hemiptera: Coreidae), froghoppers (Homoptera: Cercopidae), aphids (Homoptera: Aphididae), pit scale insects (Homoptera: Asterolecaniidae), and gall-making chalcids (Hymenoptera: Eurytomidae and Ceraphronidae). In many cases, both adults and nymphs of these insects, which have highly modified piercing-sucking mouthparts, feed on the sap of leaves, injure by egg laying, inject toxic compounds into the plant, and transmit diseases. The results are defoliation, wilting of young shoots and branches, and even death of the culm (Fig. 6.9). A heavy outbreak of suckers also can cause huge economic losses.

Galls, induced mainly by chalcid wasp species, are a common sight on bamboo twigs (Fig. 6.10). Galls cause abnormal growth and shedding of leaves on the

Fig. 6.9 Damage symptom of bugs and froghopper. (**a**) *Hippotiscus dorsalis*; (**b**) *Brachymna tenuis*; (**c**) *Aphrophora horizontalis*

Fig. 6.10 Damage symptom of gall maker. (**a**) Gall; (**b**) *Aiolomorphus rhopaloides* larva in the gall; (**c**) *Aiolomorphus rhopaloides* adult

affected twigs and thus probably affect the photosynthesis. The impact of galls on the productivity of bamboo stands however remains to be evaluated.

6.2.6 Bamboo Flower and Seed Pests

Seed pests, which affect seed production, may have some impact on the establishment of new plantations. Up to now, only few insects including seed bugs

(Heteroptera: Pentatomidae), lygaeid bugs (Heteroptera: Lygaeidae), and grain moths (Lepidoptera: Gelechiidae) are presented to damage bamboo flowers and seeds. Both adult bugs and nymphs feed on the seeds during the formative stage and also after they have fallen to the ground, thereby destroying the means for natural reproduction. The seed bug, *Udonga montana*, on natural bamboo stands occurs occasionally and causes serious damage in India (Mathew and Sudheendrakumar 1992; Choudhury and Ahktar 2007).

6.2.7 Postharvest Pests of Bamboo

Bamboo under storage, either as culms or as finished products, is very susceptible to damage by insects. More than 50 insect pests have been reported to attack felled culms and products made of bamboo timber (Garcia 2005). They include shot-hole borers (Coleoptera: Bostrichidae), powder-post borers (Lyctidae), long-horned beetles (Coleoptera: Cerambycidae), and termites. However, the most important species are mainly from the families of Cerambycidae, Bostrichidae, and Lyctidae. The adults and larvae cause direct damage by boring and continuously feeding inside the culm. This attack may result in severe damage or complete destruction of raw materials or finished bamboo products. In storage yards, stacks with immature culms become the starting point of attack, and the bamboo is often converted to dust.

6.2.8 Bamboo Diseases

A total of 440 fungi, three bacteria, two viruses, one phytoplasma, and one bacteria-like organism have been reported as associated with these diseases and disorders (Zhou et al. 2010). In China, a list of bamboo diseases was published including 208 species including 183 species of fungi, 1 bacteria, 2 fastidious prokaryotes, 1 virus, 3 nematodes, and 18 mites (Xu et al. 2006a, b, 2007). In India, a total of 36 species of pathogenic fungus have been reported. Only a few among these diseases are recognized as potentially serious ones, affecting the bamboo industry as well as the rural economy as a whole. Potentially serious diseases of bamboos in Asia include culm blight caused by *Sarocladium oryzae* in the village groves in Bangladesh and in the coastal areas of Orissa state, India; rot of emerging and growing culms of industrially important bamboo species in India caused by *Fusarium* spp.; witches' broom incited by *Balansia* spp. in different species of bamboos in Japan, China, and Taiwan and in reed bamboos in India; little leaf of *Dendrocalamus strictus* caused by phytoplasma (MLO) in the dry tracts of southern India; culm mosaic caused by bamboo mosaic virus (BaMV) in Taiwan; culm rust caused by *Stereostratum corticioides*; and top blight of *Phyllostachys* spp. caused by *Ceratosphaeria phyllostachydis* in China.

6.2.9   Control of Bamboo Pests and Diseases

Various control methods including cultural control, biological control, physical control, and chemical control against bamboo pests and diseases have been reported in bamboo plantation area, but the application of chemical pesticides is the most predominant measure used very often. Besides polluting the environment, the excessive use of broad-spectrum insecticide also killed the natural enemies which result in resistance and resurgence of pests. Therefore, IPM program for bamboo is required.

(1) Cultural Controls

Cultural control is the basis of IPM program for bamboo, and the aims of cultural control are to kill insects directly, create favorable environment for the natural enemies, and improve the insect-resistant ability of bamboos. These cultural controls include management of the culm population, such as removing or reducing the culm age population most favored by the pest or by adjusting the culm density to modify the grove's environment to make it unfavorable for the most prevalent pest. Another method used is to turn over the soil within and around the grove during fall and winter to expose buried pupae to the cold and bird predation. For example, removing weed in *Pseudosasa amabilis* forest decreases 20 % of the damage rate of shoot-boring noctuids (Huang et al. 2003).

(2) Physical Control

Physical control is a method of getting rid of insects by removing, attacking, or setting up barriers that will prevent further destruction of bamboos. These methods are designed according to common or specific behaviors of target insects. A large number of lepidopterous pests of bamboo have strong phototaxis action toward black light. Light trapping is very effective at controlling moth pest species during their adult stage. Mass trapping in this manner has proven to be effective and successful in control leaf rollers, tussock moths, and bamboo cicadas in China (Zheng et al. 1992; Xu et al. 2001; Liang et al. 2004). Figure 6.11 shows the black fluorescent light used in bamboo forest.

Some bamboo insects are attracted to a particular food and odor, and bait trapping, either through the use of baited traps or poisoned bait, is used to control some bamboo pests. The bamboo shoot fly is baited with fresh bamboo shoot pieces treated with insecticides. The yellow-spined bamboo locust (*Ceracris kiangsu*) adults are known to visit and feed on human urine, so poison baits with blend of human urine and insecticides were applied in fields to suppress its population. And 507 *Ceracris kiangsu* adults killed daily per bait containing 30-day-incubated urine with bisultap (Fig. 6.12, Shu et al. 2013).

Physical barriers are effective at controlling insects that need to crawl up the culm to get to their feeding/egg laying sites on the plant. Typically these barriers are a sticky band applied around the base of the culms. It is most effective and successful to set sticky or adhesive barriers around the basal part of culms in early April in the control of the stink bug, *Hippotiscus dorsalis*, in China

Fig. 6.11 Black lights used in bamboo forests

Fig. 6.12 Mass tapping of locusts with attractive toxic bait

(Fig. 6.13, Wang et al. 1998; Xu et al. 2006a, b). And it is useful to package shoot with plastic bag to prevent the bamboo shoot weevils laying eggs and feeding on the top of *Phyllostachys heterocycla cv. pubescens* shoot (Fig. 6.14, Cai et al. 2008).

Of course, the old standby handpicking is useful for controlling the larger, slower-moving bamboo pests in small groves and on specimen plants. The damage caused by bamboo shoot weevils can be significantly reduced by removing adults through handpicking before egg laying. Some insects, which feed in groups at basal

Fig. 6.13 Stick barrier for the stink bugs, *Hippotiscus dorsalis*

Fig. 6.14 Shoot packaged with plastic bag against bamboo weevils

portion of culms or shoots or leaves of lower branches, such as aphids and stink bugs, can also be physically removed.

(3) Biological Controls

Almost all insect pests are attacked or infested by a number of other living organisms which are their natural enemies. The natural enemies often include parasitoids, predators, pathogens, bacteria, and virus. In natural bamboo stands with high biodiversity, natural enemies play an important role in regulating insect pest populations. More than 50 species of natural enemies are reported to be helpful

Fig. 6.15 Shoot wireworm infested by *Metarhizium anisopliae*

in the control of bamboo insects (Xu et al. 2003). It is an important strategy for the conservation of natural enemies to create environment conditions favorable to natural enemies, such as retaining the surrounding vegetations and reducing the use of broad-spectrum pesticides. Moreover, *Metarhizium anisopliae* has been used to kill bamboo shoot wireworms in China. In 30 days, a conidia concentration gradient of 107 conidias/g soil can cause 100 % mortality of bamboo shoot wireworms (Wang et al. 2010, Fig. 6.15). *Beauveria bassiana* and *Bacillus thuringiensis* are used widely to control bamboo leaf defoliators, such as tussock moths, puss moths, and leaf rollers, in China (Li 2006; Sun et al. 2007; Wu 2009).

(4) Chemical Controls
Chemical control is highly effective, easy to use, and low in cost, and a large number of pesticides are available. Pesticides can be applied to standing bamboo plants by dusting, spraying, injecting, and smoking. Systemic pesticides are effective against most insect pests including leaf feeders, sapsuckers (including those with waxy coverings), shoot borers, and gall makers, but can't be used during the shoot harvesting season. One of the methods used for applying systemic pesticides is culm cavity injection, which is effective and safe to the environment and natural enemies (Zhen et al. 1999; Sun et al. 2005). And many botanical pesticides (rotenone, matrine, nicotine, azadirachtin) and biochemical pesticide (avermectin) have been applied to control bamboo pests. Sex pheromone is very effective and a safe method to control the insect population by mass trapping, to disrupt mating and to prevent further egg laying. And now mass trapping of sex pheromone has been applied to bamboo shoot noctuids in China. Figure 6.16 shows that about 70 moth adults are trapped by the mixture of Z11-16: Ac with Z11-16, OH per trap (Shu et al. unpublished data).

In conclusion, the long-term strategy for bamboo pests control is to develop IPM programs based on the fully understanding of the effects of various environmental conditions on insects and their interactions.

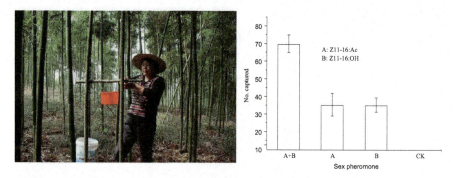

Fig. 6.16 Mass trapping of bamboo shoot noctuids with sex pheromone

References

Anonymous (1960) Population dynamics of bamboo locust and integrated control measures. Entomol Knowl 6(2):65–67

Anonymous (1979) Culm cavity injection of systemics for control bamboo leaf-rollers. For Sci Technol 3:19–20

Balakrishnan B, Chandrasekharan M, Das L (1988) Some common diseases of bamboo and reeds in Kerala. Bamboos current research. Proceedings of the international bamboo workshop, 4–18 Nov 1988, Cochin, India

Cai FC, Mou JJ, Wu ZM, Jin S, Wu ML (2008) Preliminary study on control of Otidognathus davidis by shoot bagging. J Zhejiang For Sci Technol 28(5):37–39 (in Chinese with English abstracts)

Chang YZ (1986) Insect pests of bamboos in Taiwan. In: Higuchi T (ed) Bamboo production and utilization. Proceedings of the congress group 5.04, production and utilization of bamboo and related species, XVIII IUFRO World Congress Ljubljana, Yugoslavia, 7–21. Kyoto University, Kyoto, Japan, pp 246–252

Chen Z (1989) Notes on Hemipterian insects feeding on bamboos in Guangdong, China. J Bamb Res 8(3):58–60 (in Chinese with English abstracts)

Choldumrongkul, S (1994) Insect pests of bamboo shoot in Thailand. In: Bamboo in Asia and the Pacific. Proceedings of the 4th International Bamboo Workshop, Chiangmai, Thailand, 27–30 November 1991. International Development Research Centre, Ottawa, Canada; Forestry Research Support Program for Asia and the Pacific, Bangkok, Thailand, pp 331–335

Choudhury RA, Ahktar MS (2007) Insect pests of bamboo in Aligarh, India. J Entomol Res 31 (4):369

Dayan MP (1990) Survey, identification and pathogenicity of pests and diseases of bamboo in the Philippines. Paper presented at the first national bamboo symposium workshop, 27 Feb–1 Mar 1989. Sylvatrop 13(1–2): 61–67

Deng S, Shu JP, Wang HJ (2010) Investigation of host range of wireworms (*Melanotus cribricollis*) and their spatial distribution in soil. Chin Bull Entomol 47(5):983–987 (in Chinese with English abstracts)

Fang ZG, Wang YP (2000) The hemipterous pests endangering bamboo in China. J Zhejiang For Sci Technol 20(3):54–57, 61. (in Chinese with English abstracts)

Garcia CM (2005) Management of powder-post beetles, *Dinoderus minutus* F. in freshly cut bamboo

Huang HH, Shao SF, Chen T, Tong GJ, lai YM, Ql L (2003) Study on the integrated control technique of the noctuid on the bamboo shoot of Pseudosasa amabilis. For Res 16(4):444–448 (in Chinese with English abstracts)

Huang QY, Shu JP, Zhang AL, Xu TS, Wang HJ (2009) Niche of three dominant shoot-boring noctuids and their interspecific competition. For Res 22(5):647–651 (in Chinese with English abstracts)

Kim KC, Lee TS (1986) Studies on the host plant, bionomics, and damage of bamboo leaf rollers in Chonnam Province area. Korean J Plant Prot 25(2):85–92

Li KS (2006) Effects of Sendebao and Bacillus thuringiensis on controlling Algedonia coclesalis. J Zhejiang For Coll 23(4):445–448 (in Chinese with English abstracts)

Liang SX, Qian SS, Hei ZH (2004) Life history of the moth Pantana phyllostachysae and its control. Entomol Knowl 41(5):464–467 (in Chinese with English abstracts)

Mathew G, Sudheendrakumar VV (1992) Outbreak of Udonga montana (Distant) (Heteroptera; Pentatomidae) on bamboo in natural forests and adjoining plantations in Wynad, Kerala. BIC-India Bull 2(2):17–18

Mohanan C (1997) Diseases of bamboos in Asia. India Brill Academic

Shu JP, Teng Y, Chen WQ, Shi J, Liu J, Xu TS, Wang HJ (2012) Control techniques of Melanotus cribricollis (Coleoptera: Elateridae). For Res 25(5):620–625 (in Chinese with English abstracts)

Shu JP, Teng Y, Liu J, Xu TS, Wang HJ (2013) Behavioral responses of the yellow-spined bamboo locust, Ceracris kiangsu towards the human urine incubated for different days. Chin J Ecol 32 (4):946–951 (in Chinese with English abstracts)

Song Y (2009) Study on the occurrence mechanism and monitoring and management techniques of the bamboo wireworms. Chinese Academy of Forestry, Beijing

Stapleton CMA (1985) Noctuid shoot borers in Dendrocalamus and Bambusa species. Nepal For Tech Inf Bull 11:26–31

Sun PL, Chen WM, Huang ZG, Xie WY, Li RH, Lu G (2005) Injection of imidacloprid into bamboo cavity against Tetramesa bambusae. For Pest Dis 24(1):20–22 (in Chinese with English abstracts)

Sun PL, Lu G, Chen WM, Huang ZG, Xue WY (2007) Field trials of emulsifiable suspensions of Beauveria bassinaconidia for control aphid in shoot bamboo. J Zhejiang Univ (Agric Life Sci) 33(2):197–201 (in Chinese with English abstracts)

Wang HJ, Varma RV, Xu TS (1998) Insect pests of bamboos in Asia: an illustrated manual. International Network for Bamboo and Rattan, Beijing

Wang P, Zhang YB, Shu JP, Deng S, Wang HJ (2010) Virulence of Metarhizium anisopliae var. anisopliae to the Larvae of Melanotus cribricollis (Coleoptera: Elateridae). Chin J Biol Control 26(3):274–279 (in Chinese with English abstracts)

Wu JQ (2009) Experimentation of various application methods of Beauveria bassiana on control-ling Pantana phyllostachysae. China For Sci Tech 23(4):101–105

Xu TS, Wang HJ, Ruoqing LU (1993) Revised list of bamboo insect pests in China. Zhejiang For Pest Dis 7(4):4–34

Xu TS, Wang HJ (2004) Main pests of bamboo in China. Chinese Forest Publish House, Beijing, pp 45–46

Xu TS, Wang HJ, Xu QR, Huang ZY, Lin CC (2001) Biological characteristics of Platylomia pieli. For Res 14(4):396–402 (in Chinese with English abstracts)

Xu HC, Luo ZD, Wang XP, Zhang H (2003) A list of natural enemies found on shoot oriented bamboo pests. Nat Enemies Insects 25(3):119–124 (in Chinese with English abstracts)

Xu MQ, Dai YC, Fan SH, Jin LX, lv Q, Tian GZ, Wang LF (2006a) Records of bamboo diseases and the taxonomy of their pathogens in China(I). For Res 19(6):692–699

Xu ZW, Hua BS, Guo XH, Liang XM, Cheng TS (2006b) Control methods of Hippotiscus dorsalis. East China For Manag 20(3):47–51 (in Chinese)

Xu MQ, Dai YC, Fan SH, Jin LX, lv Q, Tian GZ, Wang LF (2007) Records of bamboo diseases and the taxonomy of their pathogens in China(II). For Res 21(1):45–52

Yoshimatsu S, Kusigemati K, Gyoutoku N, Kamiwada H, Sato Y, Sakamaki Y (2005) Some lepidopterous pests of bamboo and bamboograss shoots in Japan. Jpn J Entomol (New Ser) 8 (3):91–97

Yu HP, Shen K, Wang ZT, Mu LL, Li GQ (2011) Population control of the yellow-spined bamboo locust, *Ceracris kiangsu*, using urine-borne chemical baits in bamboo forest. Entomol Exp Appl 138(1):71–76

Zhang FP, Chen QL, Chen SL (2002) Research advances on the pests that eat leaves of *Phyllostachys heterocycla cv. Pubescens*. J Bamb Res 21(3):55–60 (in Chinese with English abstracts)

Zhen JC, Lin YY, Cao QY, Nie JQ, Zheng ZF (1999) Study on new control technique against *Pantana Phyllostachysea* pests. J Fujian Coll For 19(1):61–64 (in Chinese with English abstracts)

Zheng JJ, Yang GR, Chen JY, Jiang P (1992) Techniques for control of bamboo leaf roller Algedonia coclesalis (Walker). J Bamb Res 11(3):32–36 (in Chinese with English abstracts)

Zhou YE, Bai HQ, Shu JP (2008) Study on biological characteristics of Melanotus cribricollis. J Zhejiang For Sci Technol 28(4):28–32

Zhou CL, Wu XQ, Ji J, Ye JR (2010) Research advances of bamboo diseases. China For Sci Technol 24(5):8–13

Chapter 7
Harvesting Techniques

Ratan Lal Banik

Abstract The felling practices of bamboos both from forests and homesteads of different countries of tropics are discussed. Bamboos are felled on the basis of selection of age above 3–4 years. However, clear-felling system is also practised in some special occasions mainly in the homesteads cultivations. A detail of felling process and steps with ongoing Mohal system and Permit system of harvesting the bamboo resources including transportation through waterways from the hills of the subcontinent has been described. The methods of felling bamboos from a clump and guidelines of harvesting bamboos after gregarious flowering in the forests are outlined.

Keywords Bamboo harvesting • Felling age • Felling intensity • Clearfelling • Selection felling • Mohal system • Water ways transportation • Rafting • Felling after flowering

7.1 Introduction

Bamboos in the forests usually cover a large tract of land from valley, slopes, and tops of the hills. The management of bamboo forests on a scientific basis is beset with many difficulties owing to a variety of reasons, chief among which is the difficulty of supervision as the resource is mostly in remote hilly areas and the consequent failure to ensure correct treatment to each individual clump. In most of the countries, the management of bamboos in the natural forests is only the harvesting of resource; very seldom any cultural practices are followed.

R.L. Banik (✉)
National Mission on Bamboo Applications (NMBA), New Delhi, India

INBAR, New Delhi, India

BFRI, Chittagong, Bangladesh
e-mail: bamboorlbanik@hotmail.com; rlbanik.bamboo@gmail.com

© Springer International Publishing Switzerland 2015
W. Liese, M. Köhl (eds.), *Bamboo*, Tropical Forestry 10,
DOI 10.1007/978-3-319-14133-6_7

193

7.2   Factors Considered During Harvesting of Bamboo

The important factors for consideration in harvesting of bamboo forest are felling cycle, intensity of felling, method of felling and transportation of felled bamboo to the marketing centres and/or factories. However, transportation is not so difficult for the homestead and village-grown bamboos due to the presence of accessible roads.

Management of bamboos is based on the physiological development of the clumps. On the basis of photosynthesis, in the course of the morning, bamboo starts transporting starch from the roots into the leaves. During the height of the day, this process is at its peak making this the least ideal time of day to harvest. Therefore the best time to harvest bamboo is **before sunrise** (between 12 pm and 6 am), when most of the starch is still in the rhizomes and roots. Bamboo harvested in this manner has three advantages: they are less attractive to insects, are less heavy to transport and will dry faster. As culm production takes place through rhizomes, it is essential to ensure that the rhizomes are supplied with sufficient amount of food materials to keep them fully vigorous to give out new culms.

7.2.1   Season and Harvesting Age of Culm

Bamboos are usually not harvested in growing season, i.e. from later part of spring to the end of rainy season. The best season for harvesting is after the rainy season when starch (sugars) content in the bamboo sap is low. Bamboo possesses large amounts of starch which are the favourite food for pests (goon attack). When carbohydrates are reduced (at the end of growth period), the bamboo culm will be more naturally resistant to those biological degrading organisms. Bamboos harvested during summer to rainy period are generally more rapidly destroyed than those felled in autumn and winter. Harvesting of bamboo stands is done from the end of the growing season, over a period of about 6 months in most of the countries of tropical Asia. The Javanese in Yogyakarta avoid the harvest of bamboo culms during the early shooting period in the belief that at that time, the mother culm is taking care of its young shoots (Sulthoni 1996). Immature culms of 1–2 years have a very high water content and shrivel up when cut—this makes them useless for construction. There is a belief that if bamboos are felled in the first day of new moon, bamboos will be more durable and resistant to pests. However, for temporary use bamboo can be harvested in off season also. Bamboos are found not durable if harvested in March/April/May (spring–summer).

Bamboo harvesting takes place at different ages of the plants' growth according to the product requirements. Age of a culm is an important consideration in fixing the felling cycle. Some morphological characters such as presence or absence of culm sheath, culm colour and texture, bud break and branching nature on the culm and presence and absence of root ring on culm nodes are found useful in determining the culm age in a bamboo clump (Banik 1993). The harvested bamboos are useful for making different products at different ages. For example, the recommended age to harvest for products to use in the construction industry is 3–5 years, yet for some craft uses, they may be harvested at a younger age depending on the species. Midmore (2009) observed that the strength properties were still improving in culms older than 3 or 4 years in *B. blumeana*, but in *D. asper* those of 1–2-year-old culms were equivalent to those of 2–3-year-old culms growing in the same site in the Philippines. *Dendrocalamus asper*, if it were to be used for construction purposes, could be harvested at close to 2 years of age, whereas culms of *B. blumeana* should be at least 3 years old and ideally older.

However, 3–5-year-old culms are durable and strong so they are usually cut from a bamboo clump for utilisation. Bamboo older than 5 years is harder, and the inner culm wall becomes impermeable to the treatment solution. If the culms are older than 6 years, they may have been subjected to insect damage on the interior of the plant, but it will be difficult to assess.

7.3 Type of Felling

7.3.1 Selection Felling

It is always important to consider the health of the bamboo grove as a whole, and selection of felling may be made from an area that would benefit and not harm the clump structure. To select culms that have the greatest strength for building purposes, it is important that the culms are at the right age. The culms attain their maximum size in one season. The younger culms, 1 to less than 2 years of age, have youngest rhizomes with viable culm buds that are capable of producing new culms in a bamboo clump. Thus it is mostly the younger culms that are only productive than other culms in the clump. The new culms are produced from the rhizomes mostly along the periphery of the clump. One-year-old shoots are also not economically very useful. So these culms should not be felled to maintain the vitality and productivity of a bamboo clump. In fact 1-year-old culms are not strong and durable so they are not useful in any construction works and have little demand in the market. The silviculture, therefore, is to leave all the 1-year-old culms and also leave an equal number of older culms and to cut the rest. However, for pulping and papermaking, 1-year-old bamboo culms are harvested when fibre quality is superior and lignin content is less to that of older culms (Fu and Banik 1996).

7.3.2 Clear-Felling

A 10- to 12-year felling cycle is adopted with clear-felling of clumps; it is found that the regeneration of bamboos is inadequate and the clump does not attain its normal size within this period. During 1–2 years after clear-felling, very thin and tender culms emerge from the underground weak rhizomes; and simultaneously dormant buds on the nodes of stumps sprout and produce profuse number of big-size leaves attract animals for grazing and remain vulnerable to fire damage during dry season. Thus clumps require initial 5–7 years to produce merchantable size culms, if not disturbed further. Clear-felling is not a common practice in bamboo forest. In India it is followed in Orissa for *Bambusa bambos*. It is found that the normal sizes of culms are produced after about 10 years of clear-felling (Tewari 1992), but whenever clumps have been partly felled, the regeneration obtained was of better quality.

The clear-felling method is not generally recommended for harvesting due to slow recovery and low above-ground biomass production.

7.4 Felling Practices in Forest

7.4.1 Harvesting from Natural Forests

The existing system of felling practices in forest bamboos is being worked out on a 3–4-year rotation. The bamboo forests are divided into 3 or 4 blocks, and cutting is restricted to one block each year. Even in areas where unregulated fellings have been carried on for years, rest for a few years allows the bamboo clump to recoup, hence the need to follow silviculture requirements for sustainable development of a clump.

7.4.1.1 Harvesting of Bamboos from Natural Forests in Eastern Part of the Subcontinent and Their Transportation

The bamboo forest in eastern part of the subcontinent extends throughout the northeastern part of India to Sylhet, Chittagong Hill Tracts (hereafter CHTs), Chittagong and Cox's Bazar of Bangladesh to Myanmar. The CHT, Chittagong and Cox's Bazar are forest areas occurring on the hills of Arakan Range in the southeastern part of Bangladesh, bordering the Arakan and Chin states of Myanmar and Mizoram and Tripura states of India. Sylhet is another prominent area for forest bamboos, bordering the area to the Tripura and Assam states of India. In this vast tract of land, bamboos often formed continuous vegetation like 'Bamboo Sea' in some part of Tripura, Mizoram, Manipur, and Assam states of India and also some part of CHTs and mainly of Arakan and Yoma of Myanmar.

Morphologically, most of the forest bamboo species are comparatively thin walled and smaller in size to that of village bamboos. The important commercial bamboo species of these forests are *Melocanna baccifera* [local name, Muli (Assam, CHT, Tripura, Sylhet), *Turiah* (Nagaland), *Mautak mau* (Mizoram), *Watrai* (Garo Meghalaya)], *Bambusa tulda* [L.n, *Mitinga*, Wandal (CHT, Tripura), Rawthing (Mizoram), *Saneibi* (Manipur), Thaik-wa (Myanmar)], *Dendrocalamus longispathus* [L.n, *Orah, Khag* (CHT, Sylhet*), Rupai, Wamlik* (Tripura), *Rawnal* (Mizoram), *Siejlong* (Khasi, Meghalaya), *Wanet* (Myanmar)], *Schizostachyum dullooa* [L.n, *Dalu, Wadrow* (Garo Meghalaya), *Thaikwaba* (Myanmar)], *Gigantochloa andamanica* [L.n, *Kali, Kalyai* (CHT, Sylhet, Tripura), *Wasut* (Garo Meghalaya], etc. Among all the species, *M. baccifera* constitutes 70–90 % of the total bamboo forest. The natural bamboo vegetations are present along the river courses on the hill slopes in moist deciduous forests as pure secondary moist bamboo brakes. There are usually, but by no means always, scattered trees typical of the semi-deciduous climax standing singly or in groups over the bamboo forests.

The existing felling and transportation practices of bamboos from some of the natural forests in eastern part of the subcontinent are unique, and the management of bamboo resources in these hilly areas is interesting to learn.

In the above mentioned forest regions harvesting of bamboos is not allowed for 3 months from June to mid-September. These 3 months are the maximum shoot emerging period and therefore closed to harvest. The local ethnic people harvest bamboos in a specific time of the year. They cut bamboo during November to March, the winter and dry season of the year.

In the forestry practice only 3-year-old or older culms are harvested from the clumps as *selection felling*. In some occasions, especially in '*jhummed*' (slash and burn farming or shifting cultivation) areas, *clear-felling* of all the culms are practised by the local hill people. In case of gregarious flowering when clumps die in large numbers, clear-felling is carried out to remove the dead and dry bamboos. The bamboo forests are generally worked on a 3–4-year cycle, and felling rules prohibit cutting 1-year-old, 2-year-old and a few older ones left scatteredly in the clump. It was observed that in the clump of *M. baccifera* 3-year-old culms possessed less (0.60 Kg) amount of leaves than 1- and 2-year-old culms (Ueda 1960). It was observed in *D. longispathus* that fresh weight of leaf biomass was at its maximum in the 2-year-old culms (per culm number of leaves 6,975, oven dry weight of leaves 1.15 kg) and then it decreased about 30 % and remains somewhat static in the third and fourth years (Banik 2000; Banik and Islam 2005). In 5th and 6th years, the leaf production was drastically decreased (per culm number of leaves 907, oven dry weight of leaves 0.14 kg). Subsequently fewer amounts of leaves remain in a 4- to 5-year-old culm, and therefore it is likely that it has little contribution in photosynthesis and overall health of the clump. Thus it appears that in both *M. baccifera* and *D. longispathus* clumps felling of culms may be started after 3 to 4 years of age.

Felling Processes and Steps

• Felling is purely manual using a sharp tool (bill hook or locally called as *dao*) for cutting bamboos from the clump. The *dao* is a 0.3–0.6 m-long and 4–5 cm-wide flat iron-made sharp knife. They also use *dao* for constructing their house, as agriculture implement and also for protecting themselves from any attack of wild lives.

The different tribes (*Chakma, Bawm, Marma, Murong, Lushai and Tipra* from CHT; *Khasi and Garo* from Meghalaya; *Naga, Kuki and Mizo* from Nagaland, Manipur and Mizoram; and *Reangs, Deberma* and a limited population of *Munda* and *Molson* from Tripura) who inhabit in the upstream's hilly areas are actively involved in bamboo extraction that fall within the bamboo resource catchments. Both men and female are involved in the bamboo extraction. Among the abovementioned naturally grown bamboo species, *M. baccifera* bamboo constitutes the major portions of all extracted bamboo which is about 80–90 %. The major livelihood in the upstream is *jhumming* for subsistence and bamboo extraction for generating cash money. The maximum extraction is done near the water courses for ease in transportation. Only the mature bamboos are cut from the clump. The local indigenous people can identify the ripe/mature bamboo by hearing the sound by beating the back side of *dao* on the bamboo stem. Young bamboos are left in the bush. The felling ***processes and steps*** may be summarised as follows:

Step I—Bamboos are harvested from the natural forests on the hills and along the stream bank. The bamboo stems are felled 0.3–0.45 m above the ground, followed by trimming and cutting into pieces for making bundles and then extracted by shoulder load. From the hilltops bamboos are collected by making a walking path. Bamboos are cut at the hilltops and thrown towards the base of the hills through sliding on the walking paths. Thus bamboos are collected and stacked at the bottom of the hills. Bamboo cutting is generally carried out in two phases—roadside cutting and cutting on the hills and rolling the felled culms to the nearby waterways (river/streams). On average, a labourer can cut, trim, carry and stack 40–50 bamboos per day that is equivalent to about 100–120 kg of *M. baccifera* bamboo poles up to the river or any waterways (before rafting). So just for extracting 1 t, about 10 man-days is required. In roadside cutting, cutters cut bamboo along the road extending to an average lead of 90–150 m.

Step II—After felling, the culms are pulled out of the clump, then limbed and trimmed for handling and cut into pieces of 1.7–3.0 m in length for road transport. Generally, 5.5–6.5 m- and sometimes 8.0 m-long pieces are transported through waterways/streamlets running through the hills to the main rivers in the regions (the river *Kachalang, Sangu, Matamuhuri and Karnafuli* in CHT; *Khowai, Gomuti, Manu, Feni and Muhuri* in Tripura; *Jiri, Makru and Barak* in Manipur and lower Assam; *Twlang, Serlui and Chhimtuipui* in Mizoram; *Khowai, Surma and Juri* in Sylhet; and *Chindwin, Irrawaddi and Sittoung* in Myanmar).

Bamboo pieces are then tied with [newly emerged soft (*doga*) bamboo] strings made out of bamboo into bundles of 5–10 pieces each, so that each of the bundles can be carried on the shoulder. For waterways transportation, the bamboos are commonly bundled in 12, 16 or 26 poles (to round up 96) of 12, 14, 15, or 18 ft length. These are tied to make a bundle. The practice of bundling 96 culms has its roots in the days of the local tribal kings. The four culms from 100 are used to be paid as royalty to the king and left uncut in the forest of Tripura. This practice has been a remarkable ethnic way of conserving the bamboo resource.

The bamboos are tied into rafts by the person who rafts it downstream usually employed by the trader who buys the bamboo and ties them into challis (a *chali* is usually made of 300 bamboo poles). When the smaller bundles of bamboo poles harvested from different parts of the forest reach the main river, they are retied into challis, and usually about 10–15 challis form a raft by tribes who are skilled and only engaged in rafting bamboos. Thus, a raft will usually have 3,000–5,000 poles (Fig. 7.1).

Step III—The bamboos are rafted up to a local river bank market nearer to the roads and highways. Bamboo transportation from different sources to a river bank stockyard market point usually takes 2–5 days depending on the water level and water current in the river. One tribal person can do rafting 1,000 poles through the river. Some (may be 10 %) of the total harvested bamboos decay in 1 year in the forest stacks, and 10–20 % get lost and damaged in the monsoon floods in narrow streams or rain during waterways transportation (Banik 2000). The bamboo resources have been extracted from various collection points located inside the forest usually accounts for about more than 80 % of all.

It is estimated that around 16.1 million man-days per annum of employment is generated on the account of management and extraction of bamboos in Tripura (Tripura State Bamboo Policy 2001, Forest Department, Government of India).

Step IV—After arrival of bamboo rafts at a major bamboo market on the river bank, the bamboo challis are untied in the river and each bundle is carried to the roadside of the highway (Fig. 7.2a, b) and then assembled by the labourers. Usually in a particular day of the week, large amounts of bamboo rafts arrive here, and from this point the bamboos are marketed to the plain land traders.

The extracted bamboos are often sold to the bamboo traders at major river transportation points near major roads and highways.

Step V—The bamboo bundles are untied and rebundled, consisting of a right mix of bamboos (species and size of pole length, diameter, etc.) again in 12 numbers for loading in trucks by the labourers.

Step VI—The bamboos from this point are loaded in the truck and transported to the major secondary and retail market at nearby towns and other distant parts of the country.

Fig. 7.1 *Melocanna baccifera*: a raft usually has 3,000–5,000 bamboo poles transported through the river Khowai to Chakmaghat depot, Tripura

Fig. 7.2 (**a**) In the river the bamboo challis are untied, and each bundle is carried to the roadside of NH 44 and then assembled by the labourers in Chakmaghat, Tripura (India). (**b**) The bamboos from Chakmaghat area are transported in trucks to the retailer in Agartala town of Tripura for secondary and retail marketing

Hill communities themselves have been conducting the whole process of harvesting and transportation. But in the marketing they have little role to play; rather in most cases the plain land people act as middleman and control the marketing. The age-old practice of transportation of harvested bamboo through waterways from the forests of northeast India, CHT and Myanmar is an example of intelligence and wisdom of local indigenous people. In this region every year several million of bamboo poles are being transported through waterways and marketed without any cost of fuel and machines. In some cases bamboos are transported for more than 150 km in 4–5 weeks' time; the rower and raftsman stay on the raft full time and navigate safely along the water current to the destination market. Such water transportation practices is environment-friendly as it is manual, and bamboo poles are treated while conveying on water for a few weeks and become durable (not much food remains for pests; thus there is less 'goon' or pinhole borer attack) (Banik 2010).

The **harvesting system** of bamboos from different forest areas of some of the major bamboo-producing areas of the above region are narrated below:

- **Mohal system**—follows 3–4-year felling cycle with definite intensity of felling. Felling operations are carried out in a number of steps. Transportation of harvested bamboo poles are mostly through water courses (hilly streams and rivers).
- **Permit system**—*shoulder load transportation*

(a) ***Mohal system:*** *The Mohal system* is the main system of selling forest department bamboos. Under *Mohal system* the bamboo resources are identified by the local forest division and demarcated in areas as basis. Harvesting rights are sold annually to Mohaldars (bamboo contractors) from Cachar part of Assam and Mizoram, India, and Sylhet, Bangladesh. Mohaldars have rights to remove any quantity of bamboo (above 1 year old) from the forest.

The state of Mizoram has dual modes of leasing the bamboo resources for extraction by bamboo traders. The harvesting of bamboos from the forests has been practised by *Mohal* and *Permit systems*. There are about 20 Mahals in Mizoram covering an area of 1,772 km^2. Under *Mohal system* the bamboo resources are identified by the forest division and demarcated in area basis; in Kolasib forest area of Mizoram, mostly on the 800 m of either side of the river (called as Kolasib Mohal) are given to a Mohaldar (selected bidder of the Mohal) for a specified duration of the year. *Melocanna* accounted for 95–98 % of the total standing stock with 725,684 million culms over 1 year old. Other bamboo species (*Dendrocalamus hamiltonii*, *D. longispathus*) only accounted for 2–5 % of the total bamboo stock. Although culms mature after 3 years, regulations exist to limit cutting to 4-year-old culms to provide some safeguards for future culm availability. The steep terrain makes the harvesting very difficult. Bamboo forest covers 12,54,400.00 ha out of the total area of 21,090 km^2 of the Mizoram state.

Local people have no rights to harvest bamboo in the forest area allotted to Mohaldar and must pay the Mohaldar if they wish to take any culms.

In Sylhet (Bangladesh) harvesting of bamboos (of which 70–90 % is Muli) from the forests has also been practised by Mohal system. The Sylhet Pulp and Paper Mills (**SPPM**) Ltd. conducts major harvesting operation in the Sylhet forest. Both pure and mixed bamboo vegetations are found throughout the southern part of the Sylhet Forest Division in Rajkandi, Patharia, Harargaj reserve forests (RF) and Prithimpassa acquired forests (AF). These reserves and AF are divided into *bamboo Mahals that represent the area units under bamboo extraction every 4 years*. Bamboo Mahals identify the catchment areas of the water courses (or perennial streams, locally called as *chara*) that are used for bamboo extraction and after which they are named as *Surmachara Mahal* and *Dholaichara Mahal*. The SPPM allotment is 12,150 ha in the

Sylhet bamboo forests. Every year after getting allotment of bamboo Mahals from the Bangladesh Forest Department (BFD), the mill authority appoints a harvesting contractor through tender. The permission for harvesting bamboos remains *valid from 1 January to 31 December, with a gap of 3 months from 16 June to 15 August.* This gap period *is the closed season for harvesting.* The size and number of the felled bamboos are then checked and verified by the SPPM and Forest Department officials. These are then transported by water-ways on a particular day of a week to the base depot on the bank. A long distance of over 300 km is to be covered to the mill site at Chattak. Part of this distance, bamboo has to be transported against the river current.

Bamboo extracted by purchasers in the Sylhet Forest Division is often sold to the bamboo traders at the major transportation points, mostly located near the railway stations. Long-distance transportation even up to Dhaka, Barisal, Patuakhali and Khulna is carried out mostly through rivers and taken to these areas in big rafts. However, nowadays rail and truck also transport a substantial amount of bamboos. Sometimes bamboos are converted into tarjas (flats) and transported by bundles in trucks or on railway wagons.

In **lower Assam** (Barak Valley—Cachar, Karimganj and Hailakandi bamboo forest area), bamboo resource is harvested from the hills and brought down to *Badarpurghat* on the banks of the Barak River by leaseholder who sells lengths of 6–7 m of bamboo in bundles of 1,000 culms to a contractor. The contractor arranges to float the bamboo down the river to the cottage industry. A number of bundles of bamboo are tied together like a raft, and a couple of men live on the raft and look after the bamboo till it reaches the destination (Badarpurghat). The bamboo is brought out of the river, and each culm is split in half along the length after the external nodal rings are cut off. The half culm is then flattened by hitting it along the length, particularly at the nodes. This is done with a wooden mallet or the back of a *dao*, a broad bladed knife, which causes the semicircular cross section of the culm to creak at various places, so that it lies flat. The internal nodal walls are chopped off, and then the board is split in two through the thickness so that one board has the outer skin of bamboo, while the other has the inner surface of the culm. These are stored separately.

In Assam, large coarse bamboo matting is made from flattened bamboo boards. These are used as prefabricated walls and roofs, particularly in the bamboo frame Assam-type house common in the state. Small labour-intensive cottage industries make this type of matting, and two or three such industries are located at Badarpur. At the cottage industry, a number of people employed by the contractor convert the culms in to flattened board. The single largest user of bamboo resources in the area is the Cachar Paper Mill, a unit of M/S Hindustan Paper Corporation (HPC) Ltd., Govt. of India undertaking. The unit is located at Panchgram in the Hailakandi district, and its raw material capacity is 250,000 MT AD annually.

In **Myanmar** the bamboo areas are usually divided into sections, and felling is rotated so that a section is cut and is left for 2, 3 or 4 years, depending on the felling cycle. Among many different bamboo species, *Melocanna baccifera* is one of the major forest bamboo resources in Myanmar. Bamboos are classified according to their age. Young Bamboo (*Wa-nu* in Burmese) is 1 year old, middle (or Tanyin) is 2 years old and matured (Wayint) is 3 years old and above. The mature bamboos are cut and accepted for utilisation if they are 18 ft high and 1 in. in diameter at the tip. Bamboos which are to be used as housing materials should be extracted in the cold season but not in the rainy season. The sharp blades (locally called *Dah*), something like machete, are used to cut bamboos at about 1.5 ft. from ground. Usually they cut some of the *Tanyin* (2 or 3 Nos.) and matured *Wayint* (3–4 Nos.). The young are left to grow with other two types, while very young shoots are also harvested for edible purpose from some bamboo species. About 2.0 m from the tip is also cut as that portion is also useless. The bamboo pole is then cut in required length for export which is usually about 4.0 m. Htun (1999) reported that bamboos are harvested at 3–5-year cycle from the bamboo forests and transported mainly through riverways. The overall system of extraction is more or less similar to the practices followed in Chittagong Hill Tracts.

Bamboos are extracted by the villagers, which are collected by local brokers in the villages and sold to the local markets and merchants in towns and big cities for local use or export.

(b) **Permit system**: In the **Chittagong Hill Tracts (CHT)** and Chittagong, harvesting of bamboos (of which 70–90 % is Muli) from the forests has been practised by **Permit system** issued by the local forest officers, specifying the quantity, area and time limit. The royalty for the quantity of bamboos is paid at the time of issuing the permit. *Karnafuli Paper Mills* (KPM) located at the hilly town Chandraghona under the district Kaptai of CHT purchases a substantial quantity of the **bamboos extracted by the permit holders.** This is the cheapest method of bamboo harvesting, but, due to the change of hands, the prices go up to accommodate the profit of the different parties. This method only harvests bamboo in the accessible areas along the stream, perennial streams (locally called as *chara*) and river banks. The KPM also carries out extraction of bamboos by employing contractors using hired labourers. The major working areas for KPM are in the Kassalong and Rankhiang Reserve Forests of Chittagong and Chittagong Hill Tracts. Harvesting of bamboos by KPM include a series of operations broadly divided into three phases as follows (Flow chart 7.1).

The harvesting of bamboos is done on 3-year rotation using selection felling. Normally, felling starts in October and lasts for about 120 days in a season. The 3 months from 16 June to 15 August is the closed season for harvesting.

Bamboo harvesting is generally carried out in two phases—roadside cutting and ropeway cutting. In *roadside cutting*, cutters cut bamboo along the road extending to an average lead of 90–150 m. *Shoulder load transportation* up to the river bank extends 3.5 km in some cases and is done by the labourers due to the absence of extraction forest roads. In *ropeway cutting*, cutters cut bamboo

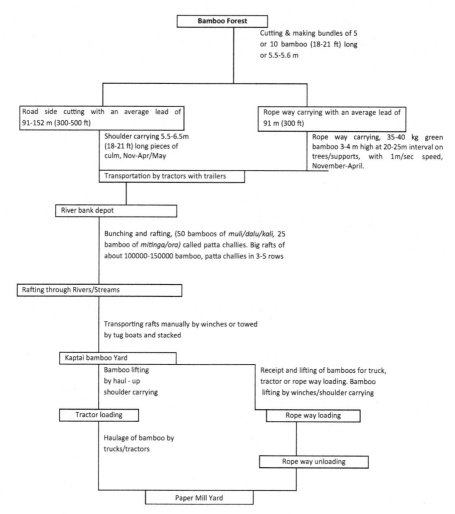

Flow chart 7.1 A schematic diagram showing the steps involved in harvesting bamboos from forests of Chittagong Hill Tracts in Bangladesh (*Source* Banik 2000)

extending to an average lead of 90 m on each side of the ropeway. A ropeway system consists of endless moving cables which are suspended at about 2.5–3.0 m above the ground at an interval of 18–25 m on trees or other supports. The cable is operated at a slow speed of 1.0 m/s by a 20–25 hp petrol/diesel engine. The maximum load given is 30–35 kg bamboo poles (green) at an interval of 12–15 m. Tractors with trailers are used for the transportation of bamboos extracted by ropeways as well as those cut along walking paths. This experience of ropeway harvesting would be useful for efficient way of extracting bamboos in the hilly terrain of northeastern part of India.

After felling, the culms are pulled out of the clump, then limbed and trimmed for handling and cut into pieces of 1.7–3.0 m in length for road or rail transport. Generally, 5.5–6.5 m- and sometimes 8.0 m-long pieces are transported through waterways. After that, bamboos are tied with *doga* bamboo, strings made out of bamboo, into bundles of 5–10 pieces each, so that each of the bundles can be carried on the shoulder. In case of *Muli/Dalu/Kali* bamboos 10 poles are tied to make a bundle. But *Orah* and *Mitinga* bamboos are comparatively bigger in size, and, therefore, usually five stems are used for a bundle. On average, a labourer can cut, trim, carry and stack 100 bamboos per day which is equivalent to about 210–250 kg. Bamboo cutting is a piece rate job, irrespective of age, size and quality. Felling is purely manual using a sharp tool (bill hook or *dao*) for cutting bamboo stems 3 years old at or above 15–35 cm from the ground level.

After the completions of the harvesting, rafts are checked by the forest officer and royalties realised for excess numbers of bamboos taken out. The entire system is a manual operation and is mostly located along the bank of rivers Sangu, Karnafuli, and Matamuhuri or stream bank for ease of extraction. Under this system bamboo is rarely cut beyond 1.6–3.2 km of the floating stream banks. Such bamboos are rafted or taken by boat by the permit holder to important selling centres. Bamboos are mostly felled, processed and rafted by Chakma, Lushai and Morong tribes and sold at Kaptai, Dohazari and Chiringa and the local markets near the forests of the Chittagong and Cox's Bazar Forest Divisions.

On an average 11 % of the total harvested bamboos decay in 1 year in the forest stacks and 7–19 % get lost in the monsoon floods in narrow streams or rain. In wider watercourses, the bamboo rafts may be torn apart by storms and high waves (Banik 2000). Thus bamboos are being transported from Kassalong and Rankhiang forest reserves of CHT to Chandraghona KPM site.

In some area like the Kolasib Forest Division in Mizoram, the bamboo resources are managed by both *Mohal system* and *Permit system*. Under the *Permit system*, in Mizoram a permit fee of Rs. 8 per 100 poles and another 100 % monopoly fee of Rs. 8 (Year 2003–2004) are collected. The allotted bamboo resource mostly along the Bairabi-Kolasib National Highway (NH-54) are leased under this system.

According to local people of **Manipur (Tamenglong) state**, October to January (autumn–winter) is the best season for bamboo harvest. Harvested bamboo culms (mainly of *M. baccifera, Dendrocalamus hamiltonii)* are hardly transported through the river. It is transported to local markets in Imphal, Jiribam and Tamenglong by truck through temporary road inside the deep bamboo jungle. Sometimes in the villages people carry bamboo as head load to nearer local markets. The villagers themselves cut bamboo from the forest; sometimes labourers are also engaged. A labourer can cut 100 15–20 ft- or 4.6–6.1 m-long bamboos per day. In one head load 5 big-size (12″ or 30 cm girth) or 10 medium-size (8–9″ or 23 cm girth) bamboo poles are carried.

Permit system for locals: As per rule, bona fide householders and cultivators in Tripura who are the inhabitants of villages entirely surrounded by reserved forest are allowed free permits for harvesting bamboos to the extent of 250 numbers per family between the months of January and March (winter to early spring).

Extraction from the village forest land or private land is based on the permission from the village council or the individual owner.

Auction purchasers: Bamboos from the non-licenced area of Sylhet, Chittagong and Cox's Bazar Forest Divisions are sold in *auctions*. The cutting cycle is 3–4 years, and the yield is regulated by area. The extraction of bamboos by the *auction purchasers* is a manual process. Generally the bulk of the bamboo extracted in Sylhet by the auction purchases is utilised outside the district. However, for local consumption there is a provision that the auction purchaser should allow a certain percentage (25–30 %) of total bamboo of the auction coupe to the local permit holders. The local demands are also met by issuing permit of bamboos in the clear-felling reforestation coupe areas. Some of the auctioned bamboos are also sold to SPPM. Shoulder load transportation up to the river bank extends 3.5 km in some cases and is done by the labourers due to the absence of extraction forest roads. In general, bamboo is transported from the forest to collection sites on foot by collection teams usually comprising 3–5 members.

The bamboo-based economy is mainly on the intensity of the bamboo extraction, which forms a major livelihood option of the tribal people of the hills.

7.4.1.1.1 Harvesting in Other Parts of the Subcontinent (Eastern, Northwest, South and Central) and Some Parts of Tropical Asia and Africa

The following factors have been, generally, considered in management of bamboo forests in these regions—felling cycle, intensity of felling, season of felling and method of felling:

(a) **Eastern, northwest, south and central part of India**: *Dendrocalamus strictus* and *Bambusa bambos* are the two major caespitose (clump) bamboo species occurring in the natural forests and also widely cultivated throughout the villages of the abovementioned region of Indian subcontinent, excepting northeast India. Therefore time to time studies were undertaken on these species to develop scientific and systematic harvesting techniques of bamboo. More often experiments were conducted on *D. strictus* than thorny *B. bambos*. The harvesting age varied from 4 to 14 years, according to locality, for the same species. A series of pioneering experiments was commenced as early as in 1910–1911 and continued up to 1915, on *Dendrocalamus strictus* by Troup (1921) dealing with 57 clumps at Ranipur (rainfall 50 in.) and one in a fairly moist locality (rainfall 70 in.) near Kotdwara of Uttar Pradesh (UP) forest dealing with 98 clumps. The study revealed that:

(1) Annual working, whether cutting high or cutting low or whether removing all old culms or only half of them, led to rapid reduction of clump size, number and diameter of new culms.
(2) When all the culms excepting the current year were cut, the clump deteriorated. This was true for 1-, 2- and 3-year rotation and for cutting at height of 1 node, 3 nodes and 5 nodes.
(3) When only half of the old culms were cut, the clumps under a 2-year rotation were in much better condition than those under a year rotation.
(4) Whatever the rotation, some old culms should be left for the mechanical support of new shoots and to maintain the rhizomes in full vigour.
(5) The effect of differences in height of cutting up on the health of clump was nil. But cutting high produces number of twigs at the top of the stumps which impedes working.
(6) Production of new shoot was not affected by the height of cutting.
(7) A 3-year felling rotation, some of the old culms left standing, gave better result than a 2-year rotation. (8) Clear-felling of all culms including shoots of the current year nearly but not quite killed the clumps.

These findings guided the preparation of bamboo working plan in different forests of the subcontinent and future plan of operation.

The clumps of *D. strictus* containing less than 8 matured culms are not harvested as clump is not considered as developed and matured.

As regards the felling cycle, a 3- or 4-year felling cycle appears to be the most suitable for subcontinent bamboos. This is the actual practice throughout India except in Punjab where a 2-year cycle is followed. In *Orissa*, a 10–12-year felling cycle is adopted with clear-felling of clumps, but 3 or sometimes 4 years are practised for selection felling. It has been found that in clear-felling system, the regeneration of bamboos is inadequate and the clump does not attain its normal size within this period. The current practice of bamboo harvest in *northern India* (Uttar Pradesh and other states) is to retain 3–6 culms per clump. The working season normally is in November to July. The culms are to be cut low, about 15–30 cm from the ground level and just above the septum of a node. Young culms of less than 1 year are not felled. The practice is to retain a minimum of 6–8 or 55 % of older culms in each clump. The clumps containing less than eight matured culms are not harvested. Felling is made on the site containing the least number of new culms. The maximum cutting height is 45 cm from the ground, leaving at least one internode next to the ground. Extraction of rhizomes is prohibited.

In *south India* (Tamil Nadu), the bamboo area is worked under the culm selection system on a 3-year felling cycle. Bamboos are generally worked on a 3-year cycle except in some forest division of *Andhra Pradesh* where a 4-year cycle is adopted. Lakshmana (1994) also suggested that for optimum production of bamboo in *B. bambos*, natural forests of Karnataka culm has to be extracted at a 4-year felling cycle. In *Maharashtra*, harvesting is prohibited in rainy season (between 15 June and 30 September). A mature clump with more than 8 culms is considered as developed and fit for harvesting. No culms below the age of 2 years are felled, and the cutting height is between 15 and 45 cm.

Bamboos and reed bamboos (*Ochlandra travancorica*) are worked on a 3-year felling cycle in Kerala. The felling rules prescribe (Manoharan and Trivedi 2008) that (1) bamboos and reeds adjacent to the stream banks and located on slopes above 30° gradients are not worked, (2) culms of age more than 2 years are only felled, (3) the felling of culms is done on a horseshoe pattern, (4) no felling is done during the regeneration period (rainy season) of June to August and (5) all dead and malformed culms are removed irrespective of age.

Falling production from overharvested stands and inability to harvest bamboo during the monsoon season have left the industry soured on the potential for further expansion of bamboo into an industrialised agriculture format (privately owned plantations).

(b) **Some parts of tropical Asia and Africa**

1. **Bhutan:** In Bhutan, most of the bamboos are of sympodial type but also include two genera (*Arundinaria* and *Chimonobambusa*) of monopodial type. Bamboos have been harvested through selective system, but the cutting dates of different species are little different from each other (Mukhia 2005). Most of the species of *Bambusa* genus is cultivated for farm household use, and the species like *Bambusa multiplex* is cultivated as an ornamental hedge. All the bamboo species under this genus have same cutting season except for the ornamental ones. Selective system is being practised by the farmers. But in this selective system, only the good, clean, straight and biggest culm is harvested unlike the trees. Farmers use patang, khukuri and any other knife to cut the stem, and they usually cut from 15 October till 15 May (dry winter). While harvesting from *Borinda grossa* matured bamboo culms are being cut starting from December till March. All the bamboo species belonging to *Dendrocalamus* have been harvested by the farmer from southern Dzongkhags who collects culm from mid-October till mid-May, whereas the eastern Dzongkhags' farmer collects from December till mid-April. Both region farmers practised selection system for harvesting the bamboo culm. The farmers cut the culm at 0.61–0.91 m above the ground level if bamboos are collected from the wild. And for the cultivated ones, farmers cut the culm at about 10 cm

above the ground level in order to maximise the utilisation of the bamboo culm.

Usually the cutting of *Cephalostachyum capitatum* and *C. latifolium* bamboo culm starts from October till mid-May.

The harvesting of culms of a monopodial bamboo species *Arundinaria racemosa* is very minimal in Bhutan because it is allowed to grow for grazing of livestock and for wildlife. Occasionally people use them for making arrows, brushes or pipe straws. Usually villagers use *geetshu* or any other small knives to cut the bamboo culm and harvest them as and when necessary. The farmers do not have the season for harvesting of this particular bamboo, *Chimonobambusa callosa*. The collectors cut the bamboo with the help of patang or khukuri and collect as and when required. Cutting this bamboo is usually done at about 0.61–0.91 m above the ground level, but thorny nature of culms makes it difficult. The bamboo culms *of Thamnocalamus spathiflorus* and *T. bhutanensis* are not usually harvested because these bamboo provide food for red pandas and bears. If necessary, farmers collect them as and when required.

2. **Nepal:** Local management practices seem to vary according to the end product required. In the case of *Bambusa* and *Dendrocalamus* spp. [*bans*], demand for poles is often sporadic and the harvesting of culms is not always at the optimum age for maximum productivity. As a general practice in all large-size bamboo species, culms above 3–4 years of age are felled during winter time. The smaller *Arundinaria* and *Drepanostachyum* species [Nepalese name,*nigalo*] are (according to local practice) harvested by removing the shoots of a lighter green colour, which are readily distinguishable in March (Thompson 1986). Harvesting age should be 16–20 months (Stapleton 1987). This leaves the clump uncongested and produces culms of consistent size and quality.

3. **Indo-China:** *Dendrocalamus membranaceus* is one of the most frequently occurring, clump-forming woody bamboos (with pachymorph rhizomes) in Southeast Asia (Laos, Myanmar, Northern Vietnam and Northern Thailand in addition to China's Yunnan Province, especially along the Lancang-Mekong River Valley); natural bamboo vegetation along the bank of Mekong River also covers the hilly terrain of Thailand, Laos, Cambodia and Vietnam. Other major bamboo species occurring in the region are *Dendrocalamus asper*, *Thyrsostachys siamensis* and some species of *Gigantochloa*.

Lao-PDR: Preliminary assessments were carried out on the extent of bamboo stands within the designated areas of upland agriculture and production forests within village boundaries and along the lower Namtha River. It was estimated that each charcoal kiln with an annual production of

30 t of charcoal would require about 150 t of green bamboo, which in turn could be harvested annually on a sustained basis from about 30 ha of bamboo stands (Dransfield and Eijada 1995). It was estimated that approximately 20 % or 11,000 ha of this area is covered with exploitable bamboo stands. Harvesting operations will be restricted to the dry season of only 8 months (RAP 2007).

Due to the difficulties in bringing individual bamboo culms to the ground, felling is done in teams of at least two persons. The culms are carried out with axes or straight-bladed machetes at a height of about 0.8–1.2 m above ground, due to the dense growth of culms at the base of the clumps.

The costs might be slightly lower in situations where whole fields may be harvested in clear-felling operations.

Manual transport of bamboo culms is presently the most common form of forwarding. This is mainly done for domestic consumption over distances of up to several kilometres. The optimal range for manual transport is between 50 and 150 m. For commercial-scale operations, the maximum range considered economically viable would be around 500 m for downhill transport. Due to the relatively high friction of long bamboo culms, manual transport is most suitable for this terrain condition. For transport over level or uphill terrain, the acceptable range diminishes considerably. Manual uphill transport on slopes steeper than 35–40 % is limited to distances of less than 50 m.

The use of bamboo rafts for transport on the Mekong River and its tributaries is common. Such rafts are typically constructed for the transport of culms themselves, whereas transport of other commodities on such rafts is rarely found with the exception of transporting tourists, in the north of Thailand. Rafts are normally built in the rivers from 40 to 60 culms, in a crosswise double-layer arrangement, with a total bamboo weight of about 600–900 kg (RAP 2007).

Performance of the *horse logging is also being tested* along relatively narrow foot paths and skidding trails over distances of between 200 and 3,000 m with a total of 180 harvested bamboo culms. Two horses the size of large ponies (shoulder height of 115 cm) were trained by certain ethnic groups (Hmong, Yao) as horse-keeper in bamboo logging, using self-locking skidding chains and ropes.

Specific rules were set up to regulate the harvesting of bamboo in a sustainable manner (Ekvinay et al. 2011):

– Permission and quota must be granted before harvesting of bamboo poles.

Forbid to cut young bamboo pole age below 1 year for selling.
Forbid to cut bamboo other than the areas of permission and zoning.
Forbid to harvest bamboo shoot higher than 50 cm for selling.
Forbid to harvest bamboo shoot over 1/3 of the clump.
Allow to harvest the bamboo pole with age over 2–3 years and do not allow to cut for the whole clump.

There is a traditional rule that whoever cuts bamboo unlawfully in the spirit forest will be fined 5,000 kip for each bamboo pole.

1. **Thailand**: Bamboo is found throughout Thailand, mostly in mixed deciduous forests. It covers about 810,000 ha (5.5 % of the forest area). Thirteen genera, with more than sixty species, are found in Thailand (Subsansenee 1994). The best known species are *Thyrsostachys siamensis* and *Dendrocalamus asper*. *Thyrsostachys siamensis* is mostly collected from natural forests. The species is tolerant of drought and saline soil. It was the primary raw material for pulp and papermaking in Thailand from 1939 to 1984. Each year, over five million culms were required for the pulp and paper industry. Owing to the strong demand for bamboo stalks, this species is diminishing.

   Bamboo harvesting is carried out by selective cutting. The 1-year-old culms are not harvested in order to maintain growth. Cutting is generally done by using a small axe, machete, bill hook or saw. The first harvest is between the third and fifth year of growth. The 2- to 3-year-old culms are cut for bamboo stalks, poles, construction work and wicker work. The culms should be cut at the bottom close to the ground. Quality decreases if overaged clumps are left uncut. These clumps become brittle while the immature ones are not durable. Cutting is easier from November through March (Subsansenee 1994). Studies have indicated the suitability of a 3-year cutting cycle for *Thyrsostachys siamensis* in natural forests conditions. Consecutive cuttings 3 years apart each yielded more than 10,000 culms per hectare with no reduction in stem quality.

   De-branching of the culms is done immediately after cutting. The culms are then cut to the desired length. Bundling may or may not be done before the poles are transported to the roadside or the yard.

   Bamboo shoot harvesting is done from May to October (the rainy season). Shoots can be collected from the clumps daily or twice a week. In bamboo plantations, 1- or 2-year-old stalks of *Dendrocalamus asper* each yield about 5 or 6 shoots per year. For export, the average weight of shoots should range from 0.4 to 2.0 kg (Subsansenee 1994).

2. **Vietnam**: Vietnam is a country which has a greatest reserve of the bamboo materials in the world. More than 464 species were founded in the country which belong to 15 genera. The practice of overexploitation and harvest in wrong season have been leading to degraded bamboo forests (e.g. small stem diameter, short stem and few number of stems per clump). Depending on the bamboo species character and utilisation purpose, the first harvest is done 3–5 years after planting with harvest rotation every 2–4 years. Up to 2010, Vietnam plans to produce 2–2.5 million tons of paper/year, of which 30 % of raw materials are from bamboos (i.e. 3–4 million tons/year; 5–6 kg of bamboos = 1 kg of paper pulp).

3. **Indonesia**: The Javanese bamboos are traditionally harvested during the 'old season' (January–June) from the pure natural forest of *Gigantochloa apus* in Yogyakarta and Central Java. The people avoid the harvest of bamboo culms during October to December; during this time young shoots are still in the

process of elongation and development. In the months March to May, the young culms are considered to be strong enough to not suffer damage and the threat from powder-post beetle is at low. After 4 years of age, culms are mature and harvested (Sulthoni 1996).

4. **Malaysia**: The natural bamboo stand composed of *Gigantochloa scortechinii* occurs in abundance throughout the central and northwest areas. The exploitation of this resource has been unsystematic and haphazard, and the practices include the felling of bamboo culm annually (Mohamed 1996). According to Azmy et al. (1997), natural stands of *G. scortechinii* especially in Kedah, northern peninsular Malaysia, can be managed with recommended felling intensity of 70 % per clump; only mature culms of 3 years and above can be felled. However, 80 % felling intensity for shoot production has been suggested. Later the X-shaped harvesting technique was recommended by Othman et al. (2012) for natural stands because this method produced a higher number of new culms and greater total biomass as compared with the horseshoe-shaped harvesting technique. Furthermore, only bamboo clumps with a density of 26 culms per clump and above were suitable for this harvesting technique. Smaller clumps should not be harvested until they reach the limit of more than 26 culms per clump of bamboo. The clear-felling method was not recommended for harvesting of natural stands of *Gigantochloa scortechinii* due to slow recovery and low above-ground biomass production.

5. **Philippines**: The harvesting of bamboo culms for the commercial species (*Bambusa blumeana*, *B. vulgaris*, *Dendrocalamus merrillianus*, *D. latiflorus*, *Schizostachyum lumampao* and *Gigantochloa levis*) is selective cutting. The matured culms of 1 1.5 years are cut first leaving the younger culms for future harvests. The matured culms are cut at approximately 30 cm above ground level. The culm is crosscut into 5 m poles as specified for the banana props. Poles are then piled at the clump site, counted and quality checked by the supervisor before the minor transport. Minor transport from the clump site to the roadside over an average distance of 175 m is carried out by carabao (water buffalo) with a sled (Cruz 1989).

   Schizostachyum lumampao, locally known as *buho* mainly used for making woven material, occurs extensively in Luzon. The species is durable and from thin strips of culms. Due to unregulated exploitation, the *buho* stands have been very much depleted and the supply continues to decline. Through a number of studies, it has been recommended that application of moderate thinning followed by harvest cutting of culms having 3-year and above age with a 2-year felling cycle is optimum for managing clump sustainability with maximum culm yields and harvest cut from *buho* natural stands (Virtucio and Tomboc 1994).

6. **China**: Four- to five-year-old culms are being harvested for large-size bamboo (*Bambusa spinosa*, *B. stenostachya*, *B. lapidea*), while 3 years old bamboos are felled from the clumps (*B. textilis*, *Lingnania chungii*, *Schizostachyum funghomii*, etc.) for making thin strips. (Qiehui 1994). For large-size bamboo, 3-year-old culms are retained, while those over 4 years old are felled, except for those necessary for maintaining the required canopy density (Guoging 1987).

It is also reported that in Japan, felling are selective and carried out in autumn with cutting cycle varying from 3 to 5 years for *Ph. reticulata* and 5–10 years for *Ph. edulis* (Sharma 1982). In the case of noncaespitose bamboos, although clear-felling again would be most practical and is more frequently used, it seems desirable even under this practice to retain a few stems per acre, regularly spaced, so as to ensure maximum vigour and productivity of rhizomes (Huberman 1959).

While timber quality of culms of some species such as *P. pubescens* improves with increase in cell-wall density until 6–7 years (Huang et al. 1993), culms of most bamboo species 'mature' and can be harvested after 3–4 years. At this stage, they have attained their maximum static bending and compression strength (Huang et al. 1993; Espiloy 1994; Liese and Weiner 1995), physical properties that usually deteriorate beyond that age.

7. **Taiwan**: Chu-Shan forest has been famous and the most important bamboo-producing region in Taiwan. One of the main commercial bamboo species is *Phyllostachys makinoi*. At 3 or 4 years old, the culm reaches maturity, and this is the suitable harvesting age.

8. **East Africa**: Little is known about the harvesting of bamboo in Africa. In East African countries such as **Kenya, Uganda and Tanzania**, a highland bamboo species *Yushania alpina* occurs naturally in mountain areas of the country between 2,200 and 3,500 m above sea level. The uncontrolled exploitation of this resource has resulted in reduced productivity and yields as well as deterioration in quality. The coverage of this species in Ethiopia was roughly estimated in 1997 to be about 130,000 ha. In recent years, the resource base has been significantly reduced because large areas of indigenous bamboo forests have been cleared for conversion to agriculture. The CFC/UNIDO/INBAR while making cultivation manual for *Yushania alpina* in Eastern Africa (Ethiopia and Kenya) has also made a practical guideline for harvesting of *Y. alpina*. As per field guide a newly established plantation of *Y. alpina* should normally be ready for first harvesting after 5–6 years from the time of planting. A technical document has been prepared entitled 'Guidelines for Growing Bamboo' by the Kenya Forestry Research Institute (Bernard 2007) as a guideline for managing the bamboo forest in Kenya. Thereafter, cutting of mature culms can be done annually or at predetermined intervals of years, according to the management plan and the end use of culms. The clump should never be overharvested or clear-cut. New culms as well as 1–2-year-old culms should not be harvested. A few 3-year-old culms should also be left standing so that the clump remains robust and so that harvesting can be performed annually. Following this method, culms are left standing on the clump until they mature, after which, they may be harvested selectively according to the age and maturity of the culms. Harvesting is a labour-intensive operation, and it is necessary to make good arrangements with plantation workers so that harvesting operations are not delayed (Anonymous 2009).

The introduction of bamboo in South Nyanza region in **Kenya** as an alternative source of income initially met pessimism among the farmers due a lengthy wait, of at least 3 years, before harvesting. To convince the farmers to plant bamboo, they needed to know the expected quantity of harvest, its timing and consequently its associated income. The harvest is determined by the height and multiplicity of bamboo clumps. This varies from place to place due to diverse environmental conditions and other growth factors. A Markov model for forecasting bamboo harvest is introduced. Since the growth measures were random in nature, the stochastic modelling approach by Markov chains was applied. The resultant estimates from the model were 2,640 poles of *Bambusa vulgaris* and 1,188 poles of *Dendrocalamus giganteus* per acre of land per year. The model was tested for validity during the actual harvesting and found to be a good estimate of the actual values (Arori et al 2013).

According to the 'Bamboo Cultivation Manual Guidelines for Cultivating **Ethiopian** Lowland Bamboo (Eastern Africa Bamboo Project 2009)', a newly established plantation of *Oxytenanthera abyssinica* should normally be ready for first harvesting after 5–6 years from the time of planting. Thereafter, cutting of mature culms can be done annually or at predetermined intervals of years, according to the management plan and the end use of culms. Harvest culms only during the dry season. Harvesting should be selective: only mature culms which are 3–4 years old should be harvested. Do not cut young culms unless congestion in the clump prevents the cutting of mature culms.

9. **South America**: *Guadua angustifolia* is one of the major bamboo species in this region. Bamboo is considered mature between 4 and 7 years, after which they slowly start to deteriorate. The best time to harvest bamboo is at the end of rainy season—beginning of the dry season.

Reviewing the existing harvesting and transportation methods of bamboos, the following issues are highlighted:

7.4.2   Major Issues in Harvesting Forest Bamboos

(a) *Overexploitation in the easily accessible natural forest:* As the cutting and extraction is done on a piece rate basis, the workers have tendency to cut more bamboos in the accessible areas and also immature (less than 1 year old) bamboos ignoring the prescribed cutting rules. Labourers are always tempted to harvest bamboos from the easily accessible areas, usually from the banks of perennial streams (*chara*). Therefore bamboos are overexploited in these accessible areas. The overuse causes a gradual degeneration in health and size. Bamboos are seldom harvested from the steep slopes or inaccessible areas in the forest, and as a result bamboo clumps in these areas remain undercut (under exploited), and congested conditions develop.

(b) *Incomplete utilisation in the unexploited areas of natural forest:* Due to hilly terrain and difficulty in cutting and carrying, the bamboos are usually not harvested from the inaccessible areas. Thus the bamboo resources in many areas of northeast India and Myanmar and in other forests remain unexploited and not utilised for years together. Development of ropeways in the hills, as practised in CHT, for easy extraction of bamboo from inaccessible areas is very important. Paper mills in association with respective forest departments may initiate the work. The deportable condition of national highways and district roads including narrow roads on the hills should be maintained and improved for the better transportation of bamboo resources.

(c) *Resource wastage during harvesting and transportation from the forest:* Often bamboos are cut 1.0–1.5 m above ground level and chop off the upper narrow portion 1.5–2.0 m, only harvesting the mid-portion. The basal 1.0–1.5 m is heavier and has more biomass, and the upper portion of the culm has longer fibre length much desired for pulp production. In most of the cases, these portions are left in the forest (which is productive raw materials for pulp and paper mill) and not harvested. It is estimated that about 50 % of the total length of the harvested bamboo is left on the forest floor.

(d) *Uneconomic harvesting for pulping:* The pulp and paper mill authority purchases bamboo culms by weight, not by number because the utilisation is dependent on cellulose content. Like other plants the fresh weight of a culm includes its total moisture, cellulose and other chemical content. The amount of moisture in a culm may vary with the age, season of measurement and type of species. A study on five naturally grown bamboo species of Bangladesh shows that a culm may contain 60–87 % moisture at 1 month of age and 50–72 % at 6 months and up to 9 months; it varies within 49–57 % (Table 7.1).

After 12 months of age, the moisture content reduces to 35–47 %, and then in subsequent years, it becomes more or less constant, 22–40 %. If the culms are harvested before 9 months of age, they may contain more than 50–70 % moisture by weight. So the pulp and paper mill authority should not purchase bamboo culms less than 9 months of age harvested at March of the next year. Since the maximum fibre content in a culm is already achieved by the time the culm is 1 year old and since lignin and silicon accumulation increases with age, the culms for pulping should be harvested when they are young. It is necessary to remove old stumps and rhizomes every 2 years when clumps are grown in homesteads and farms in villages. Cutting of the top parts should be avoided since this will make the culm very fragile and is detrimental for pulping.

7.5   Felling Practices in Homestead and Small Farms

In villages, bamboos are cultivated in the homesteads of the plain land; usually the clumps are few in number, rarely covering one to two hectares of land owned by the families.

Table 7.1 Moisture content (per cent) of freshly felled culm of five naturally grown bamboo species as affected by age and season at Chittagong (*Source* Banik 2000)

Species	Culm age (month)/month of data collection										
	1 month (June)	6 months (December)	9 months (March)	12 months (June)	18 months (December)	21 months (March)	24 months (June)	30 months (December)	33 months (March)		
B. tulda	68.7	67.4	53.7	35.8	36.8	49.8	26.6	39.5	33.5		
D. long	58.5	46.3	39.3	32.8	27.7	25.2	22.3	25.2	21.6		
M. becc	86.5	71.5	57.2	43.5	36.2	38.5	37.6	39.6	34.2		
O. nigr	76.5	61.4	54.3	47.3	44.3	42.9	41.5	39.2	35.2		
S. dull	83.2	66.3	50.2	48.5	38.8	39.2	39.5	34.4	35.4		

Species: *D. long = D. longispathus, M. becc = M. beccifera, G. adm = Gigantochloa andamanica* (syn. *Oxytenanthera nigrociliata*), *S. dull = Schizostachyum dullooa*

Fig. 7.3 Most of the immature culms growing in the peripheral zone of a bamboo clump (*Bambusa balcooa*) are felled, while the older ones in the centre not harvested, as the harvesting is easy and quick at the peripheral zone. This practice ultimately destroys the clumps by gradually decreasing the rhizome vitality

7.5.1   Demand Basis Harvesting

The farmer has been harvesting bamboo mostly on his and market demand basis; so very rarely any management system is followed. Due to the price hike, the rate of felling of bamboos is increasing in some villages near urban areas. Compelled by poverty and tempted by higher price, poor farmers also sell immature good-looking bamboos to traders from their homesteads. Bamboo traders mostly cut immature culms growing in the peripheral zone of these clumps, as the harvesting is easy and quick. This practice ultimately destroys the clumps by gradually decreasing the rhizome vitality (Fig. 7.3). Thus, the bamboo supply is decreasing, and less is available for house and fence construction. Decreased numbers of immature culms adversely affect the regenerative capacity of the clumps. In contrast, rich farmers and absentee landlords in some of the districts in the different states of India and Bangladesh where bamboos are cultivated as patches on small farms do not usually fully sell or harvest their bamboos in due times. As a result, the clumps are in a congested condition. In the clumps unharvested, mature culms die due to overage and start rotting in the rainy season. These unhygienic conditions attract insects and fungi initiating diseases in the clumps. As a result majority of bamboo clumps are over or under extracted. In some cases culms are not harvested, and clumps remain unattended; this situation also induced deterioration of clump health. These disadvantages are overcome under the clear-felling system with its simplicity and concentration of working and suitability for mechanisation; but such advantages are again offset by the need for a considerably longer cutting cycle (a clear-felled clump taking longer to mature into a full-sized production stage again) and perhaps greater liability of clumps to deterioration and mortality (Huberman 1959). The choice of a system is, therefore, a matter for decision by individual farm owners. However, selective cutting gives a somewhat greater yield than clear-cutting and is preferable.

7.6   Clear-Felling

When culms are not harvested for several years, it creates congestion and clumps gradually become susceptible to injury and diseases. In such a case to maintain the proper clump growth and shape, clear-felling of culms is good for clump health. Occasionally farmers do practise clear-felling in congested clump—*B. bambos, B. balcooa and D. strictus* in Uttarakhand (Kalinagar, Dineshpur and Gadarpur), Bihar (Pusa) and West Bengal (Bankura). A study conducted in Shahdol district (Prasad 1987) has shown that clear-felling of the congested clumps and allowing the new shoots to come up are more beneficial than the prevalent practice of working the congested clumps. This is because the congested clumps do not allow new shoots to come up easily; even if any shoot comes up, it becomes malformed. On the other hand, if the clump is clear-felled in one stroke, good-quality culms can be obtained in greater numbers after 4 years. Therefore, clear-felling of the congested clumps of *Bambusa bambos, B. balcooa* and *D. strictus* is occasionally a beneficial practice and seen in northwest and central part of India.

Sometimes rhizomes and roots of the bamboo, which are invaluable for the clump reproduction, are ruthlessly burnt for brick production, and these are the only source of cash money for the poor farmers. Farmers dig up the old and dead rhizomes to sell to the local brick kilns (Fig. 7.4), while doing so some productive young rhizomes are also dug out from the clump. Kiln owners purchase the bamboo rhizomes at the rate of Rs. 800–1,000/t (1 US$ = INRs 50, in the year 2009), while the cost of fuelwood is double that amount. For the more than last 20 years, bamboo rhizomes are being extensively used for brick burning in these areas. It requires about 28 metric tonnes of wood to produce about 100,000 brick. So poverty in rural areas also compels the farmers in harvesting the underground rhizomes from their homesteads to earn some immediate cash.

7.6.1   Selection Felling

Some farmers also follow selection felling system. However, dead culms and branches can be collected any time for fire wood purposes. Harvesting of bamboos is carried out during winter season (December–February). Males are usually involved in harvesting and carrying of bamboo poles from forest to their residences. The common village-grown tall bamboo species (*B. balcooa, B. bambos, D. hamiltonii, D. strictus, B. tulda, B. nutans, B. polymorpha, B. pallida, B. cacharensis, B. vulgaris, Pseudoxytenanthera stocksii*, etc.) are harvested for different purposes such as local house construction, fencing, flooring, animal cages, making mats, baskets, trays, caskets and other household items. A large knife or '*wait*' is used to cut bamboo poles from clumps. The poles are cut at a point about 0.6 m above ground on average. Straight, well-formed poles are preferred, with an average age of 3 years. Where clumps of *B. bambos* were well managed and

Fig. 7.4 While clear-felling bamboos, the farmers sometimes dug out old and dead rhizomes from the clump to sell to the local brick kilns for earning some immediate cash, while doing so some productive young rhizomes are also dug out from the clump

scientifically harvested with a felling cycle of 6–8 years, culm production was higher, poles were longer and output was larger. Due to the higher proportion of high-graded long poles, well-managed clumps fetched higher price and enhanced farm income (Krishnankutty 2005). However, overmature culms start to decay, reducing their quality and market value of the stems, and the precautions must be taken to preserve maximum vigour and productivity of rhizomes and culms, such as by cutting out overmature, defective and unmerchantable culms and securing even distribution of culms within each unit stand or clump.

Once extracted, the culms are trimmed to full length. The bamboos are cut into suitable lengths for ease of transportation depending on the mode of transportation used. In road and rail transportation, the lengths are cut to accommodate the space. In water raft transportation, full lengths of bamboos are usually transported. If, however, boats are used, culm length is cut accordingly.

For mat making, 1-year-old bamboo is most commonly harvested.

As discussed earlier, for maintaining the sustainable growth, bamboos should not be harvested every year from a clump. Felling should be stopped at least for 3 years so that culms in the clump attain maturity. Therefore, in a homestead at least four clumps need to be planted from where harvesting may be started sequentially in all these clumps to obtain bamboos every year. Thus culms will be harvested from each clump at 3-year intervals.

On the principles of selection felling system following steps of bamboo harvesting, *horseshoe* and *tunnel* methods (Sharma 1987; Banik 1992, 2000; Fig. 7.5) have been prescribed.

7.6.1.1 *Horseshoe* and *Tunnel* Methods

a. A 60–100 cm-wide path has to be made inside the clump so that one can enter into the central part to start felling and dragging out the mature culms. As the path will be made from periphery towards the centre, it is likely that a few

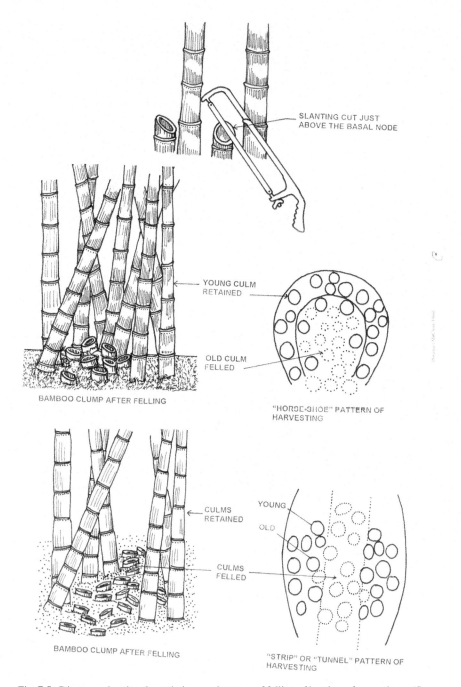

SLANTING CUT JUST
ABOVE THE BASAL NODE

YOUNG CULM
RETAINED

OLD CULM
FELLED

BAMBOO CLUMP AFTER FELLING

"HORSE-SHOE" PATTERN OF
HARVESTING

CULMS
RETAINED

YOUNG

OLD

CULMS
FELLED

BAMBOO CLUMP AFTER FELLING

"STRIP" OR "TUNNEL" PATTERN OF
HARVESTING

Fig. 7.5 Diagrams showing the technique and pattern of felling of bamboos from a clump (*Source* Banik 2000)

number of young culms may have to be cut. So one must make path from that side of a clump where minimum number of young culms are sacrificed.

b. Most of the mature culms from the central part of the clump have to be cut. As a felling tool either a knife or a pruning/portable chain saw may be used depending on the thickness of the bamboo to be harvested. The unharvested mature culms should be left scattered throughout the clump to provide mechanical support to the young immature culms against the strong wind and storm. This felling procedure involves opening from one side of a clump, and the central mature culms are cut and dragged out. This type of felling is *horseshoe* method of felling of culms (Fig. 7.5). In other method two openings are made, one opposite to other, forming a *tunnel* in the clump through which harvested culms dragged out.

c. The following care should be taken when culms are cut and dragged from the clump:

   (1) The number of harvested culms should not exceed the number emerged last year. For example, a clump has 15 culms, out of which four and six culms emerged in 1993 and 1994, respectively. Thus the clump has six 1-year-old and four 2-year-old culms in the felling year 1995. To keep the clump productive and healthy, do not cut more than five to seven mature culms. Mature culms should be left evenly distributed throughout the clump to provide mechanical support as well as nourishment.

   (2) Before entering into the clump, branches, if any, on the lowest nodes need to be trimmed thoroughly to facilitate harvesting operation.

   (3) Cut culms in a slanting manner (cuts at 45° angle) just above the lower most nodes to minimise the wastage. Due to slanting cut rainwater will not stagnate at the remaining stump portion of the felled culms and will not harbour pests and fungi. When the felling is done far above the ground, buds on the nodes of the stumps of the felled culms become activated and produce twigs and branches. Thus congested and bushy conditions are created in the clump and interfere in future felling operation.

   (4) The branches and twigs from the harvested culms from the clumps have to be cleared and trimmed out. The harvested and cleaned bamboo pole may be kept aside so that it can dry and cure.

   (5) The dead and rotten stumps of the felled culms may be carefully uprooted as a sanitary cleaning. Add soil and organic manure (crowding/rice husk/rotten water hyacinth at 3:1 ratio) to the clump after digging out such stumps. Use these stumps as fuel.

   (6) The felling operation should not be undertaken during the culm emergence period. Harvest culms from September to April. Felling of culms is more desirable in the month of November to January as culms possess less amount of starch during this time of the year. Thus harvested culms become less vulnerable to powder-post beetles (goon) attack.

   (7) Felling during the year of flowering is discussed under Sect. 7.7.

However, on the hilly terrain horseshoe cut initially is more labour-intensive than selective cutting (SC) and requires more skills. After initial greater labour

Fig. 7.6 During the first year of clear-felling, many thin tender branches are produced from the stump nodes in a clump of *Bambusa balcooa* and resulted in vigorous leafy and bushy growth and little or no emergence of new shoots

investment, harvesting operation is easier to administer, as shoots and culms are easily accessible from clump edge.

Audiovisual aids may be utilised to train and motivate the farmers about the proper bamboo harvesting technique.

7.6.2 Management of Clump After Clear-felling

The dormant buds on the stumps nodes activate and produce many thin branches resulted in leafy and bushy condition in the clump; as a result little or no emergence of new shoots are observed in first year (Fig. 7.6).

1. So tender branches need protection from browsing and trampling by cattle so that carbohydrate reserve is created in the rhizome below the ground to rejuvenate the clump.
2. Small weak shoots/branches should be trimmed; otherwise the clump will be congested.
3. In the second year lightly thin out the poor shoots, make space for new shoot emergence and protect and manage these shoots from browsing, weed and vine suppression and fire.
4. The too old rhizomes with stumps should be dug out from the clump.
5. Do mounding/heaping fresh, loose soil mixed with FYM at the base of shoots to provide moisture and protect the buds in the basal part of the clear-felled clump.
6. Mulch and if possible water in dry climate up to 2–3 years. The leaves must be left in the grove after cutting off the branches at the harvesting. Silicate (SiO_2) is contained roughly more than 5.0 % in the leaves, about 2.0 % in the branches, about 0.4 % in the culm and about 0.5 % in the rhizome (Ueda 1960). So SiO_2 is effective in increasing the culm production.

7.7   Felling of Bamboo After Flowering

The following activities may be undertaken through total collaboration of local people of bamboo flowered areas.

- Felling during the year of flowering is not desirable so that regenerating seedlings are not disturbed and can establish on the ground. Flowered bamboo is allowed to complete the flowering and then seeding. Meanwhile start preparations for harvesting the dry bamboo after the seeding is complete everywhere; this is followed by toppling down of the bamboo clumps. This may take a few months to years depending on the extent of flowered area (Maslekar 2003). In many occasion all the clumps located in an area do not flower and produce ripe seeds at a time; some clumps may remain alive for few more months in producing mature seeds. Harvesting bamboos from these clumps is to be delayed till the mature seeds are produced and fall on the ground.

- Only when the clumps start falling the harvesting is undertaken.
- The rain starts the profuse regeneration of bamboo seedlings from the fallen seeds on the ground, and during this time the harvesting activities would kill the very young germinating seedlings. So dry and dead bamboos are to be cut and carried away from bamboo areas before the start of the rainy season.
- Keep a few standing dead culms evenly distributed in the areas to provide partial shade to the regenerating seedlings. These bamboos may be harvested after 9–12 months as during this time bamboo seedlings develop sustainable rhizome system.
- However, in ecologically vulnerable sites, such as water catchment, steep slopes, wildlife protected areas, etc., the flowered bamboos are usually not cut or removed because flowering and drying and further degradation of bamboos in situ is itself an ecological event. Flowered bamboo has a higher resistance to beetles because the starch is depleted when bamboo flowers. These bamboos are brittle and, so usually, not used for any construction works.
- Among the major *cultivated bamboo species* in the villages of India and Bangladesh harvesting of culms in the clumps of *B. vulgaris*, *B. balcooa* and *Ps. stocksii* should not be stopped during flowering as these species do not flower frequently and if they at all flower they do not produce any seeds. Other cultivated village-grown bamboo species (*Bambusa bambos*, *B. cacharensis*, *B. polymorpha*, *B. nutans*, *B. pallida*, *B. tulda*, *Dendrocalamus hamiltonii*, *D. strictus*, *Thyrsostachys oliveri*, *T. siamensis*, etc.) usually produce fertile seeds when they flower, and so culms should be harvested after seed collection. Felling during the year of flowering is not desirable so that seeds and wild seedling can be collected from the ground for raising new groves.

Efforts should also be made to harvest a substantial amount from huge quantities of dead bamboos from the inaccessible areas and marketing them.

- *Good storage practices for huge amount of harvested dead bamboos:* Storage of the huge quantity of bamboo for a long period results in damage by microorganisms, and as a result bamboo deteriorates rapidly. Bamboos may be protected against deterioration by giving prophylactic treatment by spraying dilute preservatives like sodium pentachlorophenate (1 % solution) and boric acid + borax (1:1) 2 % solution.
- *Maximising utilisation of available huge dead culms:* As mentioned earlier, the dead flowered bamboos are brittle in nature and are not suitable for any construction works. Bamboo is a good substitute for fossil fuels in the form of charcoal briquettes. Such dead bamboo can be best utilised for generating electric power through a bamboo-based biomass gasifier. By this process, the energy present in the biomass is converted into a gaseous combustible or chemical energy. Gas products are easy to handle and can be used in combustion engine or gas turbines. The combustion is clean and less polluting. However, establishment of microenterprise for producing incense sticks, chopsticks, toothpicks, bamboo-made pencils and charcoal, etc. may be carried out so that the indigenous people can also earn livelihoods and tide over the difficult period till the bamboo areas regenerate, or before.

References

Anonymous (2009) Bamboo cultivation manual guidelines for cultivating Ethiopian highland bamboo-Eastern Africa (Ethiopia & Kenya)—a field guide. CFC/UNIDO/INBAR, pp 1–58

Arori WO, Kibwage JK, Netondo GW et al (2013) A Markov model for bamboo harvest forecasting in South Nyanza region, Kenya. Adv J Agric Res 1(004):45–50

Azmy HM, Norini H, Wan Razali WM (1997) Management guide-lines and economics of natural bamboo stands. FRIM technical information no. 15, Kepong, Malaysia, pp 1–40

Banik RL (1992) Bamboo: forestry master plan of Bangladesh. ADB (TA. No 1355-BAN), UNDP/ FAO BGD 88/025 (8 Appendix), Dhaka, pp 1–62

Banik RL (1993) Morphological characters for culm age determination of different bamboo species of Bangladesh. Bang J For Sci 22(1&2):18–22

Banik RL (2000) Silviculture and field-guide to priority bamboos of Bangladesh and South Asia. BFRI, Chittagong, pp 1–187

Banik RL (2010) Biology and silviculture of muli (*Melocanna baccifera*) bamboo. NMBA (National Mission on Bamboo Applications), TIFAC, Department of Science & Technology, New Delhi, pp 1–237

Banik RL, Islam SAMN (2005) Leaf dynamics and above ground biomass growth in *Dendrocalamus longispathus* Kurz. J Bamb Rattan 4(2):143–150

Bernard NK (2007) Guidelines for growing bamboo. KEFRI guideline series no. 4. Kenya Forestry Research Institute, Nairobi, Kenya

de la Cruz V (1989) Small-scale harvesting operations of wood and non-wood forest products involving rural people: part I. Harvesting a bamboo plantation and a natural bamboo stand. FAO forestry paper no. 87, pp 1–31

Dransfield S, Eijada WA (1995) Bamboos, plant resources of South East Asia Network, Nr. 7. PROSEA and Backhuys, Bogor, Indonesia

Eastern Africa Bamboo Project (2009) Bamboo cultivation manual guidelines for cultivating Ethiopian lowland bamboo.CFC/UNIDO/INBAR. Ministry of Agriculture and Rural

Development Federal Micro and Small Enterprises Development Agency, Kenya Forestry Research Institute, pp 1–61

Ekvinay S, Vilaisi P, Khongkha M et al (2011) Towards communal land title in Sangthong District: participatory development of a format for communal land titles in four villages in Sangthong district, Greater Vientiane Capital City Area. SNV/gef/UNDP, pp 1–35

Espiloy Z (1994) Effect of age on the physio-mechanical properties of some Philippine bamboo. In: Proceedings of the 4th international bamboo workshop on bamboo in Asia and the Pacific. FORSPA Publication 6 IDRC FAO-UNDP, Chiangmai, Thailand, 27–30 Nov 1994, pp 180–182

Fu M, Banik RL (1996) Bamboo production system and their management. In: Rao IVR, Widjaja E (eds) Bamboo, people and the environment. Propagation and management, vol I. Proceedings of the 5th international bamboo workshop and the 4th international bamboo congress, 19–22 June1995. Ubud Bali, Indonesia. INBAR/EBF/IPGRI/IDRC, INBAR, New Delhi, pp 18–33

Guoging L (1987) Improved cultivation techniques of bamboo in north China. In: Rao AN, Dhanarajan G, Sastry CB (eds) Recent research on bamboos. Proceedings of the international bamboo workshop, 6–14 Oct 1985. The Chinese Acad of Forest, Hangzhou, China; IDRC, Canada, pp 71–77

Mohamed AH (1996) Effect of fertilizing and harvesting intensity on natural stand of *Gigantochloa scortechinii*. In: Rao IVR, Widjaja E (eds) Bamboo, people and the environment. Propagation and management, vol I. Proceedings of the 5th international bamboo workshop and the 4th international bamboo congress,19–22 June 1995, Bali Indonesia, pp 86–95

Htun N (1999) Bamboos of Myanmar. In: Rao AN, Rao VR (eds) Bamboo conservation, diversity, ecogeography, germplasm resource utilization and taxonomy. In: Proceedings of a training course cum workshop, 10–17 May 1998, Kunming and Xishuangbanna, Yunnan, China; IPGRI-APO, Serdang, Malaysia, pp 201–214

Huang Q-M, Yang D-D, Shen Y-G et al (1993) Studies on the primary productivity of phyllostachys pubescens grove. RISF annual report 1995. Research Institute of Subtropical Forestry, Chinese Acad of Forestry, Fuyang, Zhejiang, P R China, pp 16–17

Huberman MA (1959) Bamboo silviculture. Unasylva 13(1):36–48

Krishnankutty CN (2005) Bamboo *(Bambusa bambos)* resource development in home gardens in Kerala State in India: need for scientific clump management and harvesting techniques. J Bamb Rattan 4(3):251–256

Lakshmana AC (1994). Culm production of *Bambusa arundinacea* in natural forests of Karnataka, India. In: Proceedings of the 4th international bamboo workshop on bamboo in Asia and the Pacific, 27–30 Nov 1991. Chiangmai, Thailand, FORSPA publication no. 6, IDRC FAO-UNDP, pp 100–103

Liese W, Weiner G (1995) Ageing of bamboo culms. A review. Wood Sci Technol 30:77–89

Manoharan TM, Trivedi BNV (2008) Forest policy and laws governing cultivation, harvesting, transport and trade of bamboo in Kerala. In: Choudhary ML, Salam K (eds) Proceedings on international conference on improvement of bamboo productivity and marketing for sustainable livelihood, 15–17 Apr 2008, New Delhi, pp 182–192

Maslekar AR (2003) Bamboo the life blood of the people: alarm to ecosystem (Bamboo flowering - *gregarious*, more on bamboo flowering). EPW, 13 Dec 2003. www.sos-arsenic.net/english/homegarden/bamboo.htm

Midmore DJ (2009) Overview of the ACIAR bamboo project outcomes. In: Midmore DJ (ed) Silvicultural management of bamboo in the Philippines and Australia for shoots and timber. Proceedings of a workshop Los Baños, Philippines, 22–23 Nov 2006. ACIAR proceedings no. 129, pp 7–12

Mukhia PK (2005) General guidelines for management of bamboo in Bhutan. Royal Govt of Bhutan, Ministry of Agriculture Department of Forest, Forest Resources Development Division, Thimphu, Jul 2005, pp 1–60

Othman AR, Lokmal N, Hassan MG et al (2012) Culms and above-ground biomass assessment of *Gigantochloa scortechinii* in response to harvesting techniques applied. J Agrobiotech 3:1–8

Prasad R (1987) Effect of clear felling of congested clumps on yield of bamboo (*D. strictus*). Indian For 113:609–615

Qiehui D (1994) Orientation cultivation of bamboos. Part I. Nursery and afforestation technology for sympodial bamboo. In: Fu M, Jianghua X (eds) Cultivation and utilization on bamboos. The Research Institute of Subtropical Forestry, The Chinese Academy of Forestry, Oct 1994, pp 81–100

RAP (2007) FAO corporate document repository 5 SECTION: Technical aspects:13. Appropriate Forest Harvesting and Transport Technologies for Village-based Production of bamboo charcoal in mountainous areas of Northern Lao PDR, pp 1–12. Series title: RAP Publication. ftp://ftp.fao.org/docrep/fao/010/ag131e; http://www.fao.org/docrep/010/ag131e/ag131e00.htm

Sharma YML (1982) Some aspects of bamboos in Asia and the Pacific. FAO regular programme, No Rapa 57, Bangkok, pp 1–56

Sharma YML (1987) Inventory and resource of bamboos. In: Rao AN, Dhanarajan G, Sastry CB (eds) Recent research on bamboos. Proceedings of the international bamboo workshop, 6–14 Oct 1985. The Chinese Acad of Forest, Hangzhou, China; IDRC, Canada, pp 4–17

Stapleton CMA (1987) Bamboos, *Gramineae*. In: Jackson JK (ed) Manual of afforestation in Nepal, pp 199–214

Subsansenee W (1994) Thailand. In: Patrick BD, Ward U, Kashio M (eds) Non-wood forest products in Asia. Regional Office for Asia and The Pacific (RAPA Publication 1994/28)/FAO, Bangkok, pp 127–150

Sulthoni A (1996) Shooting period of sympodial bamboo species: an important indicator to manage culm harvesting. In: Rao IVR, Widjaja E (eds) Bamboo, people and the environment. Propagation and management, vol I. Proceedings of the 5th international bamboo workshop and the 4th international bamboo congress, Bali, Indonesia, 19–22 June 1995, pp 96–100

Tewari DN (1992) A monograph on bamboo. International Book Distributors, DehraDun, pp 1–498

Thompson IS (1986) A forest management research study in broadleaf middle-hill forest of Nepal. O.F.I. occasional papers no. 30, p 31

Troup RS (1921) The silviculture of Indian trees, vol III, Gramineae. The Clarendon Press, Oxford, pp 978–1013

Ueda K (1960) Study on the physiology of bamboo with reference to practical application. Prime Minister's office, Resources Bureau Science and Technology Agency, Tokyo, Japan, pp 1–167

Virtucio FD, Tomboc CC (1994) Effect of thinning, cutting age and felling cycle on culm yield of Buho (Schizostachyum lumampao) natural stands. In: Proceedings of the 4th international bamboo workshop on bamboo in Asia and the Pacific, 27–30 Nov 1994. FORSPA Publication 6 IDRC FAO-UNDP, Chiangmai, Thailand, pp 106–112

Chapter 8
Properties of the Bamboo Culm

Walter Liese and Thi Kim Hong Tang

Abstract The properties of bamboo culms determine their possible uses. They are based mainly on the structure of the tissue, which is dealt with in some detail. According to the scope of the Tropical Forestry Handbook, the following presentation provides a general overview on the chemical compounds and physical and mechanical properties. Detailed information provides the referred literature.

Keywords Anatomy • Chemical composition • Physical properties • Mechanical properties

8.1 Structural Properties

8.1.1 Anatomy of the Culm

The structural composition of a bamboo culm provides numerous possibilities for its utilization, either in round form or split into parts. Continuous research has focused on evaluating the relationships between structures, processing, and product quality (Liese 1985, 1998, 2004a, b; Jiang 2007).

The bamboo culm is separated by nodes into internodes (Fig. 8.1). Internodal length differs considerably between species. It is longest in the middle of a culm. The strictly axial arrangement of the vascular bundles is interrupted at the nodes, so that species with long internodes, like *Bambusa textilis* with up to 60 cm long internodes, are preferred for furniture, splitting, and weaving. The culm consists of a culm wall around a hollow center, the lacuna; only a few species have a solid culm at the lower end, like *Dendrocalamus strictus* and *Thyrsostachys siamensis*. The thickness of the culm wall shows great differences between genera and species, like

W. Liese (✉)
Department of Wood Science, University of Hamburg, Leuschnerstr. 91, 21031 Hamburg, Germany
e-mail: wliese@aol.com

T.K.H. Tang
Department of Wood Science & Technology, Nong Lam University, Ho Chi Minh City, Vietnam
e-mail: kimhongtang@yahoo.com

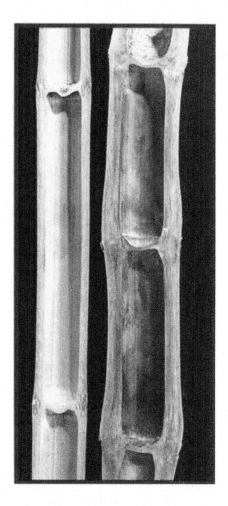

Guadua with a thick wall as compared to *Melocanna* with a very thin wall. The thickness of the wall has a great impact on mechanical properties.

Culm diameter tapers from bottom to top with differences between species. Base and middle portions with long internodes are utilized generally for construction work, furniture, mats, and boards. The reduction in diameter is accompanied by a reduced wall thickness, whereby the overall density increases. The number of vascular bundles does not change with height, while the fraction of parenchyma tissue diminishes.

The anatomical structure of a bamboo culm appears rather uniform compared with wood. Although the differences between the around 1,200 bamboo species are small, certain species are preferred for specific uses.

Fig. 8.2 Cross section of a
bamboo culm, vascular
bundles within the
parenchyma (Liese 1998)

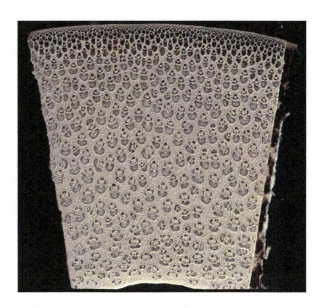

8.1.1.1   Structure of the Internode

In general, the culm consists of about 50 % parenchyma cells, 40 % fibers, and 10 %
vascular bundles (vessels, sieve tubes with companion cells). A cross section shows
the dark vascular bundles with their fiber agglomerates within the parenchyma
tissue (Fig. 8.2), which strongly contribute to the structural character of the culm.
All bamboo culms exhibit a similar pattern in the distribution of their cells. The
percentage of fibers decreases from outside to inside, while the parenchyma
increases. The culm base contains more parenchyma, and the upper part has
many smaller vascular bundles with a high portion of fibers, providing a superior
slenderness (Grosser and Liese 1974).

Parenchyma

The parenchyma cells are the ground tissue, in which the vascular bundles are
embedded. Both jointly contribute to stability and flexibility of the culm. At the
outer culm part, the parenchyma cells are small but become larger and longer
toward the inner part. There are two types: vertically elongated cells and short
cube-like ones interspersed (Fig. 8.3). The longer ones lignify in the early stages of
internodal development, whereas the shorter ones with denser cytoplasm remain
unlignified. The cell wall of the longer ones has a polylaminate structure consisting
of up to 20 lamellae with alternating fibril orientation (Fig. 8.4) (Parameswaran and
Liese 1975). The cells are connected by pits on their tangential wall. Parenchyma
cells are vital for the culms life. At the end of the season, densely stacked starch
particles fill the cell lumina (Fig. 8.5) and are mobilized before shoot formation
(Magel et al. 2005). For later processing, starch retards the setting reaction in
cement-bonded particleboards, so that soaking or chemical additives should reduce

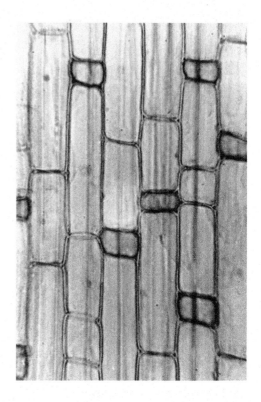

Fig. 8.3 Parenchyma
consisting of longer and
shorter cells (Grosser and
Liese 1971)

the sugar content. Parenchyma cells at the outer culm wall, the cortex, contain silica, which affects the cutting and pulping properties. It is species dependent so that the ones with a low content are preferred for furniture.

Fibers

The fibers are characterized by their slender form, long and often forked at the ends (Fig. 8.6). They are present around the vascular bundles as sheaths and as isolated strands. They amount to about 40 % of the mass and to 60–70 % of the weight of the culm. Their length follows a definite pattern across the wall and along the height of the culm. It varies considerably between species, with a factor of variation between 1.5 and 3.0 (Table 8.1), being much longer than those of hardwoods or softwoods. Content and length influence the specific gravity (0.5–0.9 g/cm^2) and strength as well as pulping properties. Fiber length is strongly correlated with fiber diameter, cell wall thickness, as well as with the modulus of elasticity and compression strength. The outer part of the culm with its denser arrangement of fibers has a far higher specific gravity than the inner, more parenchymatous part (Liese and Grosser 1972).

The fiber wall consists of the primary, secondary, and tertiary wall. Especially the secondary wall is made up by numerous layers of microfibrils with a varied orientation (Figs. 8.7 and 8.8a, b). The inner layer, also called tertiary wall, is often covered by a warty structure (Fig. 8.9) (Parameswaran and Liese 1976, 1977a, b).

Fig. 8.4 Polylamellate wall of parenchyma cell, *Phyllostachys edulis* (Liese 1998)

Fig. 8.5 Starch particles filling the parenchyma cells, *Phyllostachys viridiglaucescens* (Liese 1998)

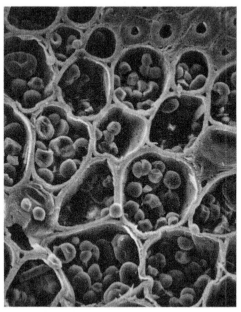

Fig. 8.6 Small and a big
bamboo fiber with forked
end; parenchyma cell (Liese
1998)

parenchyma cell

100μm

Table 8.1 Fiber length of
some bamboo species (Liese
1985)

Bamboo species	Average fiber length (mm)
Bambusa arundinacea	2.7
B. textilis	3.0
B. tulda	3.0
B. vulgaris	2.3
Dendrocalamus giganteus	3.2
D. membranaceus	4.3
D. strictus	2.4
Gigantochloa aspera	3.8
Melocanna bambusoides	2.7
Oxytenanthera nigrocilliata	3.6
Phyllostachys edulis	1.5
P. makinoi	2.5
P. pubescens	1.3
Teinostachyum sp.	3.6
Thyrsostachys siamensis	2.3

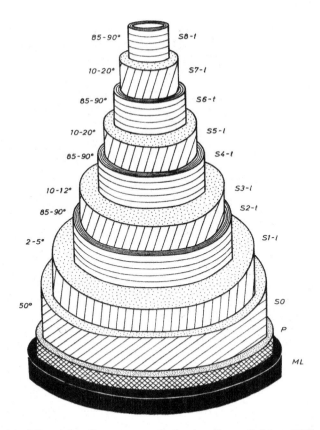

Fig. 8.7 Model of the polylamellate structure of a bamboo fiber wall (Liese 1985)

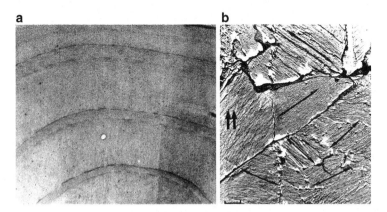

Fig. 8.8 (**a**) Cross section of a fiber wall with alternating broad and small lamellae and (**b**) surface view on lamellae with different orientation of their microfibrils, *Phyllostachys edulis* (Liese 1998)

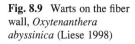
Fig. 8.9 Warts on the fiber wall, *Oxytenanthera abyssinica* (Liese 1998)

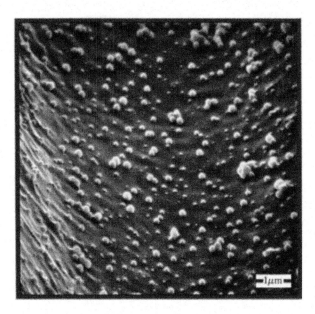

The microstructure of the cell wall contributes to the great flexibility of fibers and, in combination with the other cell types, to the excellent flexibility of the culms. It also influences its fractured appearance after breaking.

Vascular Bundles

The vascular bundle consists of the two metaxylem vessels, fibers, and the metaphloem of sieve tubes with companion cells (Fig. 8.10). They are the prominent and clearly visible components on a cross section and the most varied structures of the various types of bamboo (Grosser and Liese 1971). This is mainly due to the agglomerated fibers, which are attached as fiber sheaths or as separated fiber strands. Their form and shape is genetically determined. Generally, leptomorph species have their metaxylem vessels attached by four fiber sheaths, whereas the taller pachymorph species growing as a clump present additionally isolated fiber strands (Fig. 8.11). The large culms of the pachymorph genus *Guadua* show extensively formed fiber sheaths. Altogether six types with eight subtypes have been distinguished (Grosser and Liese 1973; Liese and Grosser 2000) (Fig. 8.12). The amount of fibers, as sheaths or additional strands, is closely related to the specific gravity. It increases within the culm from base to top and influences the strength properties.

The two large vessels in the individual vascular bundles accomplish the water transport within the culm (Fig. 8.13). They are considerably bigger in the inner culm part and smaller toward the outside. Their volume amounts to only about 6–8 % of the total tissue. Consequently, its high conductivity is vital for the transpiration of the culms leafs but later also for any axial treatment of the culm, like by the sap replacement process. Properly applied, the easiness of the water conductivity provides the best treatment result.

Fig. 8.10 Three-dimensional view of the culm tissue with vascular bundles and fiber sheaths embedded in ground parenchyma, *Oxytenanthera abyssinica* (Liese 1998)

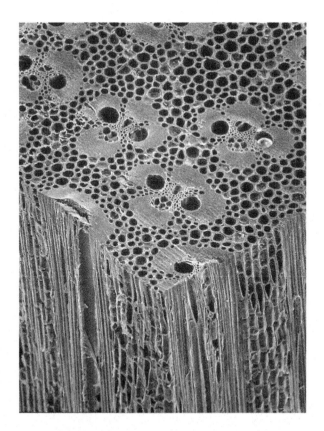

The phloem serves the downward transport of assimilates to be stored in the parenchyma. It consists of large thin-walled sieve tubes and smaller companion cells (Fig. 8.14). They are connected with each other by sieve pores. Distributed among the sieve elements are companion cells, characterized by a dense cytoplasma and a large nucleus. Mitochondria are numerous, and plastids are few. They are connected with the sieve elements by plasmodesmata.

Outer and Inner Layers of the Culm Wall

The culm wall is covered on both sides with a special tissue (Fig. 8.1). Its outer part, the cortex, presents a watertight seal to prevent any moisture loss of the living culm. The structural composition of compact fiber bundles with thick walls provides also a protection against mechanical wounding. As a consequence for processing, the compact structure hinders the loss of moisture during drying of culms, as well the penetration of any preservative liquid although the outer layer provides protection against wounding and biodeterioration.

At the inner side toward the central cavity, the lacuna, layers of parenchyma cells, forms a special tissue, the terminal layer. Its cell walls are often heavily thickened and exhibit distinct differences between species (Figs. 8.15 and 8.16) (Liese and Schmitt 2006).

Fig. 8.11 Different growth habitats reflect as different vascular bundle types (**a**) *Phyllostachys* sp., (**b**) *Guadua* sp., (**c**) *Dendrocalamus* sp. (Liese 2012)

8.1.1.2 Structure of the Node

The bamboo culm is divided by nodes (Fig. 8.1). A node consists of a sheath scar, nodal ridge, diaphragm, and the intra-node between the nodal ridges (Ding and Liese 1997). The nodes have special significance for the intercalary growth and for the function of the culm. A three-dimensional structure of the vascular system at the nodal region is shown in Fig. 8.17. Most of the main axial vascular bundles pass directly from an internode through the node into the next internode. In the peripheral zone of the culm, they bend slightly outward while branching partly into the sheath. In the inner zone, they are connected with those in the diaphragm. In the upper part of the node, the bundles become larger in diameter, and vascular anastomoses develop intensively. At the upper edge of the diaphragm, many small bundles turn horizontally and twist repeatedly.

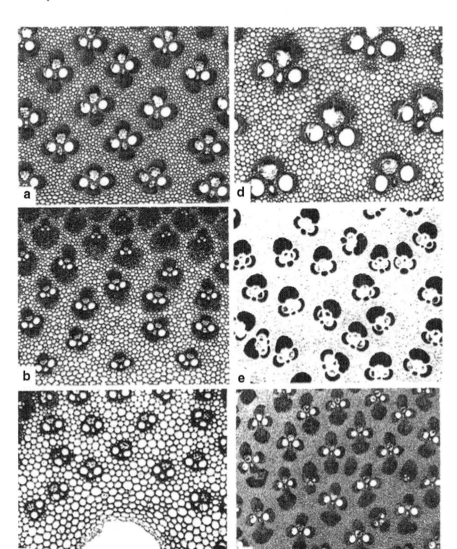

Fig. 8.12 Expanded typology of vascular bundles, for example, (**a**) type Ia (*Phyllostachys nigra*, 22×), (**b**) type Ib (*Fargesia robusta*, 30×), (**c**) type Ic (*Shibataea kumasaca*, 50×), (**d**) type IIa (*Bambusa multiplex*, 35×), (**e**) type IIb (*Guadua angustifolia*, 10×), and (**f**) type IIc (*Melocanna baccifera*, 12×) (Liese and Grosser 2000)

In the separating tissue, the diaphragm, the typical vascular bundle structure of bamboo disappears. The differences of fiber arrangement in the bundles of mono-podial and sympodial bamboos vanish. The xylem consists of only one metaxylem vessel. At the branching of vascular bundles, smaller cells develop abundantly with an intensive reticulate pitting.

Fig. 8.13 Vascular bundle
with two vessels and sieve
tubes with fiber sheaths
within ground parenchyma
(Liese 1998)

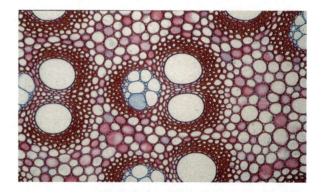

Fig. 8.14 Fine structural
details of a sieve tube and
companion cells, *Bambusa
vulgaris* (Liese 1998)

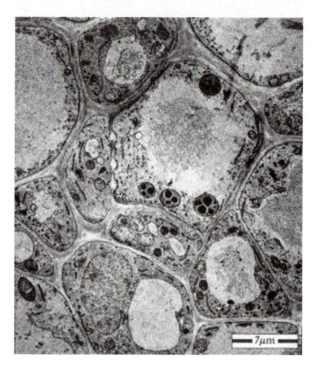

The ground tissue of the diaphragm consists of short parenchyma cells interspersed with sclerified cells. Tyloses are often present in both monopodial and sympodial bamboos.

Size and form of the various cell types in the nodal area differ considerably from those in the internode. The metaxylem vessels are much shorter than in the internodes; they are also smaller and often deformed. They possess several large simple perforations on their side walls and numerous small pits that are open for direct contact between vessels (Fig. 8.18). The sieve tubes appear generally longer than the vessels. Fibers and parenchyma cells in the nodal regions often contain

Fig. 8.15 Inner skin on the culm wall, *Phyllostachys bissettii* (Liese and Schmitt 2006)

Fig. 8.16 Inner culm wall consisting of the skin and the transition layer, *Cephalostachyum pergracile* (Liese and Schmitt 2006)

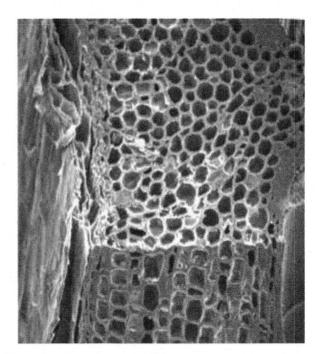

starch granules. Owing to the lack of radial conduction cells in the internodes, the vessels enable the necessary cross-transport of water and nutrients. The nodal structure is also important for the movement of liquid during drying and preservation and influences some physical and mechanical properties of the culm. Due to the shortening of the fibers in the nodal part and their simultaneous wall thickening, bamboo culms often break at the nodes when under tension. Nodes have a lower holocellulose content but more extractives, pentosans, lignin, and ash than the

Fig. 8.17 Structure of a
node with vascular
anastomoses (Ding and
Liese 1997)

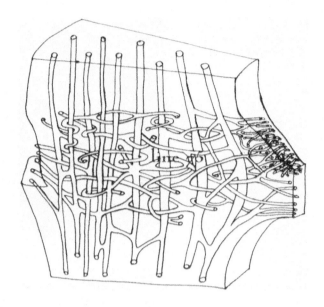

Fig. 8.18 Metaxylem
vessel with large, simple
perforations, *Pleioblastus
maculata* (Ding and Liese
1997)

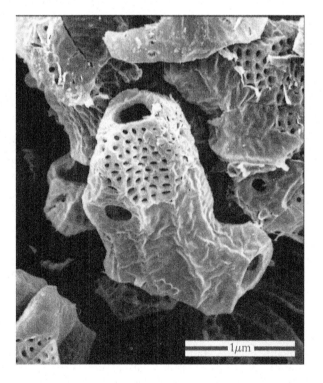

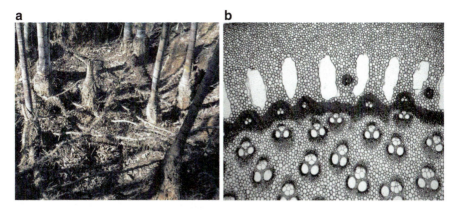

Fig. 8.19 (**a**) Rhizome of *Phyllostachys edulis*. (**b**) Cortex of rhizome with large air canals, *Phyllostachys heteroclada* (Liese 2012)

internode portions. Nodes produce pulp of lower strength but can hardly be excluded.

8.1.1.3 Structure of the Rhizome

Unlike other woody plants, bamboo has no main axis. The rhizome is one piece of the modified branch of a monopodial bamboo plant. Ding et al. (1992, 1993, 1997a) reviewed the terminology and suggested that only "running bamboo" has a genuine rhizome. It serves the uptake, transport, and storage of nutrients, as well as the vegetative reproduction. The rhizome initiates lateral buds which may develop into the culm shoots or into a new rhizome (Fig. 8.19a). The rhizome consists of internodes and nodes. From the nodes adventitious roots develop with root hairs as outgrowth of the epidermis for the uptake of water and mineral elements. However, the adventitious roots develop not only from the rhizome but also from the culm base, which bears another root system and performs the same function as the roots from the rhizome (Ding et al. 1997b).

The rhizome has a smaller diameter and shorter internodes than the culm. The anatomical structure, as revealed in a transverse section, is basically similar to the culm. Distinct differences exist regarding a thick cortex, random orientation of vascular bundles, poorly developed fiber strand, absence of a pith ring, and only a small pith cavity (Ding et al. 1992, 1993).

The epidermis of the rhizome is made up of long and short cells, like that of the culm. Abundant silica cells and stomata are present. Beneath the epidermis lies the hypodermis which consists of 2–3 layers of sclerenchymatous cells. Under the stomata, they are substituted by parenchyma cells as passage for gas exchange. The structure of vascular bundles in a rhizome is similar to that in the culm. Based on the arrangement of vascular bundles beneath the cortex and the development of air canals within the cortex, monopodial bamboos can be grouped into four types (Ding et al. 1997b).

The rhizome comprises about 62 % parenchyma cells, 20 % fibers, and 18 % conducting elements, significantly different from the culm tissue. The distribution of vascular bundles is related to the thickness of the rhizome. Form, number, and density of vascular bundles vary greatly from species to species. The inner part has mostly the lowest number, whereas the outer and center parts have a higher number but a similar distribution. Remarkable is the presence of large air canals in the cortex of several species, e.g., of *Phyllostachys heteroclada*, *P. nidularia*, and *P. stimulosa* where they are distinctly separated from the parenchyma ground tissue (Fig. 8.19b). The presence of air canals of some species shows their ability to grow along flooded river bands, contributing to their soil stabilization.

8.1.2 Structural Changes During Lifetime

During lifetime the culm undergoes an aging process, especially during its maturation period of 3–4 years, but also still later (Liese and Weiner 1997; Murphy and Alvin 1997). This process changes certain structures and, consequently, it influences properties and utilization. Fibers and also parenchyma cells exhibit a thickening of their walls by the deposition of additional lamellae on the existing wall layers with subsequent lignification (Fig. 8.20). The wall thickening is expressed by an increase of density and strength properties. Culms of 3–4 years of age are suitable for any utilization.

Younger, immature culms with lower lignin content can be more easily split. They are preferred for handicraft work. The lower lignin content is also beneficial

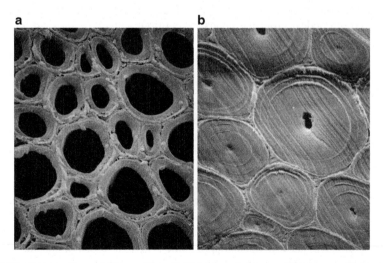

Fig. 8.20 Cross sections of fibers with wall thickening: (**a**) 1 year, (**b**) 6 years, *Phyllostachys viridiglaucescens* (Liese 2012)

Fig. 8.21 Flowering culms become brittle (Liese 2012)

for pulping. However, the harvest of young culms is detrimental for the vitality of the stand, which has to produce and store the amounts of energy for the growth of the next year's generation.

After around 8–10 years, senescence affects the functional efficiency of a culm, but not the technological properties. This natural aging occurs as blocking off the water-conducting vessels by tyloses and slime-like substances as well as the sieve tubes by callose occlusions and tylosoids. The functional inefficiency results in the dying of individual culms within a clump or grove.

Quite contrary appear the structural consequences for a dying culm after flowering. The tissue structure becomes brittle and the whole culm often bends down and breaks (Fig. 8.21). Since this phenomenon is not associated with any biodegradation, it must result from biochemical changes affecting the lignin-cellulose complex. In spite of the great impact for the utilization of the masses of dying culms, the processes are not yet fully understood (Liese 2008).

Site conditions influence more the morphological characters then anatomical parameters, which appear as rather stable (Latif and Liese 2001, 2002a). A higher fiber content may occur in drier areas and on slopes, resulting in higher density and increased strength properties. Fertilization affects shoot production, fiber diameter, and wall thickness (Abasolo et al. 2005), but apparently not the anatomical composition.

8.1.3 Wound Reactions of Culm and Rhizome

Bamboo culms and rhizomes respond to wounding in order to protect the surrounding tissue against damaging influences through the wound surface.

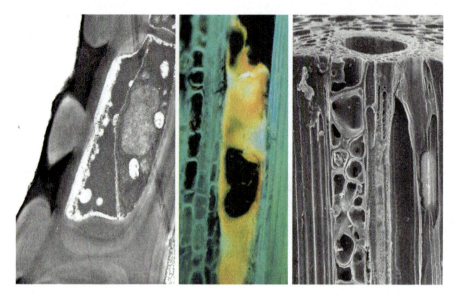

Fig. 8.22 Wound reactions for occlusion of vessels by slime and tyloses (Liese 1998; Liese and Kumar 2003)

The defense arsenal consists of quite a number of cellular reactions, like the closure of sieve tubes by callose; the formation of slime and tyloses, phenolics, and suberized wall layers; wall lignification; and also septa development in fibers (Fig. 8.22) (Weiner and Liese 1996; Liese and Weiner 1997).

The responses observed in monocotyledons are similar to those of hardwood species (Schmitt and Liese 1995; Liese and Dujesiefken 1996). However, two additional phenomena occur in injured bamboos: starch granules to be mobilized later accumulate in parenchyma cells and also in fibers. Furthermore, parenchyma as well as fibers show cell wall thickening by developing additional lamellae, at first unlignified and later lignified. Suberization is a well-known response in dicotyledons, but is hardly known in monocotyledons. A suberin layer develops in the vascular bundle parenchyma as well as in the ground parenchyma. Whereas in hardwoods, this layer appears to be a final lamella formed before cell death, in bamboo an additional wall layer is developed on the suberin layer with a peripheral cytoplasm present.

The extension of wound response is much limited laterally but increases axially with time. The latter could halt either in an internode above a node or extend even through into the adjacent internode. The extension of the wound reaction in an internode can be stopped and the functional tissue thus be protected against further inactivation.

Also for rhizomes, wound reactions occur in some kind of time schedule. Metaxylem vessels are filled with slime-like substances from the surrounding vascular parenchyma, parenchyma cells, and fibers that form additional lamella on their lignified cell wall, which become later lignified, and phenolic substances appear in sieve tubes and vascular parenchyma.

Tyloses in metaxylem vessels are not observed in the rhizome, whereas it is a general phenomenon in the protoxylem vessel of culms. It also develops in the metaxylem vessels as a reaction of wounding or aging. Underneath the wounded tissue of the rhizome, a special feature develops, as the living tissue is separated against the wounded tissue by a layer of parenchyma cells with thickened walls. Furthermore, fibers adjacent to the wounded parenchyma develop additional wall lamellae. Such demarcation is not observed in wounded culms, although as a zonation in dying branches. These differences may likely be associated with the anatomical structure of the rhizomes as a main functional organ of storage and transport of nutrients. They possess more parenchyma cells, more conducting tissue, less fibers, and a very small lacuna.

Due to the lack of a secondary meristem, bamboos as all monocotyledons do not form a barrier zone by developing a callus to close a wound. The term "compartmentalization" appears therefore not suitable for the defense system of bamboo as for conifers and hardwoods. Wounded bamboo culms also show a limitation of the wound response laterally, but not a defined axial blocking of the wound influence. These wound reactions resemble more of a gradual "fading out" of the cellular response (Ding et al. 1997a, b).

8.2   Chemical Compounds

The main constituents of the culm tissue are cellulose, hemicelluloses, and lignin; minor constituents consist of various soluble polysaccharides, proteins, resins, tannins, waxes, and a little amount of ashes. The main chemical constituents of bamboo are not too different to those of woody materials (Vena et al. 2013). The composition varies according to species, the conditions of growth, the age of the bamboo, and the part of the culm. Because the bamboo culm tissue matures within a year when the soft and fragile sprout becomes hard and strong, the proportion of lignin and carbohydrates is changed during this period. However, after full maturation of the culm, the chemical composition tends to remain rather constant. Table 8.2 gives an approximate chemical analysis for some bamboo species; small differences exist along a culm (Table 8.3).

The nodes contain less water-soluble extractives, pentosans, ash, and lignin but more cellulose than the internodes. The season influences the amount of water-soluble substances, which are higher in the dry season than in the rainy season. The starch content reaches its maximum in the driest months before the rainy season and before sprouting. The ash content (1–5 %) is higher in the inner than in the outer part. The silica content varies from 1 to 6 %, increasing from bottom to top. Most silica is deposited in the epidermis, and the tissues of the internodes almost contain none. Silica content affects the pulping properties of bamboo (Liese and Latif 2000).

Table 8.2 Chemical composition of some bamboos (Latif and Liese 1995)

Species	Holocellulose (%)	Alpha-cellulose (%)	Lignin (%)	Alcohol-benzene solubility (%)	1 % NaOH solubility (%)	Hot water (%)	Cold water (%)	Ash (%)
Gigantochloa levis	63.5–67.2	36.2–42.5	23.3–26.6	1.7–2.2	24.7–26.8	5.9–6.1	3.9–5.1	1.4–1.9
G. scortechinii	66.8–68.1	40.5–41.4	24.9–27.9	3.2–3.5	19.2–19.6	5.4–6.3	4.3–5.3	1.1–1.4
Bambusa vulgaris	67.8–69.6	37.9–43.2	22.7–23.9	3.9–4.5	20.6–23.1	5.7–5.9	3.4–5.6	1.8–2.1
B. blumeana	65.7–72.6	40.3–45.1	20.5–22.7	3.1–4.4	21.9–24.7	6.1–8.5	3.4–5.1	–
Schizostachyum zollingeri	68.8–74.3	48.7–52.6	21.1–22.9	2.2–2.7	21.8–26.8	3.7–6.5	2.7–5.4	–
Range of values for 10 Indian bamboo species	62.5–71.5	–	22.1–32.2	0.7–3.2	15.1–21.8	3.4–6.9	–	1.7–3.2

Table 8.3 Chemical composition of some bamboos at different heights

Species	Culm portion	Holocellulose (%)	Lignin (%)	Ash (%)	Alcohol-benzene (%)	Hot water extract (%)	1 % NaOH (%)
Bambusa sp. (1)	Top	78.8	25.0	2.4	3.6	12.4	25.9
	Middle	80.8	25.1	1.8	3.9	10.6	24.9
	Bottom	78.5	25.9	1.4	4.6	11.3	24.2
Bambusa blumeana (1)	Top	81.9	27.6	4.9	7.1	8.6	28.9
	Middle	81.2	27.6	5.8	6.7	6.9	28.7
	Bottom	81.7	26.3	3.4	8.7	9.1	28.2
Dendrocalamus asper (1)	Top	80.5	27.5	1.8	5.3	8.6	26.6
	Middle	78.4	26.9	1.4	11.3	11.8	29.0
	Bottom	79.7	28.2	0.9	8.5	9.1	21.8
Dendrocalamus strictus (1)	Top	79.3	27.4	2.8	6.2	12.7	27.0
	Middle	77.7	27.8	1.1	7.6	12.8	26.9
	Bottom	77.9	28.0	2.1	6.2	11.6	26.0
Gigantochloa albociliata (1)	Top	79.6	23.4	1.5	4.1	8.9	26.3
	Middle	80.8	27.2	2.4	6.2	12.4	27.9
	Bottom	79.6	27.3	1.5	4.0	8.3	23.2
Phyllostachys pubescens (2)	Top	54.1	24.7	1.2	6.0	7.0	25.6
	Middle	53.6	24.5	1.2	7.6	8.5	27.6
	Bottom	54.4	24.0	1.1	7.4	9.3	28.3

Data from: (1) Ratanophat (2004), (2) Li et al. (2007)

8.2.1 Holocellulose

Cellulose

The content of cellulose in bamboo of 40–60 % is higher than in hardwood. Vena et al. (2013) summarized for bamboo 40–53 % glucan, whereas the cellulose content of softwoods and hardwoods is 40–52 % and 38–56 %, respectively. As in other plants, it consists of linear chains of β-1-4-linked glucose anhydride units. The number of glucose units in one chain is referred to as the degree of polymerization (DP). The DP for bamboo is higher than for dicotyledonous woods where it has maximum of 15,000. Within the cell wall, cellulose is closely associated with hemicelluloses and lignin according to various models.

Hemicellulose

More than 90 % of the bamboo hemicelluloses consist of xylan (4-O-acetyl-4-O-methyl-D-glucuronoxylan) which is a characteristic feature of hardwoods. Xylan is a relatively short linear polymer (*Fagus sylvatica*: DP 200) consisting of β-1-4 linked xylose units with side chains of 4-O-methyl-D-glucuronic acid, L-arabinose, and acetyl groups (Puls 1992). Bamboo xylan accounts for approximately 25 % of the cell wall material. It contains 6–7 % acetyl groups, whereas hardwood xylan has 8–17 % (Vena et al. 2013). It is also different from the xylan found in gymnosperms with regard to the degree of branching and molecular properties. Due to the presence of arabinose, the xylan is closer to that of softwoods. Thus, the bamboo xylan is intermediate between hardwood and softwood xylans. These results indicate that the bamboo xylan has the unique structure of Gramineae (Higuchi 1980). A chemical analysis of bamboo xylan is given in Table 8.2.

8.2.2 Lignin

After cellulose, lignin represents the second most abundant constituent in bamboo, and much interest has been focused on its chemical nature and structure. Lignin is a macromolecular 3-dimensional polymer. Bamboo lignin belongs to the grass (monocotyledons) type of lignin consisting of three types of phenyl-propane units: p-hydroxyphenyl (H), guaiacyl (G), and syringyl (S) units interconnected through biosynthetic pathways in a molar ratio of 10:68:22. This shows that bamboo lignin is qualitatively but not quantitatively similar to hardwood lignin.

Bamboo grows very rapidly and completes height growth within a few months by reaching the full size. The growing bamboo shows various lignification stages from the bottom to the top portions of the same culm (Itoh and Shimaji 1981). The lignification within every internode proceeds downward from top to bottom, whereas transversely it proceeds from inside to outside. During the height growth, lignification of epidermal cells and fibers precedes that of ground tissue parenchyma. Full lignification of a bamboo culm is completed within one growing season, showing no further aging effects. No difference has been detected in the lignin composition between vascular bundles and parenchyma tissue (Higuchi et al. 1966). Bamboo has been chosen as one of the suitable plants to study the biosynthesis of lignin. Initially, these investigations were almost exclusively based on feeding experiments with radioactive precursors, and it has been found that lignin is synthesized from glucose formed by photosynthesis via the "shikimic acid pathway" (Higuchi 1969). Several key enzymes involved in the synthesis of shikimic acid were isolated from bamboo shoots (Fengel and Shao 1984, 1985).

8.2.3 Extractives

The extractives—both organic and inorganic—show a certain species relationship, but they are more influenced by age and season. The appearance of the epidermis with its apposition of cutin and wax can be species dependent, but as a taxonomic parameter, it needs additional anatomical criteria.

The starch content not only varies with season but also shows a relation to species. Bambusa vulgaris and Dendrocalamus asper are known to be rich in starch, whereas Gigantochloa atter has a lower content. Also, the ash content and, here especially, the amount of silica vary between species, influenced by age and site. *Bambusa vulgaris* has a low silica content, while *Schizostachyum* has a higher one (Tamolang et al. 1980). Although these differences influence certain properties such as the taste of shoots and pulping and cutting properties, they have no taxonomic value.

8.3 Physical and Mechanic Properties

8.3.1 Moisture Content

The moisture content of the bamboo culm and derived products influences the dimensional stability of the bamboo material. Moisture content also affects toughness, density, strength, working properties, and durability. The water contained in bamboo can be expressed as percentage of either dry weight (moisture content) or wet weight (water content). For most purposes, moisture content based on the ovendry weight is used. It is defined as follows:

$$\text{Moisture content (MC\%)} = \frac{\text{wet weight- dry weight}}{\text{dry weight}} \times 100$$

The term water content is mainly used for solid fuel material or for raw material used for pulping purposes. Water content is defined as follows:

$$\text{Water content (MC\%)} = \frac{\text{wet weight- dry weight}}{\text{wet weight}} \times 100$$

The moisture content varies within one culm and is influenced by its age, the season of felling, and the species. In the green stage, greater differences exist within one culm as well as in relation to age, season, and species. Young, 1-year-old shoots have a high relative moisture content of about 120–130 % both at bottom and top. The nodes, however, show lower values than the internodes. These differences can amount to 25 % and are larger at the base than at the top. In 3–4-year-old culms, the base has a higher moisture content than the top. The moisture content across the culm wall is higher in the inner than in the outer part (Latif and Liese 2002a, b; Islam et al. 2002; Kamruzzaman et al. 2008).

The season has a great influence on the moisture content of the culm, with a minimum at the end of the dry period, followed by a maximum in the rainy season. During this period the culm can double its moisture content. The variation due to the season is higher than the differences between base and top as well as between species. Among species the moisture content varies even in the same locality. This is mainly due to the variation in the amount of parenchyma cells, which corresponds to the water holding capacity (Liese and Grover 1961). The considerable differences in the moisture content of freshly felled culms have to be considered when determining the yield of bamboo expressed by its fresh weight.

Being a hygroscopic material, bamboo absorbs or loses moisture until it is in balance with the surrounding climate conditions. The amount of moisture at this point of balance is called the equilibrium moisture content (EMC). The EMC depends mainly on the relative humidity and temperature of the surrounding air.

Many studies on physical properties (Sulthoni 1989; Sattar et al. 1994; Hamdan et al. 2007; De Vos 2010) showed that the EMC of bamboo is very similar to wood. Thus, the computed data of wood, relating EMC to temperature and relative humidity, can also be used for the bamboo-moisture relationship.

8.3.2 *Fiber Saturation Point and Shrinkage*

The fiber saturation point (FSP) is defined as the moisture content at which the cell walls are fully saturated without any free water in the cell cavities. In bamboo, the FSP is influenced by the composition of the tissue and the amount of chemical constituents. Older references state that the mean FSP of bamboo in general is around 17–25 % (Ota 1955; Kishen et al. 1958; Sharma 1988; Hamdan et al. 2007). But, due to the low content of hydrophobic extractives found in bamboo, one would expect an FSP between 28 and 32 % similar to wood species which do not form heartwood. The low FSP values reported in literature are probably a result of the problem of maintaining a 100 % relative humidity atmosphere over longer periods of time while determining the EMC at such condition.

Shrinkage of bamboo is the basic cause of many problems that occur during drying of culms and during their service life. Unlike most wood species, shrinkage of bamboo starts to become apparent in a decrease of both cell wall thickness and cell diameter which is due to capillary forces leading to cell collapse as soon as moisture begins to decrease. But this apparent shrinkage does not continue regularly. When the average moisture content diminishes to 70–40 %, shrinkage stops, because no more fully saturated cells are present. But, as soon as moisture content of the cell walls drops below FSP, shrinkage is initiated again. Parenchyma tissue shrinks less in bamboo than in timber, while vascular fibers shrink as much as in timber of same specific gravity.

Bamboo tissue shrinks mainly in radial direction, whereas the minimum deformation occurs in axial direction. The tangential shrinkage is higher in the outer parts of the wall than in the inner parts. The shrinkage of the whole wall appears to be governed by the shrinkage of the outermost portion, which possesses also the highest specific gravity. Mature culms shrink less than immature ones.

Parenchyma tissue shrinks less in bamboo than in timber, while vascular fibers shrink as much as in timbers of the same specific gravity. When the average moisture content is low (below FSP), swelling due to absorption of water is almost equal to shrinkage due to adsorption, which explains that the apparent shrinkage observed in the high moisture content range is mainly caused by cell collapse. The percentage of swelling decreases with an increase of basic density (Kishen et al. 1958; Sekhar and Rawat 1964).

Table 8.4 Physical properties of some Malaysian bamboo species (Latif and Liese 1995)

Species	Moisture content (%)	Ovendried density (g/cm^3)	Shrinkage (%)	
			Radial	Tangential
Bambusa blumeana	57–97	0.43–0.60	5–10	6–20
B. vulgaris	79–118	0.27–0.57	6–11	10–20
B. heterostachya	92–132	0.44–0.58	19–27	6–11
Dendrocalamus asper	28–105	0.55–0.78	4–9	6–11
Gigantochloa levis	30–77	0.65–0.94	5–14	2–10
G. scortechinii	79–108	0.47–0.60	7–14	14–19

Compared to wood, the anisotropy of shrinkage (relation of dimensional change between tangential and radial direction) of bamboo is much smaller. Whereas tangential shrinkage of wood is almost double as shrinkage in radial direction, apparent shrinkage of bamboo in radial direction is often reported to be greater or equal to shrinkage in tangential direction. This can be attributed to two effects: First, bamboo does not contain any radially oriented cells, like ray cells in the case of wood, which could act as structural components restraining shrinkage in radial direction; the result is similar shrinkage in radial and tangential direction. Second, when bamboo culms dry, the main moisture transport is radially directed toward the lacuna, and collapse becomes visible mainly in the radial direction.

8.3.3 Density

The density of bamboo varies from about 0.4 to 0.9 g/cm^3 (Tables 8.4 and 8.5) depending on the anatomical structure such as the quantity and distribution of fibers around the vascular bundles (Zhou 1981; Razak et al. 1995; Qisheng et al. 2002). Accordingly, density increases from the inner layer to the outer part of the culm and along the culm from the bottom to the top (Liese 1985; Nordahlia et al. 2012).

8.3.4 Mechanical Properties

The strength of bamboo generally increases with the thickening of the fiber walls until maturity is reached after approximately 3 years, but also later on (Liese 1987a, b). The selection of suitable bamboo species and age in addition to other factors, such as site and season, influencing the strength properties, is of utmost importance.

The variation of density within a culm and between species has a major effect on the strength (Espiloy 1987; Janssen 1985; Anwar et al. 2005). With the exception of the modulus of rupture (MOR), shear strength, compression strength parallel to grain, modulus of elasticity (MOE), and stress at proportional limit correlate

Table 8.5 Physical properties of some Myanmar bamboo species (Sint et al. 2008)

| Species | Ovendried density (g/cm³) | Shrinkage (%) | | | |
		Diameter (%)	Wall thickness (%)	Longitudinal (%)	Volumetric (%)
Bambusa longispiculata	0.73	6.7	10.3	0.17	16.8
B. burmanica	0.76	12.4	14.8	0.20	25.1
Dendrocalamus calostachyus	0.77	9.0	8.1	0.26	23.5
D. giganteus	0.68	4.8	5.0	0.19	9.0
D. hamiltonii	0.72	6.9	11.4	0.12	20.3
D. maclellandii	0.83	7.3	10.2	0.12	17.7
Melocanna baccifera	0.61	7.9	11.9	0.14	26.8
Thyrsostachys oliveri	0.80	5.5	6.4	0.16	11.2
T. siamensis	0.81	7.2	9.9	0.13	20.6

$n = 108$ (6 clumps per species, 6 culms per clump, and 3 replications per culm)

positively with height increment of the culms. The negative correlation of MOR with culm height indicates that the maximum bending load of culms with a thinner wall is much lower than of the basal portion of a culm with a thicker wall due to an unfavorable moment of inertia (Latif and Liese 1995; Kamruzzaman et al. 2008).

The vascular bundle distribution correlates positively with all the strength properties except MOR. This implies that the increase in the amount of the vascular bundles will be accompanied by an increment of density and thus with an increase of strength properties. The decrease in the size of vascular bundles (higher radial/tangential ratio) is also associated with the increment of strength properties. The fiber wall thickness correlates positively with compression strength, stress at proportional limit, and MOE, but negatively with MOR.

For bamboo to be used as an engineering material in structural applications, strength data should be generated (Table 8.6). But, up to now, generally accepted and universally applicable test methods do not exist. It is worth mentioning that the strength values of bamboo obtained from tests using three types of samples (round or split, long and short) are significantly different from each other (Table 8.7). Therefore, strength data without reference to the test mode and sample size is of little help. As different bamboo species have different diameters and wall thicknesses, for relative comparison purposes, testing bamboo in split form would be more appropriate than testing bamboo in round form in short span (Gnanaharan et al. 1994).

The specification for determining the physical and mechanical properties of bamboo is provided into the two international standards: ISO/TC 165 N315 (Anonymous 2001) applying for samples from round bamboo and ISO 22157–1 (INBAR 2004) for using samples from split bamboo.

Table 8.6 Mechanical properties of air-dried bamboos (12 % moisture content, according to standard ISO/TC 165 N315) (Sint et al. 2008)

Species	Static bending		Compression parallel to grain		Shear	
	MS (N/mm^2)	MOE (N/mm^2)	MS (N/mm^2)	MOE (N/mm^2)	MS (N) (N/mm^2)	MS (WN) (N/mm^2)
Bambusa longispiculata	86.8	31,451	57.6	1,326	8.3	6.3
B. burmanica	137.0	36,096	56.9	1,946	15.2	18.2
Dendrocalamus calostachyus	71.9	24,392	48.4	2,110	10.5	10.8
D. hamiltonii	70.9	18,860	65.5	3,404	15.3	15.1
D. giganteus	32.1	6,556	60.9	2,117	13.3	14.4
D. maclellandii	126.0	41,899	60.0	1,824	25.0	21.6
Melocanna baccifera	58.9	20,330	44.9	2,124	15.2	13.5
Thyrsostachys oliveri	64.4	22,396	59.1	2,322	17.9	16.4
T. siamensis	119.6	30,033	64.0	2,349	21.7	19.0

$n = 108$ (6 clumps per species, 6 culms per clump, and 3 replications per culm)
MS maximum stress, *MOE* modulus of elasticity, *N* specimens with node, *WN* specimens without node

Table 8.7 Bending strength of Guadua bamboo (Gnanaharan et al. 1994)

Type of samples	MOR (N/mm^2)	MOE (N/mm^2)
Round long (3 m)	72.6	17,608
Round short (0.7 m)	54.0	7,363
Split at the bottom	82.1	9,194
Split at the top	113.9	11,996

MOR modulus of rupture, *MOE* modulus of elasticity

Acknowledgments Our thanks are expressed to Prof. Dr. Dieter Eckstein, Prof. Dr. Olaf Schmidt, and Dr. Johannes Welling for their review and valuable comments.

References

Abasolo WP, Fernandez EC, Liese W (2005) Fiber characteristics of *Gigantochloa levis* and *Dendrocalamus asper* as influenced by organic fertilizers. J Trop For Sci 17(2):297–305

Anonymous (2001) Laboratory manual on testing methods for determination of physical and mechanical properties of bamboo. ISO/TC 165 N315

Anwar UMK, Zaidon A, Hamdan H, Tamizi MM (2005) Physical and mechanical properties of *Gigantochloa scortechinii* bamboo splits and strips. J Trop For Sci 17(1):1–12

De Vos V (2010) Bamboo material properties and market perspectives. Thesis report of Larenstein University, pp 19–48

Ding YL, Liese W (1997) Anatomical investigations on the nodes of bamboo. In: Soc L, Chapman G (eds) The bamboos. Academic Press, London, pp 265–279

Ding YL, Grosser D, Liese W, Hsiung W (1992) Anatomical studies on the rhizome of monopodial bamboos. In: Bamboo and its use. Proceedings international symposium on industrial use of bamboo, Beijing, China, 7–11 Dec 1992, pp 143–150

Ding YL, Tang GG, Chao CS (1993) Anatomical studies on the rhizome of some monopodial. Chin J Bot 5(2):122–129

Ding YL, Weiner G, Liese W (1997a) Wound reactions in the rhizome of *Phyllostachys edulis*. Acta Bot Sin (Beijing) 39(1):55–58

Ding YL, Tang GG, Chao SS (1997b) Anatomical studies on the culm neck of some pachymorph bamboos. In: Chapmann G (ed) The bamboos. Linnaean Society, London, pp 285–292

Espiloy ZB (1987) Physico-mechanical properties and anatomical relationships of Philippine bamboo. In: Rao AN, Dhanarajan G, Sastry CB (eds) Recent research on bamboos. Proceedings of the 3rd international bamboo workshop, Hangzhou, China, 6–14 Oct 1985. Chinese Academy of Forestry, Beijing, China; International Development Research Centre, Ottawa, Canada, pp 257–264

Fengel D, Shao X (1984) A chemical and ultrastructural study of the bamboo species *Phyllostachys makinoi* Hay. Wood Sci Technol 18:103–112

Fengel D, Shao X (1985) Studies on the lignin of the bamboo species *Phyllostachys makinoi* Hay. Wood Sci Technol 19:131–137

Gnanaharan R, Janssen JJA, Arce O (1994) Bending strength of Guadua bamboo - comparison of different testing procedure. INBAR working paper no. 3. Kerala Forest Research Institute (KFRI), India and International Development Research Centre (IDRC), Canada

Grosser D, Liese W (1971) On the anatomy of Asian bamboos, with special reference to their vascular bundles. Wood Sci Technol 5:290–312

Grosser D, Liese W (1973) Present status and problems of bamboo classification. J Arnold Arboretum 54(2):293–308

Grosser D, Liese W (1974) Verteilung der Leitbündel und Zellarten in Sproßachsen verschiedener Bambusarten. Holz Roh- Werkstoff 32:473–482

Hamdan H, Hill CAS, Zaidon A, Anwar UMK, Latif A-M (2007) Equilibrium moisture content and volumetric changes of *Gigantochloa scortechinii*. J Trop For Sci 19(1):18–24

Higuchi T (1969) Bamboo lignin and its biosynthesis. Wood Res Kyoto 48:1–14

Higuchi T (1980) Chemistry and biochemistry: bamboo for pulp and paper of bamboo. In: Lessard G, Chouinard A (eds) Bamboo research in Asia. IDRC, Ottawa, pp 51–56

Higuchi T, Kirnura N, Kawamura I (1966) Difference in chemical properties of lignin of vascular bundles and of parenchyma cells of bamboo. Mokuzai Gakkaishi 12:173–178

INBAR (2004) International Standards Organization - ISO (2004) ISO 22157–1:2004 bamboo – determination of physical and mechanical properties – part 1: requirements. ISO, Geneva

Islam MN, Hannan MO, Lahiry AK (2002) Effect of age and height positions of borak, jawa and mitingabamboo on their physical properties. J Timber Dev Assoc India 48:16–22

Itoh T, Shimaji K (1981) Lignification of bamboo culm (Phyllostachys pubescens) during its growth and maturation. In: Higuchi T (ed) Bamboo production and utilization. Proceedings of the 17th IUFRO world congress, Kyoto, Japan, 6–17 Sept 1981. Wood Research Institute, Kyoto University, Kyoto, Japan, pp 104–110

Janssen JJA (1985) The mechanical properties of bamboo. In: Rao AN, Dhanarajan G, Sastry CB (eds) Recent research on bamboos. Proceedings of the 3rd international bamboo workshop, Hangzhou, China, 6–14 Oct 1985, pp 250–256

Jiang Z (2007) Bamboo and rattan in the world. China Forestry Publishing House, Beijing, China

Kamruzzaman M, Saha SK, Bose AK, Islam MN (2008) Effect of age and height on physical and mechanical properties of bamboo. J Trop For Sci 20(3):211–217

Kishen J, Ghosh DP, Rehman MA (1958) Studies in moisture content, shrinkage, swelling and intersection point of mature *Dendrocalamus strictus* (male bamboo). Indian For Rec Dehra Dun, India 1(2):11–30

Latif A-M, Liese W (1995) Utilization of bamboo. In: Razak A-O, Latif A-M, Liese W, Norini H (eds) Planting and utilization of bamboo in Peninsula Malaysia. FRIM research pamphlet no. 118. Forest Research Institute Malaysia, Kuala Lumpur, Malaysia, pp 50–102

Latif A-M, Liese W (2001) Anatomical features of *Bambusa vulgaris* and *Gigantochloa scortechinii* from four harvesting sites in Peninsular Malaysia. J Trop For Prod 7(1):10–28

Latif A-M, Liese W (2002a) Culm characteristics of two bamboos in relation to age, height and site. In: Kumar A, Ramanuja Rao IV, Sastry Ch (eds) Bamboo for sustainable development. Proceedings of the 5th international bamboo congress and 6th international bamboo workshop, San José, Costa Rica, 2–6 Nov 1998. VSP and INBAR, pp 223–233

Latif A-M, Liese W (2002b) The moisture content of two Malaysian bamboos in relation to age, culm height, site and harvesting month. In: Kumar A, Ramanuja Rao IV, Sastry Ch (eds) Bamboo for sustainable development. Proceedings of the 5th international bamboo congress and 6th international bamboo workshop, San José, Costa Rica, 2–6 Nov 1998. VSP and INBAR, pp 257–268

Li XB, Shupe TF, Peter GF, Hse CY, Eberhardt TL (2007) Chemical changes with maturation of the bamboo species *Phyllostachys pubescens*. J Trop For Sci 19(1):6–12

Liese W (1985) Bamboos - biology, silvics, properties, utilization. Deutsche Gesellschaft für Technische Zusammenarbeit (GTZ) Schriftenreihe Nr. 180, TZ Verlagsges, Roßdorf

Liese W (1987a) Anatomy and properties of bamboo. In: Rao AN, Dhanarajan G, Sastry CB (eds) Recent research on bamboos. Proceedings of the 2nd international bamboo workshop, Hangzhou, China, 6–14 Oct 1986, pp 196–208

Liese W (1987b) Research on bamboo. Wood Sci Technol 21:189–209

Liese W (1998) The anatomy of bamboo culms. International Network for Bamboo and Rattan (INBAR), Technical report no. 18, Beijing, China

Liese W, Kumar S (2003) Bamboo preservation compendium. International Network for Bamboo and Rattan (INBAR), Beijing

Liese W (2004a) Structures of a bamboo culm affecting its utilization. Part 1: bamboo industrial utilization. In: Proceedings of international workshop on bamboo industrial development and utilization, Xianning, China, 12 Nov 2003. INBAR 2004, Beijing, China, pp 1–8

Liese W (2004b) Preservation of bamboo structures. Ghana J For 15&16:40–48

Liese W (2008) The blooming of *Melocanna baccifera* in Northeast India and its consequences. Bamb Bull Bamb Soc Aust 10(1):20–22

Liese W (2012) A personal reflection on 60 years of bamboo passion and work. In: Proceedings of the 9th world bamboo congress, Antwerp, Belgium, 10–15 Apr 2012. Keynote lecture, pp 29–56

Liese W, Dujesiefken D (1996) Wound reactions in trees. In: Raychaudhuri SP, Maramorosch KK (eds) Forest trees and palms-disease and control. Oxford, New Delhi, pp 21–35

Liese W, Grosser D (1972) Untersuchungen zur Variabilität der Faserlänge bei Bambus. Holzforschung 26(6):202–211

Liese W, Grosser D (2000) An expanded typology for the vascular bundles of bamboo culms. In: Puangchit L, Thaiutsa B, Thamnicha S (eds) Proceedings of international symposium on bamboo 2000, Chiangmai Thailand, 2–4 Aug 2000. Royal Project Foundation. Kasetsart University, Royal Forestry Department, ICDF, ROC, pp 121–134

Liese W, Grover PN (1961) Untersuchungen über den Wassergehalt von indischen Bambushalmen. Berichte Deutsche Botanische Gesellschaft 74:105–117

Liese W, Latif A-M (2000) The starch content of two Malaysian bamboos in relation to age, culm height, site and harvesting month. In: Proceedings of the 21st IUFRO world congress, Kuala Lumpur, Malaysia, 7–12 Aug 2000. Poster Abstracts 3: p 261

Liese W, Schmitt U (2006) Development and structure of the terminal layer in bamboo culms. Wood Sci Technol 40(1):4–15

Liese W, Weiner G (1997) Modifications of bamboo culm structures due to ageing and wounding. In: Soc L, Chapman G (eds) The bamboos. Academic, London, pp 313–322

Magel E, Kruse S, Lütje G, Liese W (2005) Soluble carbohydrates and acid invertases involved in the rapid growth of developing culms in *Sasa palmata* (Bean) Camus. Bamb Sci Cult 19:23–29

Murphy RJ, Alvin KL (1997) Fiber maturation in bamboos. In: Soc L, Chapman G (eds) The bamboos. Academic, London, pp 293–303

Nordahlia AS, Anwar UMK, Hamdan H, Zaidon A, Paridah MT, Razak A-O (2012) Effects of age and height on selected properties of Malaysian bamboo (*Gigantochloa levis*). J Trop For Sci 24(1):102–109

Ota M (1955) Studies on the properties of bamboo stem. Part 11: on the fiber saturation point obtained from the effect of moisture content on the swelling and shrinkage of bamboo splint. Bull Kyushu Univ For (Japan) (24):61–72

Parameswaran N, Liese W (1975) On the polylamellate structure of parenchyma wall in *Phyllostachys edulis* Riv. IAWA Bull 4:57–58

Parameswaran N, Liese W (1976) On the fine structure of bamboo fibers. Wood Sci Technol 10(4): 231–246

Parameswaran N, Liese W (1977a) Occurrence of warts in bamboo species. Wood Sci Technol 11(4):313–318

Parameswaran N, Liese W (1977b) Structure of septate fibers in bamboo. Holzforschung 31(2): 55–57

Puls J (1992) α-Glucuronidases in the hydrolysis of wood xylans. In: Visser J, Beldman G, Kusters-van Someren MA, Voragen AGJ (eds) Xylans and xylanases. Progress in biotechnology, vol 7. Elsevier, Amsterdam, pp 213–224

Qisheng Z, Shenxue J, Yongyu T (2002) Industrial utilization on bamboo. Technical report no. 26. International Network for Bamboo and Rattan (INBAR), Beijing, China

Ratanophat C (2004) Sustainable management and utilization from bamboo. Final technical report. Project: PD 56/59 Rev. 1(I) Promotion of the utilization of bamboo from sustainable sources in Thailand. Royal Forest Department and International Tropical Timber Organization, Bangkok, Thailand

Razak A-O, Latif A-M, Liese W, Naron N (1995) Planting and utilization of bamboo in Peninsular Malaysia. Forest Research Institute Malaysia (FRIM), Kuala Lumpur, Malaysia. Techical pamphlet no. 118

Sattar MA, Kabir MF, Battacharjee DK (1994) Effect of age and height position of muli (Melocanna baccifera) and borak (Bambusa balcooa) bamboos on their physical and mechanical properties. In: Proceedings of the 4th international bamboo workshop bamboo in Asia and the Pacific, Chiangmai, Thailand, 27–30 Nov 1991. International Development Research Centre, Ottawa, Canada, and Forestry Research Support Program for Asia and the Pacific, Bangkok, Thailand, pp 183–187

Schmitt U, Liese W (1995) Wundreaktionen im Xylem einiger Laubbäume. Drevarsky Vyskum 40(4):1–10

Sekhar AC, Rawat MS (1964) Some studies on the shrinkage of *Bambusa nutans*. Indian For Dehra Dun 91:182–188

Sharma SN (1988) Seasoning behavior and related properties of some Indian species of bamboo. Indian For 114(10):613–621

Sint KM, Hapla F, Myint CC (2008) Investigation on physical and mechanical properties of some Myanmar bamboo species. J Bamb Rattan 7(3&4):183–192

Sulthoni A (1989) Bamboo: physical properties, testing methods and means of preservation. In: Bassali AV, Davies WG (eds) Proceedings of a workshop on design and manufacturing of bamboo and rattan furniture, Jakarta, Indonesia, 3–14 Mar 1989, pp 1–15

Tamolang FN, Lopez FR, Semana JA, Casin RF, Espiloy ZB (1980) Properties and utilization of Philippine bamboos. In: Lessard G, Chouinard A (eds) Bamboo research in Asia. Ottawa, IDRC, pp 189–200

Vena PF, Brienzo M, del García-Aparicio MP, Rypstra T (2013) Hemicelluloses extraction from giant bamboo (*Bambusa balcooa* Roxburgh) prior to kraft or soda-AQ pulping and its effect on pulp physical properties. Holzforschung 67:863–870

Weiner G, Liese W (1996) Wound reactions in bamboo culms and rhizome. J Trop For Sci 9: 379–397

Zhou FC (1981) Studies on physical and mechanical properties of bamboo woods. J Nanjing Technol Coll For Prod 2:1–32

Chapter 9
Preservation and Drying of Bamboo

Walter Liese and Thi Kim Hong Tang

Abstract Bamboo culms are a unique building material for various kinds of structures. A wider acceptance of bamboo for structural uses, however, is often hindered by its propensity to biological degradation. The preservation of bamboo structures against biological hazards is an important requirement for utilizing this valuable lignocellulose resource. Compared with the preservation of timber in tropical countries, there are certain similarities, but also considerable differences.

Drying of bamboo before use is necessary since dry bamboo is stronger and less susceptible to biological degradation than moist bamboo. Furthermore, shrinkage and swelling are directly related to the moisture content. Moist bamboo affects the processing, such as machining, gluing and painting. Dimensional changes would ultimately occur if bamboo has not been dried before being used. The bamboo should be dried to the equilibrium moisture content corresponding to the service conditions before the manufacturing process.

Keywords Natural durability • Fungi • Beetles • Termites • Marine borers • Preservation methods • Chemical treatment • Non-chemical treatment • Preservatives • Air drying • Kiln drying

9.1 Natural Durability

Bamboo is one of the strongest bio-based structural materials, but succumbs to physical and mechanical damage by some non-biotic factors and is susceptible to degradation by similar organisms which attack wood. Moreover, bamboo is more likely to biodeteriorate due to its starch content.

W. Liese (✉)
Department of Wood Science, University of Hamburg, Leuschnerstr. 91, 21031 Hamburg, Germany
e-mail: wliese@aol.com

T.K.H. Tang
Department of Wood Science &Technology, Nong Lam University, Ho Chi Minh City, Vietnam
e-mail: kimhongtang@yahoo.com

© Springer International Publishing Switzerland 2015
W. Liese, M. Köhl (eds.), *Bamboo*, Tropical Forestry 10,
DOI 10.1007/978-3-319-14133-6_9

There are quite a number of general and detailed information on causes of bamboo deterioration and methods for protection in Willeitner and Liese (1992), Kumar et al. (1994), Liese and Kumar (2003), Liese (2004a, b), Jiang (2007) and Tang (2009).

The service life of bamboo structures is considerably dependent on the rate of biological degradation. Generally, the natural durability of bamboo is very low and influenced by species, environmental conditions and nature of use. Untreated bamboo has an average life of less than 1 year when exposed to outside conditions and soil contact. Under cover, it may last 4–5 years and longer or even 'forever' under favourable conditions. Split bamboo due to an easier access to the parenchymatic tissue is more rapidly destroyed than culms. The bottom part of a culm has a higher durability than the middle and top portions, and the inner part of the culm is easier attacked than the outer one (Liese and Kumar 2003; Liese 2004a, b, c).

In tropical humid areas, enormous quantities of bamboo culms stored in forest depots and mill yards decay and deteriorate. The severity of decay and biodeterioration depends on the duration of storage, bamboo species and environmental and storage conditions. Degradation of bamboo materials by fungi is a serious problem for bamboo factories during storage, processing and overseas transport of culms and bamboo products (Figs. 9.1 and 9.2) (Tang 2013).

9.1.1   Abiotic Factors

9.1.1.1   Cracks and Splits

Mechanical and physical damage of the culm occurs rarely because of its hard skin. Due to drying stresses mainly in young culms, collapse may arise as a serious defect. Cracks and splits can take place particularly when the culms are stored outdoors, and fungal deterioration may become evident (Fig. 9.1). Cracks can develop if the culm as part of a construction component is exposed to intensive sun. Such cracks may not have much influence on the tensile strength of the culm, but can lead to subsequent deterioration by fungi and insects. Since such cracks often develop on the 'weather side' they are equally exposed to rain so that water collects inside the culm's lacuna. In such a humid chamber, fungi will thrive resulting in rapid deterioration. Splitting cannot be prevented by any chemical treatment because it is a consequence of shrinkage-induced stresses due to too rapid drying.

Bamboo is also subject to mechanical wear caused by rope/wire friction when used for fastening bamboo components. Nailing is a frequently used method for fastening bamboo. Species with thick walls, like *Guadua* spp., appear to tolerate nailing better than thin-walled species.

9.1.1.2   Weathering

Weathering of exposed bamboo structures is a result of an interaction of different atmospheric conditions, such as ultraviolet radiation, surrounding temperature and

Fig. 9.1 Cracks during storage with subsequent fungal decay (Liese and Kumar 2003)

Fig. 9.2 Blue stain on stored culms (Liese and Kumar 2003)

moisture content. Severe temperature and relative humidity changes have an extremely deleterious effect on bamboo because sudden fluctuating atmospheric conditions may produce steep moisture gradients between surface and inner layers, resulting in surface cracks due to repeated swelling and shrinkage. Direct exposure to sun also causes checks in bamboo due to unbalanced shrinkage. Subsequently, water contributes further to cracks and splits.

A major factor in bamboo weathering in outdoor settings is damage by UV and visible light radiation which causes photodegradation. Such radiation breaks down bonds of the lignocellulosic polymer causing the bamboo surface to turn grey and coarse.

9.1.1.3   Fire

Fire is a great danger to bamboo constructions. The material burns easily and the hollow culms explode loudly due to heat expansion, from which the name 'bamboo' may have originated. Since bamboo is utilized in modern constructions, corresponding building codes and regulations have to be considered. Approval to use the giant culms of *Guadua angustifolia* as constructional elements besides wooden stems for the impressive ZERI Pavilion at the EXPO 2000 in Hannover, Germany, was granted only after stringent flammability tests were passed.

Whereas the flammability of timber can be reduced with fire retardants applied by pressure treatment, such protection for bamboo is hardly possible because of its refractory nature. Fire-retardant chemicals have to penetrate into the culm tissue and are effective only at a high retention. A surface application on the culm by paint appears questionable, as these decorative coatings are non-fixing. Additionally, economic factors hinder such a fire-protective treatment.

9.1.2   Biotic Factors

9.1.2.1   Fungi

9.1.2.1.1   General

Fungi cause discolouration and decay of bamboo. There are many types that infest and attack bamboo under different environmental conditions. All fungi use the chemical components of the culm cells, either from cell contents (moulds and blue-stain fungi) or from the cell wall (rot fungi), as their energy source. Fungi originate from spores produced asexually from the mycelium (Deuteromycetes) or sexually from fruit bodies (Ascomycetes, Basidiomycetes). As spores can be assumed as omnipresent, they will germinate everywhere under suitable conditions to thin hyphae. The growing hyphae live from the bamboo substances by enzymatic decomposition and develop to mycelium. The effect will often be recognized only at a later stage when substantial decomposition has resulted in discolouration and physical damage (Fig. 9.1). Mould fungi grow on the bamboo surfaces. Blue-stain and decay fungi (soft-rot, brown-rot, white-rot fungi) grow mostly inside the substrate, and some develop mycelium on its surface especially under humid conditions. Later, fruit bodies may be formed on the outside for the release of new spores.

Although fungal spores are present everywhere, they require certain conditions for germination, further growth and for the enzymatic degradation of the substrate.

The required moisture content of the material must be from 40 to 80 % of the oven-dry mass. Dry bamboo with a moisture content below 20 %, which is well below fibre saturation, is not vulnerable to attack.

The temperature range in regions where bamboo is grown and used is well-suited for fungal activities. Direct sun exposure of the mycelium with temperatures above 55 ° C may lead to destruction of the enzyme system by protein denaturation. There are however several decay fungi whose mycelia survive 95 °C for some hours (Schmidt 2006; Wei 2014).

Fig. 9.3 Moulded bamboo materials at a bamboo factory in Vietnam (Tang 2013). (**a**) Moulded fresh culms during storage. (**b**) Surface of a table infected by moulds. (**c, d**) Moulded culm parts after processing

9.1.2.1.2 Moulds and Blue-Stain Fungi

Mould fungi can occur on the surface and at the cross ends of fresh culms in a humid atmosphere as these fungi require high relative humidity, generally above 70 %. Their hyphae do usually not penetrate into the culm but obtain their nourishment from sugars and airborne impurities on its surface. Moulds may easily develop in chip-storage piles where the inner, nutritious, culm part is exposed at the cut surface of the chips, but also on finished products, like furniture during transport in containers, resulting in considerable economic losses (Figs. 9.3, 9.4). Moulds do not influence the culm's physical properties but reduce its aesthetic appearance. Different mould species produce large quantities of spores of various colours: black, blue, green and yellow. The spores of some species cause skin irritations, respiration problems and allergic reactions for human beings, particularly in indoor environments. The mycotoxins produced by some moulds, e.g. aflatoxin from *Aspergillus flavus*, are highly toxic to humans and animals

Blue-stain fungi, on the contrary, easily enter through the cross ends of fresh/moist culms and penetrate the parenchyma. They nourish on the starch and soluble carbohydrates stored in these cells. Their pigmented hyphae cause a blue-greyish-black discolouration of the inner tissue. Such discolouration occurs also on the

Fig. 9.4 Moulded bamboo culm parts at arrival in Hamburg after shipping from Vietnam (Tang 2013)

surface (Fig. 9.2) in various shades, as spots, streaks or in a uniformly scattered pattern. It reduces the aesthetic quality, especially of split bamboo and slivers, but does not affect the strength properties, except impact strength in severe cases. Moulds and blue-stain fungi belong to the so-called lower fungi, Deuteromycetes, and in case of producing sexual fruit bodies, to the Ascomycetes.

9.1.2.1.3 Decay Fungi

True bamboo-destroying fungi belong to the group of the so-called higher fungi, the Basidiomycetes and Ascomycetes. Their hyphae penetrate deeply into the bamboo tissue. The Basidiomycetes grow within the lumen of the cells and produce different enzymes according to the specific fungal type which diffuse into the cell wall. Either the enzymes decompose only the cellulose and hemicelluloses, with the lignin remaining, leading to the brown-rot type or they decompose all wall substances resulting in the white-rot type. Bamboo is mainly destroyed by white-rot fungi. A common white-rot fungus is *Schizophyllum commune* (Fig. 9.5) easily recognized by its white fruit body with a radial lamellate underside (Ashaari and Mamat 2000).

The enzymatic degradation of bamboo leads to a loss of cell wall substance which results in a reduction of strength properties. Even before a slight colour change or weight loss become apparent, the strength properties are much reduced, in particular the impact-bending strength. The relation between mass loss and strength reduction appears more severe in bamboo than in timber as the weight loss concerns mainly the strength-giving fibres. Incipient decay is often overlooked resulting in severe consequences for the safety of constructions and the possible restoration of infested building components. At late stages of deterioration, the culm appears soft to touch, and only a fibrous or powdery mass may remain.

Besides the white- and brown-rot fungi, the soft-rot fungi are a special group, belonging mostly to the Ascomycetes. In contrast to the Basidiomycetes, their hyphae grow mainly inside the cell wall. They use cellulose/hemicellulose wall

Fig. 9.5 Fruit bodies of *Schizophyllum commune*, a white-rot fungi (Liese and Kumar 2003)

substances and produce few changes in the lignin molecule. This fungal type needs less oxygen and can tolerate a higher moisture content, even above 80 %, which is prohibitive for other fungi. Soft-rot fungi also resist the toxic concentration of many chemical preservatives. Culms in ground contact are generally attacked by this fungal type. The colour of the infested culm changes from its natural cream yellow to dark brown black, and normally no surface mycelium can be recognized.

9.1.2.2 Insects

9.1.2.2.1 General

Insects are responsible for the most destructive attack on bamboo. Warm and moist climatic conditions in tropical regions favour their development. About 50 insect pests have been reported to attack felled culms and bamboo products (Wang et al. 1998). Among these insects, two types are notable: borers, such as bostrychid, Lyctid and cerambycids, and termites. The life cycle of borers can be divided into four stages: egg, larva (caterpillar and worm), pupa and adult. The female beetle burrows into the bamboo tissue, particularly into the vessel openings at the cross ends or in wounds, and lays abundant eggs. Minute larvae develop and penetrate further, gnawing through the tissue by mechanical action. The particles are digested in the gut of the larvae to a variable extent, and the faeces or frass is excreted as pellets from the rear end. For wood-destroying insects in general, the pellets and the form and size of exit holes have characteristics that allow for species identification. After a certain time (weeks or months) the larva is transformed into a pupa which soon changes into a beetle. The beetle then chews its way out of the bamboo culm leaving exit holes on the surface. Pellets falling out of the flight holes are often the first sign of an ongoing attack (Fig. 9.6). The adult beetles themselves generally do not destroy the bamboo tissue as their only purpose in life is procreation.

9.1.2.2.2   Beetles

The so-called powder-post beetles cause the most devastation in stored bamboo culms. They can reduce the tissue to a flour-like powder, leaving only a thin outer shell. The beetle attack can commence within 24 h after culm felling. The larvae feed on the starch and the soluble carbohydrates in the parenchyma cells; thus, the attack occurs primarily at the inner part of the culm where most of the parenchyma is situated within the diaphragm. They can tunnel through the entire inner tissue leaving behind only a thin surface of the hard cortex which may give a false impression when evaluating the significance of damage for repair work. The most destructive species are *Dinoderus minutus* and *D. brevis* (Bostrychidae), responsible for over 90 % of damage on harvested culms and bamboo products. The adult beetle is blackish brown and about 2.5–3.5 mm long (Fig. 9.7). The new beetles make their exit through the hard cortex of the culm by circular bore holes of about 1 mm diameter. In bamboo species with a thick culm wall, like *Guadua* spp., the

Fig. 9.6 Powdery mass of pellets due to beetle attack (Liese and Kumar 2003)

Fig. 9.7 Beetle of
Dinoderus sp. (Liese and
Kumar 2003)

beetles will demolish the inner starchy tissue and then exit without much structural damage because the strength is maintained by the thick outer fibrous tissue.

Several factors determine the intensity of attack. The larval activity depends very much on the availability of starch and is therefore strongly influenced by the season. Beetle infestation is enhanced when culms are harvested just prior to the shooting time. Younger culms with a higher moisture content are more easily attacked than older ones which contain less starch and moisture. Flowering bamboo culms with their low moisture content contain hardly any starch as it has been used for seed production and remain therefore almost immune to attack.

The incidence of borer attacks depends also on the bamboo species. Although no systematic record seems to exist so far, it is well documented that *Bambusa vulgaris* is always heavily infested, whereas *Pleioblastus* species are seldom attacked.

Besides *Dinoderus* spp., other powder-post beetles, like *Lyctus* spp. and *Minthea* spp. (Lyctids), can attack dry bamboo and bamboo products, but they cause less damage in most bamboo countries. The common wood-boring beetles (cerambycids) can cause damage to bamboo. They form larger tunnels inside the tissue (Fig. 9.8). Wasps can also damage bamboo.

Fig. 9.8 Deterioration of
imported bamboo furniture
parts by a cerambycid beetle
(Liese and Kumar 2003)

Carpenter bees are seldom serious pests, but more a source of annoyance. Once a bamboo component, like a rafter or support, is attacked, it will be repeatedly visited and can be seriously affected.

9.1.2.2.3 Termites

Termites are the most aggressive insects to wreak havoc on bamboo and wood. As social organisms, they live in well-organized groups of several hundreds to some millions of individuals. They have a strong caste system with the so-called sterile 'workers', a smaller number of blind 'soldiers' for defence and only one pair of fertile termites, the king and the queen as the primary producers. The damage to bamboo or wood is done exclusively by the worker caste who construct galleries and tunnels in their search for food and who also care for the feeding of the queen. They are among the few insects capable of using cellulose as a source of food due to

Fig. 9.9 Termites are aggressive insects that wreak havoc on bamboo (Liese and Kumar 2003)

Fig. 9.10 Culm deterioration by termites (Liese and Kumar 2003)

symbiotic bacteria and protozoa in their gut. Their attacks lead to rapid destruction often leaving behind only a thin layer of the bamboo cortex (Figs. 9.9 and 9.10).

Among the various termite genera, a distinction exists between subterranean termites as soil-dwelling species and dry-wood termites which can live without ground contact in wood/bamboo structures.

Subterranean termites need high humidity and access to water and often build their nests as large mounds (Fig. 9.11). They live underground and extend above ground through tube-like runways made of soil and faeces in search of food. The gnawing takes place inside the bamboo culm (Fig. 9.12). Though the erection of

Fig. 9.11 Large mound by
subterranean termites, India
1957 (Liese and Kumar
2003)

galleries can be quite fast, an attack can be stopped if all outside galleries leading to
their underground home are completely and repeatedly destroyed. Termite shields
on a cement foundation prevent gallery construction.

Dry-wood termites may build their nest inside the bamboo culm parts that they
are eating. The infestation is done by flying adults who enter the bamboo though
cracks or openings at cross ends. Consequently they can attack constructions above
ground level. Often their attack is only recognized at a later stage of deterioration.
They recycle their own body moisture and can survive on a minimal amount of
moisture obtained from the bamboo itself.

9.1.2.2.4 Marine Borers

There is large-scale use of bamboo for waterfront structures, as poles in stake-net
fishing, in coastal aquaculture farms and also for bamboo-made rafters/vessels.
Bamboo fencing can run to several kilometres in length, and much bamboo is used

Fig. 9.12 Galleries by subterranean termites to maintain ground contact (Liese and Kumar 2003)

in fishing activities (Fig. 9.13). All structures in marine and brackish waters are physically damaged by marine wood borers within a short time, often within a few months. The pattern of destruction varies among the groups of marine wood/bamboo destroyers, but their economic cost is generally high. The principal culprit, Teredinidae *(Banksia* spp., *Teredo* spp.), or shipworms, bores into the culm and tunnels through the tissue leaving only a small hole of the size of a pinhead on the surface. Although more obvious, the damage, however, proceeds more slowly.

No protection of bamboo marine structures appears possible by technical means, and chemical treatment is not viable due to the refractory nature of the culm structure and the great environmental danger through leaching of preservatives.

Fig. 9.13 Bamboo fencing in seawater, Philippines (Liese and Kumar 2003)

9.2 Preservation Methods

Bamboo preservation can be divided into non-chemical and chemical methods. The selection of the appropriate treatment method depends on various factors such as the state of the bamboo, green or dry, and end use, indoor or outside exposure and ground or food contact (Moran 2002; Liese and Kumar 2003).

9.2.1 Non-chemical Methods

Several traditional methods for bamboo protection have been applied in rural areas without the use of chemicals since ages. These methods are easy to follow and can be carried out by untrained villagers, without the need for technical equipment and with little cost. Through experience, harvesting rules, like the proper season and the selection of mature culms, are considered as traditional rules of good practice.

Some traditional or advanced non-chemical methods can considerably increase the resistance against fungal and beetle attack. However, for bamboo structures in long-term use, their efficiency regarding the required safety aspects and the real cost-saving benefits need to be carefully evaluated.

9.2.1.1 Reduction of Starch Content

In bamboo culms, the cellulose (structural carbohydrates) and starch and sugar (nonstructural carbohydrates) are the principal nutrients for fungi and insects. While cellulose content cannot be reduced without negative effects on strength properties, a reduction of sugar and starch makes bamboo unattractive for discolourating fungi and many insects. Methods commonly used for lowering starch/sugar content are as follows.

9.2.1.1.1 Harvesting of Bamboo During the Low-Sugar Content Season

The sugar content in the culms varies with the season. During the growing season, the culm reduces its carbohydrates in the parenchyma to provide building material for the expanding shoot. Thus, the carbohydrates are reduced (Magel et al. 2006). Therefore, the culms are preferably harvested at the end of the rainy season and beginning of the dry season when the culms are fully developed.

9.2.1.1.2 Curing

The bamboo culms are cut at the bottom and left for some time with branches and leaves at the clump (Fig. 9.14). As respiration of the tissue still goes on, the starch and sugar contents in the culm are decreased. Thus, the infection by borers is reduced, but there is no effect on the attack by termites and less by fungi.

9.2.1.1.3 Waterlogging

Waterlogging is commonly applied in many Asian countries. Fresh bamboo culms are soaked in running or stagnant water for 1–3 months. This process is said to partially leach out carbohydrates thus resulting in an enhanced resistance of the culm. In fact, during water storage the starch content is reduced partly by bacterial action. The method might therefore improve the resistance against borers but not against termites and fungi (Sulthoni 1988; Ashaari and Mamat 2000; Nguyen 2002). Submergence in water may lead to staining and bad odour of the culms due to the bacterial action.

Waterlogging is still used for treating bamboo materials for making handicraft and furniture in many traditional craft villages of bamboo countries as well as generally for housing in rural areas (Fig. 9.15).

Comprehensive studies were recently undertaken with *Dendrocalamus strictus* in India on the traditional waterlogging in a water tank for 4 weeks with weekly changed water. Decay resistance was found better than untreated and partly comparable to chemical-treated samples. However, water leaching alone cannot be considered as long-term preservation and must be integrated with other technologies to provide viable resistance (Kaur et al. 2013).

9.2.1.1.4 Boiling

Green culms or slivers for weaving are boiled in water for about 30–60 min.

Fig. 9.14 Clump curing of *Bambusa blumeana*, Philippines (Liese and Kumar 2003)

Fig. 9.15 Water storage of culms in Thailand (Liese and Kumar 2003)

Fig. 9.16 Whitewashing of bamboo mats, Indonesia (Liese and Kumar 2003)

Fig. 9.17 Plastering of bamboo mat walls, India (Liese and Kumar 2003)

As an improved method, boiling in 0.5–1 % solution of caustic soda (up to 30 min) or sodium carbonate (about 60 min) is applied. The culms are wiped to remove the wax coating for better finish. Longer or repeated boiling can dull the colour of the material. Boiling may cause starch to leach out or denature, improving resistance to borers and stain fungi.

9.2.1.2   Limewashing

Lime or whitewashing is a traditional treatment and mainly used for ornamental effects (Figs. 9.16 and 9.17). Bamboo culms and mats for houses are painted with slaked lime ($CaOH_2$), which is transformed into calcium carbonate ($CaCO_3$) and inhibits water absorption. The surface becomes alkaline thus delaying fungal attack which requires an acid environment. Bamboo mats are also tarred and then sprinkled with fine sand. When the sand and tar have dried, it is limewashed up to four times.

Fig. 9.18 Cement plaster
on a city house, India (Liese
and Kumar 2003)

Fig. 9.19 Bamboo house
with plastered walls in
Costa Rica (Liese and
Kumar 2003)

9.2.1.3 Plastering

Plastering of bamboo mats is a common method used by villagers using mud, clay or sand mixed with lime, cement or cow dung for stability. Plastering is also a widely applied method for city houses in many countries like India or Colombia where the walls made of split bamboo are covered with mortar on both sides. The tight seal keeps the bamboo culms, splits or mats protected against rain and prevents the entry of beetles (Fig. 9.18).

House constructions with cement-covered bamboo mats on dry foundations have been in service for decades (Fig. 9.19). Social housing programmes in Colombia and other countries as well as houses constructed by FUNBAMBU, a nongovernmental organization in Costa Rica, have applied this system on a larger scale (Liese 1989, 1990; Jayanetti and Folett 1998; Gutiérrez 2000). The common bahareque technique uses coarsely woven bamboo panels whose purpose is to hold the mud or cement. The surface will be finished with a limewash (Fig. 9.20). The

Fig. 9.20 City houses with bamboo construction and plastered walls in Manizales, Colombia (Liese and Kumar 2003)

mats can be protected by boron soaking/diffusion. Plastering can be considered partly as a constructional method.

9.2.1.4 Smoking

Fresh bamboo culms in rural areas are, by tradition, still stored inside the house above a fireplace. The moisture content is thereby considerably reduced and insufficient for biological degradation. The culm darkens in colour due to heat. The build-up of deposits from smoke such as carbon and its derivatives forms a protective layer preventing physical and probably chemical contact of the culm material with fungal spores as well as beetles. Smoke drying also reduces splitting.

Besides this traditional way of conservation, green bamboo is stacked in a furnace of about 4.5 m in length and 2.5 m in height and heated at 120–150 °C. Due to the build-up of soot and other pyrolytic chemical products on its surface, the bamboo is said to be protected against beetles.

The traditional smoking method as applied in Japan has been developed for commercial operations in Colombia. A square chamber or a similar device of about 14 m length with a 4×4 m^2 cross-section, standing upright, is filled with semidry culms with a moisture content below 50 %. The cylinder is heated from below with organic combustibles for 12–20 days, depending on the culm proportions, until the moisture content is reduced to about 12 %. During heating with circulating air, partial pyrolysis of bamboo substances occurs. The pyrolyzed products drip down, are pumped up to the top and flow down the culm surface again where they eventually dry.

A strong acid odour is produced which will influence the fields of application. These culms are mainly for outdoor use. So far, an improved resistance against beetles is reported which is likely because of chemical changes in the carbohydrates. The durability against fungi and termites as well as in ground contact and for a longer exposure time has still to be proven.

9.2.1.5 Heat Treatment

Wood submitted to thermal treatments well above 150 °C shows a notable decrease of water absorption, a better dimensional stability and an improved resistance against microorganisms due to modification of the organic matter. The efficacy against beetles and termites is still under evaluation, and results so far indicate an improved resistance. Mechanical properties of bamboo, especially the elasticity and strength of the heat-treated material, are evidently reduced (Leithoff and Peek 2001). This effect is also observed in timber.

9.2.1.6 Constructional Methods

Much damage can be avoided by suitable construction methods by which the moisture content of bamboo components is kept well below fibre saturation, so that no fungal attack can occur. A long-standing tradition exists, now hundreds of years old, on the proper construction use of bamboo without chemical treatment. Common methods are to place bamboo posts or walls on either stones, preformed concrete footings or durable or pressure-treated wood blocks instead of putting them directly on the ground. The culm should be cut just below a node for better stability. Good air circulation throughout the structure is also important. Houses made from bamboo mats, as is common in rural areas, should not be placed directly on the ground, but on a cement base (Fig. 9.21). An overhanging roof will protect bamboo mats against rain.

By tradition and experience sound construction methods are often applied. The houses in the Cordillera Central region of Colombia and in the lowland coastal provinces of Ecuador with an extensive use of bamboo are impressive examples (González and Gutiérrez 1996; Gutiérrez 2000). They last for over 90 years without any treatment because the plaster on the ceilings and walls prevents beetle damage.

For roofing, halved culms allow an easy run-off for rain. Due to their water-repellent cortex, they last several years until fruit bodies indicate their internal

Fig. 9.21 Social housing programme with bamboo houses in Manizales, Colombia (Liese and Kumar 2003)

Fig. 9.22 Fruit bodies indicate advanced decay of a bamboo roof (Liese and Kumar 2003)

degradation (Fig. 9.22). Only older culms with a well-developed skin should be used, as chemical protection is not feasible for roofing.

The traditional methods outlined briefly can be effective against fungal and partly against beetle attack but hardly against dry-wood termites. The invasion of subterranean termites can be prevented by removing their earthen tunnels, by placing termite shields on the cement foundation or by using soil poisoning barriers, if permitted (Nguyen 2002).

The INBAR Technical Reports No. 15 by Jayanetti and Folett (1998), No. 19 by Gutiérrez (2000) and No. 20 by Janssen (2000), as well others (Cusack 1999; Heinsdorff 2010), illustrate the manifold of possibilities for bamboo protection by design and construction.

Traditional construction techniques as well as the new designs by architects like Simon Velez in Colombia, Jorge Arcila and Jorge Moran in Ecuador and a generation of very dedicated bamboo architects and promoters, like Oscar Hidalgo in Colombia, have now turned bamboo into a 'high-tech' material. The bamboo bridge constructed by Jörg Stamm in Pereira, Colombia, with a 52 m span and a 2.8 m width, is an impressive example (Fig. 9.23). It should be noted that the culms were given a boron treatment against beetles by a 4–5-day soak with punctured internodal walls for complete internal penetration.

9.2.2 Chemical Methods

In most cases, chemical treatment of bamboo is required. A key factor for the protection is the sufficient preservation of the culm. Unlike timber, the bamboo tissue is rather resistant towards penetration of liquids due to its anatomical structure (Liese and Kumar 2003).

Fig. 9.23 Bamboo bridges
in (**a**) Indonesia, (**b**)
Malaysia and (**c**) Colombia
(Liese 1985)

9.2.2.1   Treatability

The uptake of a preservation solution is restricted mainly to the metaxylem vessels, which run through the culm like long 'water pipes'. At the nodes, they become partly structurally modified by branching, so that the passage through a node may be hindered. Their total volume amounts to only 6–8 %, so that the remaining tissue of fibres and parenchyma has to be protected by diffusion. Ray cells, like in wood, do not exist in bamboo. As a typical wound reaction, slime extrudes from the parenchyma cells into the vessels causing a blockage of the lumina. Monopodial bamboos, like Moso (*Phyllostachys edulis*), also produce tyloses as balloon-like protrusions into the metaxylem vessels.

The culm is covered on its outer side by a special layer, the cuticula, which hinders any penetration by simple treatments, such as soaking. At the inner culm wall, suberin layers lying on sclerotic parenchyma cells also hinder any penetration, although to a much lesser extent than from the outside (Liese and Schmitt 2006).

This special anatomical make-up of the bamboo culm as well as its moisture content must be taken into consideration when choosing a suitable preservative and treatment method to be applied.

9.2.2.2 Preservatives

Most preservative formulations for wood have also been found suitable for bamboo. The preservatives are either waterborne or oil-based types; for specific purposes, some special formulations already exist.

9.2.2.2.1 Waterborne Types

Water-soluble inorganic or organic salts are dissolved in water and introduced into the bamboo. The water evaporates leaving the salts inside. They are either non-fixing or fixing to the bamboo tissue.

The application of non-fixing preservatives is restricted to bamboo used in dry conditions and under cover only, as in wet conditions or exposed to rain they are of no use as the chemicals are washed out.

Fixing-type preservatives are chemically bound in the woody tissue and can be applied both inside and outside. The type of product, expected durability and use dictate the type of preservative to be used.

Preservatives are sold as solids or as a paste. Salts, more recently, are often provided as pre-solutions for health and safety reasons and to minimize pollution by chemical dust.

Waterborne preservatives are clean and odourless and offer widely recognized advantages coupled with effectiveness and permanency. Such preservatives are more cost-effective as solvent costs are eliminated. Waterborne preservatives are divided into non-fixing types and fixing types.

9.2.2.2.2 Non-fixing Types

Non-fixing or leachable salts penetrate essentially by diffusion and can thus reach the entire tissue of a culm. They provide the best way to obtain full treatment which is important for the bamboo with its small volume fraction of vessel pathways. The salts remain mobile after treatment and continue to diffuse into the bamboo tissue after penetrating through the vessel walls. Bamboo culms or products treated with such preservatives should not be exposed to rain or ground contact.

Many single or mixed salts are in use. The important ones are as follows:

Boron-containing compounds are the most widely used preservatives for protecting bamboo, usually as a mixture of boric acid and borax. Disodium octaborate, formed by mixing boric acid and borax in specific proportion (1:1.4), which has a very high solubility in water, comes as a ready-made formulation. Boron salts are effective against a variety of fungi, insect borers and termites but not against soft rot. They are applied in 5–10 % concentration, depending on the moisture content and the treatment method. Boron salts in a high concentration also have fire-retardant properties.

Zinc chloride/copper sulphate may be used as single salts for protection of bamboo. Solutions of both these salts are highly acidic and can corrode metal fittings. It can retard fire to a certain extent due to the release of water vapour contained in the salt's crystal structure.

NaPCP (sodium pentachlorophenate) is the water-soluble sodium salt of the much-used preservative, pentachlorophenol. It has been extensively used in the past, but is now banned or restricted in several countries. It is basically a fungicide to protect freshly felled lignocellulosic materials against fungi that cause stain and decay.

TCMTB (thio-cyanomethyl-thio-benzothiazole). TCMTB + MBT (methylene-bis-thiocyanate) is a substitute for NaPCP and promoted for protection of green wood/bamboo against sap stain and fungi.

Non fixed type formulations may be used for both non pressure methods and pressure treatment and hence can be applied to bamboo at any moisture content level. The borax/boric acid formulation is the most common preservative for the sap-replacement treatment in several countries and widely used in diffusion treatments.

Non-fixing salts can only be used for bamboo under cover. In contrast to timber, the cortex of the culm better protects the treated tissue against leaching as loss of salts only occurs through the open ends and the nodal parts. These salts have no toxicity to humans.

9.2.2.2.3   Fixing Types

Such formulations no longer remain soluble in water after fixation, due to their chemical reaction with bamboo substances, and are therefore more weather-resistant.

The preservatives are mixtures of different salts in appropriate proportions which interact with each other in the presence of wood/bamboo and become chemically fixed. Chromium is responsible for fixation, copper is effective against soft-rot fungi and a third compound, which varies according to brand, acts against white- and brown-rot fungi and insects. Preservative solutions containing chromium, such as CCB, should not be heated before or during treatment as it precipitates chromium, making them ineffective due to the infiltration of sludge through the fine pit pores of the cells.

CCB (copper-chrome-boron) has excellent fungicidal and insecticidal properties. It is a good alternative to CCA but slightly less effective with a lower degree of fixation, particularly the boron component. A typical formulation contains copper sulphate 3 parts, sodium or potassium dichromate 4 parts, and boric acid 1.5 parts; required levels of retention are 4–16 kg/m^3.

9.2.2.2.4 Oily Preservatives: Creosotes

Tar products of both coal and wood have been used as wood preservatives since biblical times. There are two types of creosotes available, normal creosote from coal tar produced at high temperature and low-temperature creosote produced from tar produced at low temperature. Both creosotes have similar physical, fungicidal and insecticidal properties.

Coal tar creosote is a blend of several fractions of distillation containing more than 200 major and several thousand minor constituents. Creosote is a dark brown or black viscous liquid.

Creosote should be used exclusively for pressure impregnation or hot and cold processes. Being oily in nature it is insoluble in water and in fact imparts water repellency to the treated material. This, in itself, has an important protective effect as moisture is an essential requirement for fungal attack. Its dark brown colour and bad odour prevent its application for indoor use. Creosote is very effective against a wide variety of wood rotting fungi and insects and long-lasting at very heavy doses ranging from 50 to 200 kg/m^3 depending on the use of the treated bamboo. Loss of creosote from treated poles resulting from downward movement within the poles in combination with gravity is well known, which helps protection at ground contact level but leads to a contamination of the surrounding soil. Due to its carcinogenic character, creosote use is regulated in many countries, and its application and use is limited to licensed pesticide applicators.

9.2.2.2.5 Organic Acids

For short-term protection against moulding of the freshly harvested culms with their high moisture content during oversea transport, dipping the culms in harmless solutions of 10 % acetic acid or propionic acid can prevent mould growth totally during the first two sensitive months (Tang et al. 2012).

9.2.2.3 Methods of the Preservative Treatment

For preservative treatments of bamboo, two main treatment methods can be considered, non-pressure and pressure processes.

9.2.2.3.1 Non-pressure Methods

Steeping or Butt-End Treatment

Freshly cut culms with the branches, soon after harvesting, are placed upright in a suitable container (usually plastic buckets) containing a treatment solution. The butt end of the culm is kept immersed in the solution up to about 25 cm. The time of treatment may take 8–14 days, depending on the freshness and the length of the culms as well as on the type of preservative (Fig. 9.24).

The steeping or butt-end treatment is often applied to culms used to support fruit trees or banana plants. Here replacements are needed in great quantities due to

Fig. 9.24 Butt-end treatment of *Thyrsostachys siamensis* in a trough in Thailand (Liese and Kumar 2003)

decay at the butt end. It is a simple process applied without special skills and equipment which can result in great savings due to enhanced life span of the support materials.

Soaking/Diffusion

This method provides good protection and is simple and economical. Freshly felled bamboo or almost fresh culms are stripped of branches and foliage, prepared to size and submerged in a water-based preservative solution for diffusion. Since bamboo has a specific gravity below 1.0 g/cm^3, the material has to be placed in the container and weighed down before the solution is added (Fig. 9.25). Diffusion occurs mainly in axial direction, less in transversal whereby radial is slightly better than tangential.

For larger bamboo culms boring holes on opposite sides in each internode aids penetration but causes difficulties when drawing the solution out again. More effective is the puncturing of the diaphragm with a long stick making holes of about 2 cm; puncturing the solid nodal wall opens up the vessel system and allows access to the internodes. Scratching of the outer skin (Fig. 9.26) is sometimes applied and may help penetration, especially for slow-diffusing preservatives.

Fig. 9.25 Soaking method (Liese and Kumar 2003)

Fig. 9.26 Skin removal by sanding, Co. Bamboo Nature, South Vietnam (Tang and Liese 2011)

Half-split or quarter-split material may be treated in 7–10 days, whereas round bamboo behaves less satisfactorily due to its impervious and refractory skin and requires at least double the time. After soaking, the material has to be placed on a support to let the solution drip off.

If the treated bamboo is to be used only indoors, boron-based preservatives 8 % should be preferred because of their low mammalian toxicity and excellent diffusion behaviour. The soaking process requires larger tanks for efficient production. After soaking, the culm parts should be stored horizontally and closely together for about 1 week to facilitate further diffusion of the preservative before drying.

Vietnam Method

The 'Vietnam method' is a speciality of bamboo treatment, applied for fresh culms. Its principle is the use of the upper internode as a reservoir for the treatment solution. Its inner wall is either scraped at a depth of 1–2 mm or by a round incision with a sharp tool to disrupt the inner terminal layer. The cavity is filled up daily with the preservative solution, which diffuses into the parenchyma tissue, fibres and especially the vessels located in the inner part of the culm wall, where it flows down by gravity. Therefore, this method is also called in Vietnam the 'gravity method'. The treatment is completed, when the liquid at the culm foot has the same colour as the initial solution (Tang 2009) (Fig. 9.27).

Vertical Soak and Diffusion Method

The same principle as of the 'Vietnam method' has been used for the Vertical Soak and Diffusion method developed by the Environmental Bamboo Foundation (EBF),

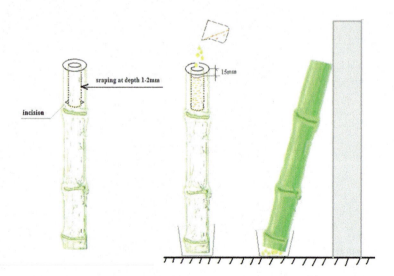

Fig. 9.27 Steps of the Vietnam method for bamboo treatment (Tang 2009)

Fig. 9.28 The Vertical
Soak and Diffusion method
for treating bamboo culms
at the Environmental
Bamboo Foundation in
Ubud, Bali (Liese and
Kumar 2003)

Bali, Indonesia (Fig. 9.28). The standardized treatment process is called 'Vertical Soak and Diffusion (VSD)' system (EBF 2003). This method does not use only the lacuna of the upper internode, but the whole culm serves as a reservoir for the solution as all diaphragms are fractured with a sharpened stick, except the lowest. The lacuna of the internodes is filled up with a borax/boric acid solution and refilled daily. After about 2 weeks the lowest diaphragm is punctured and the solution collected for further use with the required concentration (Tang 2009).

Hot and Cold Treatment

The hot and cold process is one of the most efficient production methods for large quantities of treated wood/bamboos. It can be applied without the need for any sophisticated equipment. Its principle is based on a heating of the bamboo by which the air in the cells will expand and escape. During the following cooling period, a slight vacuum effect occurs which causes the preservative to soak into the cells. Round bamboos are usually treated with two holes on opposite walls of each internode near the node or with ruptured diaphragms to allow access to the inner surface. The submerged bamboo is heated to raise the temperature to about 90 °C and maintained at this level for 2–3 h. The preservative is then allowed to cool down completely, e.g. overnight, and the oil is drained out. After treatment, the culms have to be stored to allow drainage of excess preservatives from the lacuna. This treatment is most suited for bamboo to be used as reinforcements for mud in adobe-type construction in rural and tribal areas.

9.2.2.3.2 Sap-Replacement Treatment

The sap-replacement treatment, also called Boucherie process, named after its French inventor Auguste Boucherie in 1839, has been applied since long for the treatment of wooden poles (Willeitner and Liese 1992). Its use for bamboo was first tried in India by Narayanamurti et al. (1947) and further developed to the 'modified

Boucherie treatment' by Purushotham et al. (1954), Liese (1959) and Liese and Kumar (2003).

The principle of the process is based on a pressure pump that pushes a preservative solution through the entire length of the culm, so that the sap in the vessels is replaced by the preservative. Sap replacement is a safe and environmentally friendly treatment for bamboo culms as the preservative remains entirely inside the culm. Several parameters have to be considered for a successful treatment (Fig. 9.29a, b).

Treatment time depends on the bamboo species, age and moisture content, culm length and wall thickness, the preservative and the pressure applied. In general, culms of 150 mm in diameter and 6 m in length are treated in 30–50 min and those of 9 m in length in 60–70 min with a pressure of 1.0–1.3 bar. The main advantages of the modified sap-replacement treatment are the limited need and cost of the technical equipment, the rapid treatment procedure and the complete penetration of the culm with a clean surface, thus avoiding any risks. As the sap-replacement process has been applied in several countries, some modifications have been developed for simplification and efficiency such as the system applied by Cusack (1999) with a standing treatment tank made from heavy plastic. A detailed description of the process and its modifications has been given by the EBF (1994), González and Gutiérrez (1996), Cusack (1999), INBAR TOTEM (Rao 2001) and Liese et al. (2002).

9.2.2.3.3   Pressure Treatment

Pressure treatment is the most effective method to protect bamboo against adverse conditions. Through pressure treatment, deeper and more uniform penetration of preservative can be obtained, and the retention of preservative can be more closely controlled. In addition, the time required to thoroughly impregnate the bamboo can be reduced. A sufficient demand for this value-added product is needed.

Fig. 9.29 (a) Sap replacement with roof sheath beneath by EBF, Bali. (b) Preservative solution dripping from the apex (Liese and Kumar 2003)

Fig. 9.30 Culms loaded for pressure treatment, Taiwan (Liese and Kumar 2003)

Fig. 9.31 Ruptured nodal walls by a long iron stick to ease penetration in Thailand (Liese and Kumar 2003)

The pressure method is mainly used for the treatment of dried bamboo. The principle of the process is to force the preservative solution into the bamboo tissue. This can be done by a vacuum and/or by increasing the pressure upon the preservative in the treatment cylinder. It is horizontally orientated for easy handling of the material (Fig. 9.30).

Bamboo culms are difficult to treat due to the refractory nature of their skin which hinders radial uptake. Cracks and collapse can occur, particularly with thin-walled species. To ease penetration, holes are made on opposite sides of the lower and upper part of each internode or the septa are punctured through the entire culm (Fig. 9.31). Split bamboo shows a better absorption as the preservative is pushed

Table 9.1 Vacuum/pressure impregnation schedules for bamboo culm parts (Tang et al. 2013)

Stage	Vacuum/pressure (bar)	Time (min)
Initial vacuum	−0.84	20–40
Pressure	4–10	60–120
Final vacuum	−0.6 to −0.8	15

sidewise into the tissue as well. The basal portion is less treatable than the top; the inner culm portion obtains a higher uptake than the outer one (Tang and Liese 2011). The moisture content of the culms is of more importance than impregnation pressure, vacuum and their duration.

The normal vacuum/pressure impregnation schedule in the treatment of bamboo is given in Table 9.1. The retention obtained depends on the amount of pressure applied, the duration of the pressure period, the bamboo properties and the concentration of the treating solution. For creosote it ranges between 50 and 100 kg/m^3 and for water-based preservatives (CCA or CCB type) between 5 and 15 kg/m^3 expressed as dry salt retention.

Pressure treatment of bamboo can cause environmental pollution, albeit less than for other methods. Preservative residues can remain on the smooth surface of the culm which may drip onto the soil and also dry upon the skin.

9.3 Drying of Bamboo

Drying is an important stage of the manufacturing process of bamboo products. Well-dried culms have the desired appearance, finish and structural properties to meet the requirements for a successful export into demanding markets.

9.3.1 Drying Rate of Bamboo

The drying of bamboo occurs mainly as culm parts. They are round, separated by nodes and inside mostly hollow, called lacuna. At their ends, the metaxylem vessels are the main pathways for releasing moisture. In bamboo the radial passage of moisture is slower than for wood because no ray cells exist. Generally, the anatomical structure of the bamboo culm makes drying more difficult than for wood (Kumar et al. 1994; Liese and Kumar 2003). Comparing with wood of the same density, bamboo takes a longer time to dry (Sekhar and Rawat 1964; Laxamana 1985).

The drying rate of bamboo is notably influenced by its structural features. The studies on bamboo seasoning by Glenn et al. (1954), Laxamana (1985) and Tang et al. (2013) showed a faster drying rate for species with a lower specific gravity and shorter internodes. The culm wall thickness is an important factor influencing controlling the rate of drying. The bottom part of the culm due to its thicker culm walls, therefore, takes much longer to dry than the top portion of the culm. The rate of drying of immature culms is generally faster than that of mature ones, but since the former have a higher moisture content, their rate of drying is longer. In the initial stages, drying occurs quite rapidly, but slows down gradually as drying progresses. Bamboos, from which water-soluble extractives have been removed by soaking, dry faster and take up moisture slower than untreated ones.

9.3.2 Methods of Bamboo Drying

Most drying methods applied for timber drying are also suitable for bamboo. Two major methods can be considered: air drying and kiln drying.

9.3.2.1 Air-Drying of Bamboo Culms

Air-drying is the process of removing moisture from bamboo by exposing it to atmospheric conditions. By proper stacking for air circulation, culms can be dried with no need to add energy above the capacity of the ambient air. There are two types: the horizontal and the oblique stacking (Figs. 9.32 and 9.33). In the same condition of the air-drying, culms which are stacked horizontally dry longer. They need almost double the drying time than those which are standing upright (Glenn et al. 1954; Sharma 1988).

Up to now air-drying is traditionally being used in rural areas and in bamboo factories with small capacities; however, it has some disadvantages. Drying time is long, ranging from several weeks to several months to reach the required moisture content. During air-drying, splits can occur and culms can be infected by fungi, especially moulds. The air-drying depends largely on the climatic conditions. Since the weather cannot be regulated, there is little control over the drying process. The

Fig. 9.32 Horizontal stacking bamboo culms for air-drying under cover (Montoya-Arango 2006)

Fig. 9.33 Oblique stacking
bamboo culms for open air-
drying, Colombia

air-drying conditions are difficult for reaching a moisture content below about
12 %, as required for later processing (Gandhi 1998; Montoya-Arango 2006).

9.3.2.2 Kiln-Drying of Bamboo Culms

Kiln-drying is more efficient than air-drying. By this method, bamboo can be dried
to the required moisture content in shorter time. With great demand for production
to export, kiln-drying is a better alternative for air-drying and could ensure high-
level bamboo quality.

Kiln-drying is basically a process of stacking bamboo culms or bamboo splits in
a chamber (Fig. 9.34) where air circulation and the temperature and relative
humidity are maintained and controlled so that the moisture content of bamboo
can be reduced to a target level.

9.3.2.2.1 Dry Kilns

Dry kilns are classified in some different ways and often described and named
according to its operational technique, type of heating or energy source. The kilns

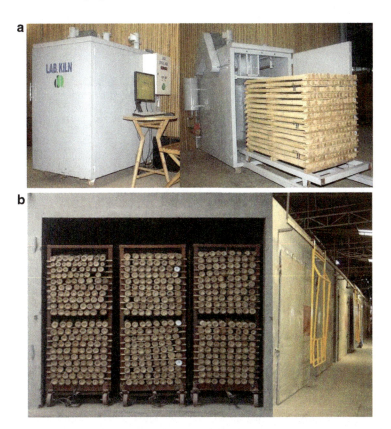

Fig. 9.34 (**a**) Pilot kiln-drying and (**b**) industrial kiln-drying at Bamboo Nature Co., South Vietnam (Tang et al. 2013)

commonly used for timber drying are conventional heat and vent kilns, dehumidification kilns, vacuum kilns and solar kilns. For bamboo drying, the conventional heat and vent kilns, either equipped with heat exchangers (hot water or steam) or directly heated with flue gas from waste incineration, and solar kilns should be considered.

9.3.2.2.2   Kiln-Drying Schedule

Kiln schedules are used to define the temperature and relative humidity needed in the kiln to dry bamboo with a minimum occurrence of degrades and in the shortest time possible. A typical kiln schedule is a series of drying conditions, expressed as temperature and relative humidity, which is used as a directive on how to operate a kiln throughout a period of time comprising the whole drying process. A drying schedule may be designed for manual, semi-automatic or fully automatic control

Table 9.2 Three drying schedules, applying for bamboo culm parts of the three main species of Vietnam (Tang et al. 2013)

Step	Moisture content (%)	Schedule No. 1 Mild		No. 2 Severe		No. 3 Highly severe	
		T (°C)	RH (%)	T (°C)	RH (%)	T (°C)	RH (%)
1	Over 90	50	80	55	80	65	80
2	90–70	50	70	55	75	65	60
3	70–50	60	60	60	65	70	45
4	50–40	60	50	65	50	70	35
5	40–30	60	30	65	35	70	30
6	30–20	65	30	70	25	75	25
7	20–10	65	20	70	20	75	15

Schedule No. 1 applying for *Dendrocalamus asper*, No. 2 for *Bambusa stenostachya*, No. 3 for *Thyrsostachys siamensis*

Table 9.3 Drying schedule for bamboo splits of *Bambusa blumeana* (Yosias 2002)

Step	Drying time (hours)	T (°C)	RH (%)
1	24	38	30
2	48	38	30
3	56	49	25

systems. It is usually formulated in a table-type format showing the drying conditions at different stages or periods of a drying process.

In kiln-drying, two general types of kiln-drying schedules (Tables 9.2 and 9.3), i.e. moisture content schedules and time-based schedules, are commonly employed. When bamboo is dried by moisture content schedules, the temperature and relative humidity conditions are changed according to the loss of moisture content, which has to be determined either by electrical moisture content measurements or by means of process control samples which are weighed at regular intervals to be able to calculate actual moisture content. When bamboo is dried by time-based schedules, drying conditions are kept constant for certain periods of time and changed to new set point values after this time has elapsed. Sometimes, also combinations of these two types of schedules are applied.

For developing a kiln schedule, culms of one bamboo species, with different diameter and wall thicknesses, initial moisture content and other factors, are dried according to several different drying schedules. The result of the kiln runs has to be evaluated and expressed in the form of using specific drying quality indicators. The schedule leading to the best results is selected and applied. Successful kiln-drying of bamboo requires an appropriate (optimized) drying schedule and good control of drying condition.

Kiln-drying of bamboo culms normally takes 6–15 days, depending on the bamboo species, the kiln and the schedule being used (Laxamana 1985; Montoya-Arango 2006; Tang et al. 2013). Kiln-drying enables to dry bamboo to any moisture content. For large-scale operations with high-level bamboo quality, kiln-drying is more efficient than air-drying. With the growing demand of large quantities of high-quality products for export, many big bamboo manufacturers have expanded their kiln-drying facilities and run batteries of kilns parallel to each other.

9.3.2.3   Drying of Bamboo Splits, Strands and Splinters

Bamboo is not only used and dried in the form of culms but also in the form of splits, strand or splinters, depending on the type of processes for further processing into a wide range of different bamboo-based products. These industrial drying processes can be differentiated into batch-type and continuous drying processes.

9.3.2.3.1   Batch-Type Bamboo Particle Driers

Bamboo splits are normally dried either in bundles or placed on special racks in batch-type drier which are similar to conventional kilns. Due to the smaller dimensions of the bamboo particles to be dried, drying time is much shorter. As for culm drying specific drying schedules will be used. Air circulation can be along the longitudinal orientation of the splits or transversal. With the aim of changing the colour of the bamboo splits, the chambers might also be used for steaming or heat treatment. When temperatures close to or even well above 100 °C are used, the colour of the bamboo split will change from light yellow to light or even dark brown.

Normally drying will be completed within several hours and certainly will not last longer than 1 or 2 days.

9.3.2.3.2   Continuous-Type Bamboo Particle Driers

There are quite a number of different drying systems in the wood sector which can also be used for drying bamboo particles. In belt dryers the material to be dried is spread in the form of thin layers on a moving belt. Hot air is circulated in the dryer and forced through the thin layers. While moving through the kiln, the bamboo particles will lose their moisture and leave the kiln in dry form. To adjust the drying conditions, the temperature, relative humidity, thickness of the material layers and speed of the belt can be changed.

9.3.3   Drying Defects

Defects may develop during and after drying of the bamboo. Some common defects are ruptures of culm tissue such as surface checks, end checks, node checks and splits (Fig. 9.35). Uneven moisture content within individual culms or moisture

Fig. 9.35 Drying defects:
end checks and splits (Tang
et al. 2013)

differences between the culms of a kiln load as well as discolouration, i.e. mould, blue staining and water staining at the nodes, also negatively affect the drying quality.

Cell collapse is a serious seasoning defect. It occurs during artificial as well as during natural air-drying processes and leads to cavities on the outer surface and to wide cracks in the inner part of the culm. Green bamboo is susceptible to collapse due to capillary tension during drying. Cells while still filled with liquid water collapse which leads to unusual high shrinkage. Collapsed tissue regions exhibit a higher density than non-collapsed regions. In the vicinity of collapsed tissue regions, small internal checks can frequently be observed. This abnormal shrinkage takes place in the early stages of seasoning while the average moisture content is still well above fibre saturation. Immature bamboo is more likely to develop collapse than mature bamboo. Collapse occurs more often during the dry season than during the rainy season, because of more severe drying conditions and the associated faster drying during the dry season. The lower portion with thicker walls is more susceptible to collapse than the upper portion. Slowly drying bamboo species are apparently more prone to collapse than others.

Changes in colour can occur during seasoning. Fresh bamboo normally looks green or rather yellowish according to the stage of maturity; it changes during seasoning to a light green shade. Immature bamboos turn emerald green and mature ones pale yellow. Culms which are slowly air-dried develop a darker yellow colour than those which are dried rapidly in a kiln. Discolourating fungi can grow during kiln-drying when bamboo moisture content is still high and kilns are operating at low temperature and high humidity.

Under very mild drying conditions, the bamboo will not collapse and will shrink almost equally in radial and tangential direction. The bamboo which has been dried carefully will exhibit volumetric shrinkage and swelling behaviour very similar to most wood species but with less anisotropy.

Different bamboo species require drying at different temperatures and different speeds to produce the best results. Each species has a different drying behaviour and, therefore, requires a specific and well-adapted drying schedule. If a correct kiln-drying schedule is used, good results in terms of drying time and drying quality can be achieved leading to a low percentage of rejects and good economic results. Laboratory and industrial investigations on kiln-drying of bamboo have led to suitable schedules for the three main bamboos of Vietnam (Tang et al. 2013), which are now applied by industry to reach high quality of dried culms and derived products for export.

Acknowledgement Our thanks are expressed to Prof. Dr. Dieter Eckstein, Prof. Dr. Olaf Schmidt and Dr. Johannes Welling for their valuable review and comments.

References

Ashaari Z, Mamat N (2000) Traditional treatment of Malaysian bamboos: resistance towards white rot fungus and durability in service. Pak J Biol Sci 3:1453–1458

Cusack V (1999) Bamboo world. The growing and use of clumping bamboos. Kangaroo, East Roseville, updated 2010

EBF- Environmental Bamboo Foundation (1994) Training manual for bamboo preservation using the Boucherie system. EBF, Ubud, Bali, Indonesia

Environmental Bamboo Foundation EBF (2003) Vertical soak diffusion for bamboo preservation. Ed Linda Garland, Ubud, Bali

Gandhi Y (1998) Preliminary study on the drying of bamboo (*Bambusa blumeana*) in a wood waste fired kiln. In: Proceedings of the 5th international bamboo congress and the 6th international bamboo workshop. San José, Costa Rica, 2–6 Nov 1998. INBAR, vol 7, pp 495–510

Glenn HE, Brock DC, Byars EF et al (1954) Seasoning, preservative treatment and physical property studies of bamboo. Bull Clemson Coll (Clemson, South Carolina) 7:27–36

González G, Gutiérrez JA (1996) Bamboo preservation at the Costa Rica national bamboo project. In: Proceedings of the 5th international bamboo workshop 1995, Ubud, Indonesia, INBAR. Technical report 3(8):121–129

Gutiérrez JA (2000) Structural adequacy of traditional bamboo housing in Latin America. Technical report no. 19. INBAR, Beijing

Heinsdorff M (2010) The bamboo architecture design with nature. Hirmer, München

Janssen JA (2000) Designing and building with bamboo. INBAR technical report no. 20, Beijing

Jayanetti DL, Folett PR (1998) Bamboo in construction. INBAR technical report no. 15, TRADA Technology Ltd

Jiang Z (2007) Bamboo and rattan in the world. In: Bamboo preservation. China Forestry Publishing House, Beijing, pp 144–150 (Chapter 9)

Kaur P, Kardam V, Pant KK, Satya S, Naik SN (2013) Scientific investigation of traditional water leaching method for bamboo preservation. Bamboo science and culture. J Am Bamb Soc 23(1): 27–32

Kumar S, Shukla KS, Dev I, Dobriyal PB (1994) Bamboo preservation techniques: a review. Technical report no. 3. International Network for Bamboo and Rattan (INBAR), New Delhi, India; and Indian Council of Forestry Research and Education (ICFRE), Dehra Dun, India

Laxamana MG (1985) Drying of some commercial Philippine bamboos. J For Prod Res Dev Inst (Philippines) 14(1&2):8–19

Leithoff H, Peek RD (2001) Heat treatment of bamboo. International research group on wood preservation. IRG/WP/01-40216, Stockholm

Liese W (1959) Bamboo preservation and soft rot. FAO report to the Government of India. Rome (1106):1–37

Liese W (1985) Bamboos-biology, silvics, properties, utilization. Schriftenr. Deutsche Gesellschaft für Technische Zusammenarbeit (GTZ), Nr. 180, TZ Verlagsges, Roßdorf

Liese W (1989) Bamboo preservation in Costa Rica. Progress UNHABITAT- report no.1, pp 37; no. 2, pp 24

Liese W (1990) Bamboo preservation in Costa Rica. Progress UNHABITAT report no. 3, pp 11

Liese W (2004a) Preservation of bamboo structures. Ghana J For 15&16:40–48

Liese W (2004b) Preservation of a bamboo culm in relation to its structure. In: Proceedings of symposio internacional guadua, Pereira, Colombia, 27 Sept–2 Oct 2004, pp 20–29

Liese W (2004c) Preservation of a bamboo culm in relation to its structure. World Bamb Rattan 3(2):16–21

Liese W, Kumar S (2003) Bamboo preservation compendium. International Network for Bamboo and Rattan (INBAR), People's Republic of China

Liese W, Schmitt U (2006) Development and structure of the terminal layer in bamboo culms. Wood Sci Technol 40:4–15

Liese W, Gutierrez J, Gonzales G (2002) Preservation of bamboo for the construction of houses for low-income people. In: Bamboo for sustainable development. Proceedings of the 5th international bamboo congress and the 6th international bamboo workshop, San José, Costa Rica, 2–6 Nov 1998, pp 481–494

Magel E, Kruse S, Lütje G, Liese W (2006) Soluble carbohydrates and acid invertases involved in the rapid growth of developing culms in *Sasa palmata* (Bean) Camus. Bamb Sci Cult 19:23–29

Montoya-Arango JA (2006) Trocknungsverfahren für die Bambusart Guadua angustifolia unter tropischen Bedingungen. Ph.D. Dissertation, University Hamburg

Moran JA (2002) Traditional preservation methods in Latin America. INBAR technical report no. 25, 70 pp

Narayanamurti D, Purushotham A, Pande JN (1947) Preservative treatment of bamboo. Part 1: treatment of green bamboos with inorganic preservatives. Indian Forestry Bulletin No. 137

Nguyen TBN (2002) Investigation on preservation of bamboo used for construction. Ph.D. Dissertation, Forest Science Institute of Vietnam, Vietnam, pp 44–51

Purushotham A, Sudan SK, Sagar V (1954) Preservative treatment of green bamboos under low pneumatic pressure. Indian Forestry Bulletin No. 178

Rao KS (2001) Bamboo preservation by sap displacement. IWST/INBAR

Schmidt O (2006) Wood and tree fungi. Biology, damage, protection, and use. Springer, Berlin

Sekhar AC, Rawat BS (1964) Some studies on the shrinkage of *Bambusa nutans*. Indian For 90(3):182–188

Sharma SN (1988) Seasoning behaviour and related properties of some Indian species of bamboo. Indian For 114(10):613–621

Sulthoni A (1988) A simple and cheap method of bamboo preservation. In: Proceedings of the 3rd international bamboo workshop, Cochin, India, 14–18 Nov 1988, pp 209–211

Tang TKH (2009) Bamboo preservation in Vietnam. International Research Group on Wood Protection. IRG/W/40457, pp 1–11

Tang TKH (2013) Preservation and drying of commercial bamboo species of Vietnam. Ph.D. Dissertation, Hamburg University

Tang TKH, Liese W (2011) Pressure treatment of bamboo culms of three Vietnamese species by boron and CCB preservatives. J Bamb Rattan 10(1&2):63–76

Tang TKH, Schmidt O, Liese W (2012) Protection of bamboo against mould using environment-friendly chemicals. J Trop For Sci 24(2):285–290

Tang TKH, Welling J, Liese W (2013) Kiln drying for bamboo culm parts of the species *Bambusa stenostachya, Dendrocalamus asper* and *Thyrsostachys siamensis*. J Indian Acad Wood Sci 10(1):26–31

Wang H, Varma RV, Tiansen X (1998) Insects of bamboos in Asia. An illustrated manual. INBAR, Beijing, Technical Report No. 13

Wei D (2014) Bamboo inhabiting fungi and their damage to the substrate. Dissertation, Department of Biology, Hamburg University, Hamburg

Willeitner H, Liese W (1992) Wood protection in tropical countries. A manual on the know-how. GTZ Schriftenreihe No. 27

Yosias G (2002) Preliminary study on the drying of bamboo (*Bambusa blumeana*) in a wood waste–fired kiln. Bamboo for sustainable development. In: Proceedings of the 5th international bamboo congress and the 6th international bamboo workshop, San José, Costa Rica, 2–6 Nov 1998. INBAR 2002, New Delhi, India, pp 495–510

Chapter 10
Utilization of Bamboo

Walter Liese, Johannes Welling, and Thi Kim Hong Tang

Abstract There are more than thousand uses of bamboo as a plant and especially as a culm, of which only a few can be briefly outlined. The broad spectrum of applications will be divided into uses of the living culm and of the harvested culm.

In tropical countries, bamboo was one of the first materials used by mankind to increase comfort and well-being. Its wide distribution and availability, rapid rate of growth, superior technological properties, and easy handling make bamboo an ideal material for countless uses. Bamboo is sometimes still called "the poor man's timber," which is not fair, as in certain fields of application, bamboo has superior technological qualities. Since the species vary in their properties, their suitability for different products also varies. The manifold uses of bamboo have been intensively dealt with in numerous comprehensive books, partly listed under references.

Keywords Food • Beverages • Structural application • Housing • Bridges • Interior work • Furniture • Handicraft • Musical instruments • Engineered products • Panel products • Charcoal • Pulp

10.1 Utilization of the Living Culm

The use of bamboo as living plant is valued in many fields of utilization, such as vegetables and also as garden ornamentals, just to mention.

W. Liese (✉)
Department of Wood Science, University of Hamburg, Leuschnerstr. 91, 21031 Hamburg, Germany
e-mail: wliese@aol.com

J. Welling
Thuenen-Institute of Wood Research, Hamburg, Germany
e-mail: johannes.welling@ti.bund.de

T.K.H. Tang
Nong Lam University, Ho Chi Minh City, Vietnam
e-mail: kimhongtang@yahoo.com

© Springer International Publishing Switzerland 2015
W. Liese, M. Köhl (eds.), *Bamboo*, Tropical Forestry 10,
DOI 10.1007/978-3-319-14133-6_10

Fig. 10.1 Rice and fish cooked in fresh culm parts, Thailand (Liese)

10.1.1 Shoot and Culm

10.1.1.1 Shoot as Food

Bamboo shoots are an important daily food in Asian countries, especially in China, Taiwan, Japan, and Thailand. Their price is low, and they are easily available. The sprouts come from commercial plantations as well as from natural forests but are also imported in larger quantities. They are a staple food and mostly sold fresh at local markets and along the roads. The special flavor of a fresh culm is used for cooking rice and fish in culm parts (Fig. 10.1). In factories, they are processed into pickles in tins or in dried condition (Fig. 10.2).

Since each sprout taken terminates the growth of the respective culm, harvesting must be regulated in order to avoid the depletion of a bamboo stand; it is done once or twice a year. The management of bamboo stands for good quality shoots requires soil dressing and application of straw litter and farmyard manure or fertilizers.

Different countries prefer different bamboo species. Young shoots of some species contain lethal amounts of cyanogens, which may be toxic to cattle but are destroyed by cooking. In Thailand, most of the local species produce edible shoots, preferably *Dendrocalamus asper, D. giganteus, D. merrillianus, Gigantochloa albociliata,* and *Thyrsostachys siamensis.* Research on the nutritive value of different bamboos revealed that *Bambusa spinosa* has the highest amount of protein, calcium, and phosphorus.

10.1.1.2 Bamboo Wine

None of the thousand possibilities of using bamboo creates as much gaiety as the bamboo wine produced in Tanzania (Liese 2003). Bamboo wine is a 100 % product of the plant, while bamboo beer and liqueur made in China gain their specific taste from a low additive of flavonoid bamboo leaf extract.

Of about 1,200 species of bamboo, only one is known to be used nearly exclusively for the production of alcohol, *Oxytenanthera abyssinica.* The

Fig. 10.2 (**a**) Bamboo shoot for harvesting, (**b**) Different species on the market (**c**) Shoots as staple food, (**d**) Industrial production, Thailand (Liese)

plantations are owned by families or villages and are managed according to long-established agricultural methods. If the culms have grown about one meter at the beginning of the rainy season, the tip will be cut by a sharp knife in a slanting direction and after 7 days, a slice of 5 mm is removed. After 8 days, the dripping sap, called "Ulanzi," will be collected in an attached internode in the mornings and evenings, whereas in the afternoons the surface of the wound is newly cut to keep the sap oozing. The daily harvest consists of about 0.5–1 L. The harvest time can last a few weeks up to a month, depending on the rainy season. The sap is transported to the collection points in 20 L plastic containers and sometimes cleaned with filters.

Soon enough, the product is being enjoyed as a "social affair," in the family, circle of friends, or in a local bar. On the first day, the product tastes sweet and is enjoyed by women and children, and on the second and third day, it becomes sour dry and turns more alcoholic due to the fermentation up to about 7 % alcohol.

Fig. 10.3 Bamboo wine: collecting, tasting, transporting, and enjoying, Tanzania (Liese 2003)

During the harvest time, all work stops and on the streets one has to be especially careful (Fig. 10.3).

Due to the relatively low alcohol content, the pure bamboo drink is basically a beer, but known as "bamboo wine." The high value put on this cheap delight by about five million consumers in Tanzania has also led to an industrial commercialization. The fermented juice is filled in bottles and tins, sold within the country and also exported. This use of bamboo saves at least the usual fermentation of corn and millet for the wanted alcohol, which are important foodstuffs.

Harvesting and utilization of bamboo sap have also been reported by the Korea Forest Research Institute by taping culms of *Phyllostachys pubescens, P. bambusoides,* and *P. nigra.*

Fig. 10.4 (**a**) Larva of the bamboo borer (**b**) larvae collected at the node (**c**) bamboo larvae roasted with grasshoppers on the market in Bangkok (Liese)

10.1.1.3   Bamboo Borer

Bamboo borer is generally a pest leading to the deterioration of the culm wall. However, one beetle, the caterpillar *Omphisa fuscidentalis*, is of great value. The female lays a mass of 80–130 eggs mostly in the sheaths at the base of a culm. The eggs are inoculated into culms of *Dendrocalamus, Bambusa,* and others and develop within about 10 months into big larvae, which are collected from the cut culm. They are roasted and sold at the market as a delicacy, in comparison to the meager larvae of the grasshoppers, e.g., *Locusta* sp. (Fig. 10.4). The delicious "bamboo worms" are also available in tins.

10.1.1.4   Tabasheer

Tabasheer is the common name for the amorphous siliceous deposit in the lacuna of some bamboo culms, especially in sympodial taxa such as *Melocanna baccifera*

Fig. 10.5 Tabasheer above a node and its crystals (Liese 1998a)

and *Bambusa arundinacea*. It is also called "Pearl Opal" as a native remedy to South India. It appears as loose lumps of porous silica lying on the top of the diaphragm of the node. Before it becomes solid, a thick liquid is present. The phenomenon by which tabasheer develops in only a few old culms within one stand and in only some bamboo species remains still unknown.

Tabasheer has always been attractive because of its silica content, which can amount for more than 85 % of the inclusion. It also contains iron, calcium, and chlorine. Bamboo tabasheer appears in different colors; it can be white, bluish-white, transparent, or chalky. In general, the deposits in the pit cavity are loosening grains of the size of peas (Fig. 10.5). The formation of tabasheer in the pit cavity is an age-related process since the material is found only in old bamboo culms. The reason for the origin of such enormous amounts of silica in the lacuna remains obscure.

The extraction process starts with the identification of culms that contain the deposits. The telltale rattling method consists of shaking the culms vigorously to detect the noise. The loose tabasheer grains in the pit cavity of the culm are immediately manually collected.

Tabasheer has been used for centuries in traditional medicine. It is thought to clear away heat, eliminate phlegm, and reduce fiver and is also used as cooling tonic, to draw poison out of a wound and as an aphrodisiac. Because of its porous structure, tabasheer absorbs large quantities of fluid and adheres strongly to the tongue with a severe taste.

Tabasheer has also interesting technical properties as a catalyst. It is highly priced by collectors (Liese 1998a).

10.1.2   Leaves

Leaves amount to 5–10 % of the bamboo plant biomass. Their function is the photosynthesis for the production of biomass by transferring the atmospheric CO_2 into glucose with the release of O_2. Morphology and structures of leaves vary between species, so that they are used for identification. The evergreen leaves of bamboo are worldwide a characteristic part of the landscape as well as for the home garden.

Bamboo leaves are an important food for animals. Well known are the Pandas, which need 30–50 kg bamboo per day of which only about 20 % are digested. Also cows and goats belong to the leave consumers, by which the shape of the plant becomes domelike (Fig. 10.6). Leaves are also collected as a supplement fodder during the dry season. They contain up to 10 % hard-to-digest substances such as silicium dioxide, flavonoids, and various phenolic acids.

The use of dried bamboo leaves, like from *Sasa tessallata* to deodorize fish oil, has been patented in Japan (Liese 1998b).

The sheaths strengthening the expanding culm can be of magnificent size and structure to be used for many ornamental designs. Even in small size, they make an attractive envelope at sweetie shops (Fig. 10.7).

Bamboo sap is obtained by extraction of leaves. It contains flavonoids, amino acids, phenolases, peptides, mangan, etc. and serves as additive to soft drinks, shampoo, toothpaste, and medicine. In powder form containing 10–15 % flavonoids, it is used as medicine and as antioxidant (Fig. 10.8).

Bamboo beer is widely distributed in China and also available in western countries (Fig. 10.9). It is made with addition of flavonoids from leave extract, contains 5.5 % alcohol, and shows multiple health benefits.

Bamboo tea (Fig. 10.10) is known since ages in the bamboo countries as a delicious and healthy drink. It is now also spreading to the western countries. The youngest leaves of broad-leaved species, like *Sasa* sp., are cut during the first 5 weeks, cleaned, dried, and roasted. Bamboo tea contains neither theine nor caffeine and is rich in protein, calcium, iron, and magnesium. It is recommended for various pharmaceutical applications, especially stomach pain and can be enjoyed warm or cold.

10.1.3   Bamboo Seeds and Fruits

Bamboo flourishes after about 20–40, up to 70 years, dependent on the species. All its individuals with the same genetic resource flower simultaneously, also in distant countries and continents. Since bamboo is a grass, the flowers of almost all bamboo species produce masses of small inconspicuous seeds, like the grain of corn (Fig. 10.11). Tribal farmers can collect at such occasion 20–30 kg seeds per day to be used like rice.

Fig. 10.6 Cattle like bamboo leaves and demolish new growth, Ethiopia (Liese 1998b)

Fig. 10.7 Culm sheaths make an attractive packing for sweets (Waitkus)

However, there is one bamboo species, *Melocanna baccifera*, which instead of little seeds develops big pear-shaped fruits of about 35 mm up to 45 mm in diameter, distributed on the branches (Fig. 10.12). The fruits are edible and contain a rich pulp with approximately 50 % starch and 12 % protein. Their collection is enhanced since a virility capacity is been attributed to the fruit. While hanging on the branches, the fruit already develops strong rhizome strands as well as a tip for a new culm, a spear. This birthing living entity is also called viviparism. Once dropped, the fruit develops into a new plant through further rhizome and spear

Fig. 10.8 Toothpaste with bamboo leaves extractives (Tang)

Fig. 10.9 Bamboo beers made with extractives from bamboo leaves (Tang)

development due to the good nutrient supply by the fruit. Thus, it could grow vigorously, if left for live, which however, seldom occurs, due to its special fruits.

Melocanna baccifera is the dominant bamboo species in the Indian State Mizoram, growing also at other locations. In Mizoram, it started flourishing last before about 45 years and again since around 2000, as observed in 2004 in India, Indonesia, and Columbia (Liese 2008). As soon as the fruits began to develop on the branches, rats climbed up to start eating them. The fallen fruit were devoured quickly.

The frightening biology of rat reproduction is to be mentioned. The animals reach their reproductive age after 3 months and have a lifespan of up to 3 years. Since the female can give birth to 4–12 young every month, she can produce theoretically up to 15,000 a year under good feeding conditions. From May to October, the bamboo fruits are all eaten by the rats. After finishing this limited

Fig. 10.10 Bamboo tee made of bamboo leaves (Tang)

Fig. 10.11 Bamboo seeds, also a food resource (Waitkus)

resource, the meanwhile enormous hordes of rats move to the fields to eat everything, like grain and potatoes. They move further into the villages due to their existential gluttony. The previous blooming periods are historically known due to subsequent famines and diseases such as cholera, malaria, and typhus; also political unrest was reported.

Fig. 10.12 (**a**) Fruits of *Melocanna baccifera,* Bangladesh, 1969 (**b**) Fruits with rhizome strands and spear for a new culm Colombia, 2004 (**c**) Fallen fruit ready for life, South India, 2004 (**d**) fruits attacked by rats (Liese 2008)

10.2   Utilization of the Harvested Culm

10.2.1   *Building and Structural Applications*

Evolution has designed bamboo culms with the following features: efficient raw material use (hollow tube-like structure), extreme flexibility associated with high strength, high axial permeability (good for impregnation) protective layer with extremely low radial permeability (build-in protection against moisture uptake). Mankind has learnt to take advantage of these outstanding properties. Since ages, whole bamboo culms are used for constructing all types of buildings, e.g., houses, shelter, halls. Here, bamboo culms may serve as structural components (pillars, posts, columns, roof trusses, or stringers) or in split versions as cover, shingles, wall cladding, and many other applications. Bamboo culms are used as load-bearing elements for building bridges and towers. Many authors consider bamboo as one of the most environmentally friendly building materials. Its fast growth and low prize in conjunction with well-adapted building techniques make bamboo structures competitive (Van der Lugt et al. 2005). Yu et al. (2011) reported about the energy use and carbon emissions (LCA) of bamboo-structure residential building prototype in China.

Bamboo is not a durable building material. Without appropriate protection by design combined with good chemical preservation (see Chap. 9), bamboo buildings will not last very long. But with proper design and protection, bamboo buildings may stay in service for several decades.

Due to its round shape, bamboo culms require specific connectors not only when large structures are designed but also when small houses are built. In the different regions of the world, specific connection techniques have been developed. It is far beyond the scope of this book describing in detail all the different types of connectors nor is it possible to give a full overview about all types of building applications. Therefore, only some examples explaining the broadness of the possible applications will be given. For more detailed information, the literature has to be visited, e.g., Janssen (2000), Minke (2012), CAN (2013).

Single- and multifamily houses made from bamboo culms are the most frequently found application of bamboo culms. South American dwellings may be different from South-east Asian dwellings, but the principle building concepts are quite the same. Resting on a mineral or rock-based foundation, strong bamboo culms are used as the principal structural (load-bearing) elements. Transverse bracing is needed to stabilize the wall and roof structure. Smaller diameter culms or split culms (halves) may be used as in-fill material for outer and inner walls and roof cover material. Clay and other locally available building materials are used to close the wall structure.

Due to the light weight of the hollow culms, good insulation properties are achieved. Walls of bamboo dwellings are normally diffusion open which results in a good and healthy indoor climate. Bamboo buildings are earthquake resistant due to their lightness and flexibility (both are inherent properties of the bamboo culm and the connectors). Local availability of the raw material and the need for skilled labor generate employment and income. Therefore, building with bamboo supports the rural development and helps to mitigate migration into cities. On the following pages, examples for simple buildings and structural applications of bamboo culms are given (Figs. 10.13 and 10.14).

Bamboo is used as a building material, mainly in rural areas. The load-bearing structure of the small houses (Figs. 10.15 and 10.16) for farm workers which were built in Columbia after the terrible earthquake in 1999 consists almost completely of Guadua culms (Tistl and Velásquezgil 2001). The decision for a reconstruction in bamboo was based on the fact that only the bamboo-based dwellings had survived the earthquake. Local craftsmanship and the local-grown bamboo raw material are visible in many subtropical areas of the bamboo belt all over the world.

Bamboo is used not only for small houses but also for larger buildings such as garages, sheds, protected yards and canopies, smaller and larger industrial buildings, and stables for cattle, horses, and other animals. Small and wide span bridges have been constructed using bamboo as the principal load-bearing material. In such cases, the proper protection of the bamboo material, both by design (good cover and protection against rain) and by means of chemical treatment, is of utmost importance.

Fig. 10.13   Small dwellings in Columbia (gtz)

Fig. 10.14   Small family houses in Columbia (gtz)

There are some architects, who gained worldwide recognition for their specta-
cular buildings in bamboo (Figs. 10.17 and 10.18). The pavilions of Markus
Heinsdorff earned worldwide reputation for their exceptional design in bamboo
(Heinsdorff 2010). The Colombian architect Simon Veléz designed and built the
famous ZERI pavilion for the EXPO 2000 in Hannover, Germany (Fig. 10.19a, b).
For getting permission to build the pavilion in bamboo, Veléz had to prove that
bamboo can fulfill the requirements of the strict building regulations in Germany.

Fig. 10.15 Small business unit or imbiss station (gtz)

Fig. 10.16 Kiosk (gtz)

For this reason, a prototype of the pavilion was erected in the park "Recinto de Pensiamento" in Manizales, Colombia. The roof structure was loaded with enormous weights to show that the pavilion structure would stand safe and not endanger visitors. While the EXPO pavilion was demolished after a bit more than 1 year, the prototype in Manizales was still in use in 2013 and could be visited. Another famous building designed by Simon Veléz was the temporary cathedral (Fig. 10.19), which was erected in Pereira, Columbia, to provide an alternative

Fig. 10.17 Look-out platform, Universidad Technologica de Pereira (UTP)

Fig. 10.18 ZERI pavilion (architect Simon Veléz) in the park "Recinto de Pensiamento" in Manizales, Colombia (UTP)

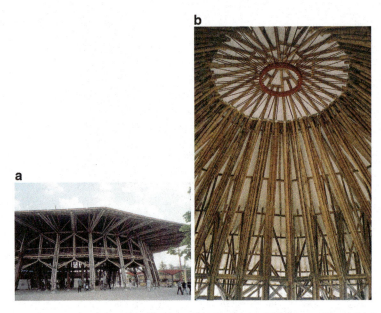

Fig. 10.19 (**a**) Pavilion in EXPO 2000 Hannover (Liese) (**b**) Ceiling structure of the Pavilion (Liese)

Fig. 10.20 Temporary cathedral in Pereira (architect Simon Veléz), Columbia, during the restoration of the ancient cathedral (UTP)

place for the religious community while the old cathedral was restored (Von Vengesack and Kries 2000) (Fig. 10.20).

When building with bamboo some similarities with building with wood can be observed. But, due to the round shape of the culms, special solutions had to be found. Lattice and skeleton framing are common. Wall structures may be filled with

Fig. 10.21 Bamboo wall: above bamboo structure, below structure filled with mortar (gtz)

mortal or cladded with panel-type products or even with split bamboo culms (Fig. 10.21). For roof constructions with bamboo (Fig. 10.22), special connectors are needed. Timmerman skills, some engineering knowledge, and a good under-standing of the bamboo material behavior are essential for long-lasting and safe buildings. Multi-material solutions are good options for cheap and sustainable building with locally available materials.

However, the use of bamboo in building applications is not limited to the bamboo culm. A wide range of applications have become possible due to the transformation of the bamboo raw material into engineered bamboo products (see Sect. 10.2.3). Bamboo-engineered products can be applied in most of the areas where wood-engineered products already have found their markets. The use of

Fig. 10.22 Roof constructions: consisting of several culms with integrated connectors (gtz)

engineered bamboo products has just started one or two decades ago. The rapid development of the novel type of bamboo products already now leads to competition with wood products, especially in regions where the wood recourses are limited and bamboo is available.

10.2.2 Interior Work, Furniture, Handicrafts, etc.

In the construction sector, bamboo culms and bamboo-based products can be used for load-bearing and non-load-bearing applications. For safety reasons and to guarantee a long life span, bamboo products must be treated with preservatives under endangered conditions. For indoor use, such safety measures are less important. Nevertheless, at least in the tropical and subtropical countries, bamboo used for interior design and furniture should be protected against mold, mildew, and fungus growth and destroying insects. Appropriate measures are described in this chap. 9.

10.2.2.1 Interior Work

In the bamboo belt all over the world, bamboo has been used for interior work since centuries, not only by the poor part of the population. Bamboo can be used in a wide range of forms including small diameter culms, split culms, processed bamboo in the form of woven sliver mats, or bamboo panels composed of rectangular four-side planed splits. Due to its unique visual appearance, bamboo and bamboo products used for interior work can easily be distinguished from wood and wood-based products. Most probably, customers and consumers will decide at early stages in the decision chain to use bamboo or wood for interior work.

While the use of bamboo for interior work is quite common in the bamboo belt, bamboo is considered as a quite unusual and "exotic" material in the rest of the world. Therefore, bamboo products are normally chosen by people who want to attract the attention of their visitors or customers. In industrialized countries, the bamboo look is often associated with holiday feelings, relaxation, and sun and tropical warmth. Bamboo-based interior design is found in bars, showrooms, expositions, wellness resorts, etc. (Fig. 10.23).

Using bamboo for interior work requires specific skills, tools, and parts to be used for fitting, binding, and mounting the bamboo parts together and/or attach them to walls, floors, and ceilings. Wall cladding, partitioning of rooms, and flooring are sectors where bamboo and bamboo products gain more and more attention. Bamboo processing companies and importers have recognized that bamboo products for interior design target growing market with a huge potential. A good marketing, quality-controlled products, and strong efforts towards standardized base products are essential to put into practice this chance for augmentation of the use of bamboo.

10.2.2.2 Furniture

According to the Oxford dictionary, "furniture are movable articles that are used to make a room or building suitable for living or working in, such as tables, chairs, or desks." The term furniture covers a very wide range of different products. There are two main types of bamboo furniture making (a) traditional bamboo furniture, which use bamboo culms as a primary material and (b) modern bamboo furniture made of prefabricated bamboo products, such as laminated bamboo, veneers, and various types of engineered bamboo panels (Fig. 10.24).

Traditional bamboo furniture uses natural round or split bamboo. The specification of products ranges from chairs, sofa sets, and beds to kitchen cabinets and gazebos. Examples are shown in Fig. 10.25 a, b. They are wholly made of bamboo culm parts and splits, which must often be treated to ensure their quality. Many of the manufacturers prefer culm parts without skin. The general process of bamboo furniture production is shown in Fig. 10.26.

Using bamboo in round form requires special skills and techniques, whereas the application of bamboo-engineered products demands more or less the same expertise which is needed in manufacturing wooden furniture.

Techniques to build bamboo furniture are described by a number of authors. A very comprehensive collection of examples covering all types of articles including furniture was presented by Rubio-Luna (2007). The book is specially designed for craftsmen. All important issues are illustrated by simple drawings and described in simple words. Another comprehensive handbook on bamboo furniture making was presented by Annonymus (2003).

Fig. 10.23 (**a**) Bamboo-based interiors Room partitioning (www.515store.de) (**b**) Inner sections of bamboo culm used as decorative wall cladding (**c**) Interior design with bamboo culms in a wellness resort (Rainer Sturm/pixelio.de)

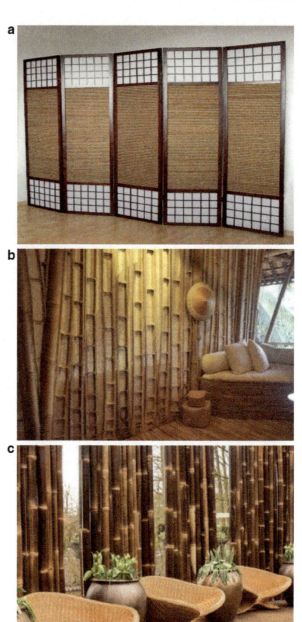

Using bamboo in its round form does not really comply with large-scale indus-trial manufacturing techniques, because the bamboo culm is not a standardized raw material. Culm diameter and culm wall thickness may vary in wide ranges. Variable

Fig. 10.24 Furniture made of nature bamboo culm parts (Bamboo Nature)

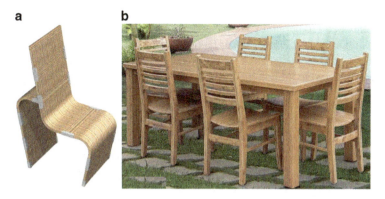

Fig. 10.25 Furniture made of laminated bamboo: (**a**) chair (Melanie Leferink); (**b**) Table chair set (Bamboodecor, Vietnam)

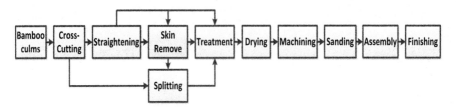

Fig. 10.26 General process of bamboo furniture production (Tang)

internode length and node structure are specific features of bamboo. Therefore, the manufacturing of furniture from bamboo automatically will include a lot of manual work and individual decisions on how and where to cut the culm for yielding the parts needed to make the furniture. Each of the resulting pieces of furniture will have its own characteristic.

The round form of the bamboo culm requires special cutting and shaping techniques for connecting bamboo furniture parts. There are quite a number of techniques available including interlacing with split bamboo/natural fibers/ropes, bolting, pinning, or using metal connectors. Glueing, being a widely used connection method for wooden furniture, is hardly ever used for manufacturing bamboo furniture, because of the round form of bamboo and the fact that the skin of bamboo does not allow good bonding even when high tech gluing systems are used. A further difficulty arises from the need to connect the open-structured and very permeable cross-sectional bamboo tissue with extremely dense and impermeable bamboo skin, while at the same time, the variable round shape does not allow perfect fit.

Furniture made of bamboo with skin still in place normally is not painted. The hard, smooth, and impermeable skin provides sufficient protection against dirt and user wear. Nevertheless, a certain protection is needed to prevent infections by insects. Due to the high sugar and starch content, mold easily develops on bamboo furniture, especially on cross sections and inner surfaces when high relative humidity and elevated temperatures are prevailing. This is normally the case in regions where bamboo grows. For export of bamboo furniture, moisture content has to be kept in tight limits. For container transport, conditions inside the containers must be controlled to avoid mold growth on surfaces during transport from origin countries to customers. Mold infection is normally associated with an unpleasant smell. While superficial mold can often be wiped off, the smell normally will not vanish, what leads to considerable or even total loss of value.

Furniture manufactured from engineered bamboo products have properties and behave like wood-based furniture. Unlike the traditional design, this furniture may be shipped in compact flat packs, to be assembled on the spot by the buyer without sophisticated tools.

Engineered bamboo furniture has to be preserved and surface treated to prevent damage from fungi, insects, and daily activities of the users. A wide range of oil, waxes, lacquers, and paints is applicable. Gluing and connecting techniques are similar or equal to those used for wood-based furniture. Special attention must be given to the types of glue. Water or weather-resistant glues should be preferred to avoid failure of glue lines. New designs overcome many of the problems of traditional bamboo furniture, such as high labor and transportation costs, low productivity, instability, varying quality, and susceptibility to insects and fungi. At the same time, it retains the distinct physical, mechanical, chemical, environmental, and esthetic features of bamboo.

Exports of furniture made of engineered bamboo have exploded during recent years due to the dramatic improvement of the production technology of engineered bamboo products and the enhanced production capacities. Nowadays, bamboo furniture manufacturing is no longer craftsmen work but rather industrial-scale series production at high-speed and high-quality level.

10.2.2.3   Handicrafts

Bamboo crafts are traditional products in China, India, Indonesia, Malaysia, the Philippines, Thailand, and Vietnam, also in Colombia, Peru, and other South American countries. In Ethiopia, as an African country, bamboo handcraft is widespread and simple furniture is sold directly along the roads (Fig. 10.27a, b). The techniques have been known for 1,000 years. The diverse products have become an indispensable part of daily life, literature, and art. Due to the special characteristics of the bamboo, culm handicraft made from this raw material can easily be recognized. The round shape, the smooth skin, the characteristic color, and its texture originating from the vascular bundles embedded in parenchyma tissue and the node/internode sequence are characteristic features of bamboo handicrafts.

Bamboo handicrafts comprise all sorts of useful items, e.g., kitchen tools and accessories to assist cooking and presentation of food, pens and pencils, boxes, bowls, mugs, bird homes, and countless other items. Bamboo handicraft also includes countless useless items, the so-called knickknack, and of course souvenirs.

The local-grown bamboo is easily accessible to small-scale enterprises and individuals who earn their living by adding value to a cheap raw material. Bamboo-based handicraft is normally sold locally to visitors and tourist or sold to wholesale buyers who provide the products to retailers in the tourist resorts or to export traders. Bamboo handicraft is bought by tourists as a reminder for holiday and relaxation in tropical counties. There is hardly any plant that is more characteristic and typical for tropical regions than bamboo.

10.2.2.3.1   Bamboo Weaving Products

There are nearly 20 categories of woven bamboo products in Asia, including fruit baskets, trays, bottles, jars, boxes, cases, bowls, fans, screens, curtains, cushions, lampshades, and lanterns (Fig. 10.28).

Bamboo weaving is the method for making two-dimensional or plane products such as woven mats, which can be of varying lengths and widths. Woven bamboo is

Fig. 10.27 In Ethiopia, bamboo handicrafts is widespread, and simple furniture are sold along the road (Liese)

Fig. 10.28 Bamboo weaving products (Liese)

found in the form of three-dimensional woven products such as baskets, caps, lampshades, and figurines. Evidently, woven bamboo products also include an additional spatial dimension, which depends on the thickness of the bamboo strips. But for all intents and purposes, plane bamboo products are distinguished from three-dimensional woven bamboo products because they are designed for flat surfaces.

Fig. 10.29 Coiled bamboo products (Tang)

Principal steps in the primary processing of materials needed for weaving bamboo include (1) Scrape off the green skin of the culm; (2) flatten the edge of the culm; (3) divide the culm into slivers of an equal width; (4) separate the slivers; (5) cut the slivers into fine layers; (6) separate the layers into the trips; and (7) classify and group the strips by color (UNIDO and INBAR 2008).

10.2.2.3.2   Bamboo-Coiled Products

Coiled bamboo products are hand coiled using small strips of bamboo which are then hand shaped and finished. Producing coiled bamboo items is a multistage process that extends over months. First, culm parts are soaked in water. After drying, the artisan cuts the stalks into thin, narrow strips and coils them around a wood or bamboo form, working inward to create the desired product. The product is then soaked in glue and placed in the sun to dry. As a next step, the artisans sand or shave the piece, then apply lacquer and varnish. Finally, the product is washed and sanded several times until it is smooth.

The bamboo items produced using coiling techniques, mostly originating from Vietnam (Fig. 10.29), were considered trendy during recent years in Europe.

10.2.2.3.3   Musical Instruments

Among all the natural materials used worldwide in the manufacture of musical instruments, one stands out in its versatility, is the grass bamboo. Bamboo is the only material whose instruments of all classes like wind, percussion, string instruments, and even the strings themselves can be made exclusively from this one material. The hollow structure of a bamboo culm provides a characteristic sound to relate to the music of many ethnic communities. When and what kind of musical instruments were made first from bamboo is unknown. The bamboo list is endless, and new instruments incorporating bamboo are invented all the time.

Details about the many bamboo instruments, their properties, and manufacture are presented in detail by Cusack (1999) and Wegst (2008), richly illustrated. The following can only give a brief touch to this wonderful world of special music.

Fig. 10.30 Bamboo organ, Las Pinas, Phillipines (Liese 1985)

Of all the musical instruments, the bamboo organ at Las Pinas in the Philippines is the most famous. It was constructed in 1818 from 950 bamboo culms (Fig. 10.30). Since then it has been repaired several times, preserving its original form. Presently a bamboo organ is built for a concert house in Pingtung, Taiwan, by a German Company.

In West Java, there are more than 20 musical instruments made from bamboo, like Angklung (a rattle diaphon, Suling and Celempung as hammer, aero, or string instruments, mostly from *Gigantochloa* or *G. apus* (Widjaja 1980). Lengths of the culms are used as resonators below the bronze keys of xylophone-like instruments. Bamboo flutes have existed for time immemorial. They are easily made due to the hollowness of the culm, and the Japanese flutes Shakuhachi and Shinobue are classical instruments as also is the Pan flute Zampano. The bottom part of smaller culms requires only a few finger holes and a mouthpiece to produce an excellent flute. One would need to write an encyclopedia to describe all the traditional and modern instruments made from bamboo.

A few of the better known were listed by Cusack (1999): Didgeridoo (Aboriginal drone tube) and the Pan flute (a series of tubes of different length and diameter); Flute (cross-blown version); Indonesian flute (end-blown flute with a bamboo ring mouthpiece); Whistle flute (an end-blown flute with a whistle-like mouthpiece); Shakuhachi flute (Japanese end-blown flute) Jegog (the Balinese bamboo xylophone Fig. 10.31); Click sticks (solid bamboo makes a clear, sharp, and resonating sound); Angklung (an Indonesian instrument one could compare to a bamboo handbell); Reed flute (an end-blown instrument with a reed like a clarinet); and Bamboo skin drum (made from the largest diameter culms possible). For each of them, the detailed manufacturing is described.

An extraordinary musical sight was the bamboo "brass" band performance at the Fourth International Bamboo Congress at Ubud, Bali, in 1995 (Fig. 10.32). It consisted of about 50 members with every instrument made of bamboo. The traditional bamboo instruments were to be expected, but imagine a tuba made completely of bamboo! They also had saxophones, clarinets, trumpets, French horns,

Fig. 10.31 Bamboo xylophone (Liese)

Fig. 10.32 Bamboo orchestra, Bali (Liese)

drums of all sizes, trombones with culm "slides", piccolos and flutes, with bright red uniforms, hats, and the conductor's baton all made of bamboo. The sound was superb, far softer than a true brass band, and the quality of the sound with its slightly woody trumpet and tuba tones seemed to add something unique to the music.

The craftsmanship and imagination to make these instruments were extraordinary. Culm sections had been carefully split and cut into tapered wedge pieces. They were glued together in hundreds to create the increasing diameter or bend and then carefully sanded inside until the required shape was achieved.

10.2.3   Engineered Bamboo Products

During the last decades, quite a number of engineered bamboo products have been developed. Therefore, some general information about these novel types of products seems to be necessary. The term engineered bamboo products covers a wide range of bamboo derivative products. Common to all engineered bamboo products is that the culm is broken down into smaller units (e.g., veneers, lamellas, splits, strands, slivers, particles, or even dust) which are further processed by binding or fixing the units together with adhesives, plastics, or other methods of fixation to form a composite material with uniform properties. A wide variety of different forms for composing the units is possible, and the type of composition normally is the basis for nomenclature and classification (Jiang et al. 2002; Zhang et al. 2002). Common to all these products is that they are engineered to precise design specifications. They can easily be tested in relation to national or international standards. Engineered bamboo products are used in a variety of applications, including handicraft, furniture, interior design, flooring, and structural and automotive applications.

10.2.3.1   Laminate Bamboo Lumber

For producing laminate bamboo lumber, the round-shaped hollow bamboo culm must be converted into rectangular elements. To achieve this task, the culm is first split into segments which are preplanned to remove the skin and the inner layer (Fig. 10.33). The splits are watered and bleached (if necessary), dried and planed on four sides to get uniform rectangular lamellae, which can be further processed into laminate bamboo lumber. After application of resin, the lamellae are assembled and placed in presses. Laminate bamboo lumber with a wide range of dimensions can be produced by gluing together smaller units which had been produced in previous steps. Laminate bamboo lumber is used for window scantling, structural building material, and joinery. Whenever bamboo laminate lumber is applied, appropriate protection against moisture, fungi, and insects has to be considered.

10.2.3.2   Bamboo Veneer

Bamboo veneer can be produced on a peeler lathe by rotating culm sections against a knife and nose bar assembly. Due to the small diameter and the hollow tube-like structure, peeling of bamboo culms is quite difficult compared to wood. More widely used is the slicing technique. Here, blocks or planks of laminate bamboo lumber are sliced on horizontal or vertical lathes perpendicular to the grain direction or on longitudinal lathes along the grain direction. Whenever bamboo slices or veneers are glued together to form the basis for larger pieces for veneer

Fig. 10.33 Cross section of laminate bamboo lumbe (Li Xiao 2011)

manufacturing, the resulting veneer is called reconstituted veneer. By assembling different colors of bamboo, a very wide range of decorative veneers can be produced.

Bamboo veneers have thicknesses between 0.15 (micro-veneer) and 1.5 mm. Due to its low tensile strength perpendicular to the grain direction, bamboo veneers are very fragile. For improving tensile strength in perpendicular to grain direction, very thin nonwoven cloth-type support layers may be attached to the backside.

Bamboo veneer is a product with high added value which is used as surface layer in coating processes for other bamboo- or wood-based panels. By the coating process, the visual appearance is improved and a solid bamboo composite impression is provoked.

10.2.3.3 Ply Bamboo

Wood-based plywood comprises veneer sheets which are assembled in such a way that adjacent plies have their grain direction rotated relative to adjacent plies by 90°. Thickness of veneers can vary in wide ranges. Large plies can be produced by gluing smaller sheets edgewise together.

For producing ply bamboo materials, not only veneers but also splits, lamella, or slivers are utilized for the formation of middle layers. Depending on the inner structure and alignment of the inner layers, a large variety of different ply bamboo products can be produced (Fig. 10.34).

For making the ply bamboo, middle and surface layers of the bamboo culm are normally split into sections (Fig. 10.35), which are roughly planed to remove the skin. The splits are then boiled, bleached, and anti-mold or anti-pest treated.

Fig. 10.34 Different forms of ply bamboo panels used as container boards (Schülke 2009)

Fig. 10.35 Ply bamboo based on splits, bleached or natural color, and darkened by heat treatment (Waitkus)

Color can be adjusted by heat treatment (often, but not correctly, termed as carbonizing) which is carried out during or after the drying process. The dry splits are then calibrated to equal width and thickness, sorted, resinated, and assembled to form layers, which are glued together in side presses or plate hot presses. After thickness calibration, several sheets or plies (with identical or different structure) are assembled crosswise similar to plywood forming panel-type ply bamboo products with low anisotropy (Fig. 10.35).

Another alternative for producing middle layers of ply bamboo is to use curtains or mats. Curtains are formed by connecting splits by means of thin filaments. In this case, the edges do not have to be resinated. For producing bamboo mats (Fig. 10.36), splits have to be stripped into thin slivers (0.8–1.5 mm). For high-quality ply bamboo panels, the thickness variation of the slivers should be as low as possible, certainly not more than 1 mm. Slivers have to be woven into mats which are cut to predetermined dimensions before being used to form middle layers in ply bamboo panels.

Many different forms of ply bamboo are on the market now. They can be applied as heavy duty container panels, flooring materials (Fig. 10.37), and decorative material for furniture production or for interior design purposes.

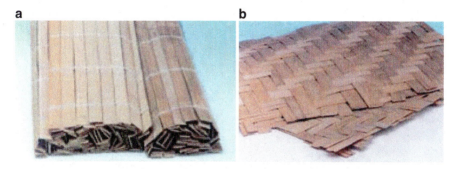

Fig. 10.36 Materials for producing ply bamboo: (**a**) bamboo curtains, (**b**) woven bamboo mats (Qisheng et al. 2001)

Fig. 10.37 Different types of bamboo parquet products (Waitkus)

10.2.3.4   Oriented Strand Board

For making bamboo Oriented Strand Board (OSB), the culm must be converted into thin strands. The culm is cut into sections which are introduced into a flaker. Preferred cutting direction is along the radial axis so that the outer dermis is always located at the thin side of the strands. Dermis located on the wide side of the strand

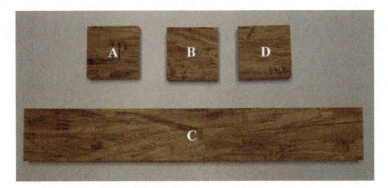

Fig. 10.38 Bamboos OSB (Malanit 2009)

cannot be resinated properly and will result in poor internal bond of the bamboo OSB. Strands are dried in rotary drum driers and resinated in mixing units. After resination, the strands are scattered onto a moving belt. Normally, strands are aligned by the scattering unit and by rotating the next scattering unit by 90°. With several adjacent layers plywood like panel properties can be achieved. A normal mat forming station comprises three or five scattering units. The strand mat is either transported to a continuous hot press or after separated into sections, to an 8–15 h daylight hot press. Malanit (2009) reported about the production and properties of OSB made from bamboo (Fig. 10.38).

10.2.3.5 Bamboo Particleboard

Worldwide, the larger part of the particleboard production is based on virgin wood and recycled wood as raw material resource. But, bamboo can also be used for producing particle boards. Special attention has to be given to the fact that bamboo, depending on age and time of harvest, contains considerable amounts of sugar and starch. This means that the raw material may be negatively affected by fungi and insects during storage. But also the finished product may be endangered by fungal attack when used under humid conditions. Therefore, precautions have to be taken. Bamboo particleboard may need to be protected by preservatives.

Normally, culm sections are cut into smaller pieces which are fed to a knife ring flaker. Here the raw material is transformed into preferably longish particles. Particles are separated in a sieving station into coarse middle layer particles and fine surface layer particles. After drying to rather low moisture content, the particles are resinated in a blending unit. Here the moisture content of the surface layer particles is adjusted to approximately 10–14 %, whereas the moisture content of the core layer particles is kept at 6–8 %. The particle mat is formed by scattering first the lower face layer, followed by the coarse core layer and an upper face layer. The mat is then transported to a continuous hot press where the mat is compressed and heated until the resin is cured. During hot pressing, the typical U-shaped density profile is formed, with high density in the face layers and lower density in the core layer. During hot pressing, part of the moisture contained in the face layer particles is converted into vapor which travels to the core where it assists the heating-up process. At the end of the pressing cycle, the moisture distribution across the panel thickness should be more or less uniform.

The endless particleboard belt is cut into sections when it leaves the press. Sections are introduced into a cooling star to allow panels to cool down to a temperature close to ambient temperature. Finally panels are calibrated in a sanding process. Panels are then stacked to form piles for transport to customer or further cut into smaller units according to customer's order.

Due to its properties, particleboard is normally used for the manufacturing of furniture. Its high density face layer allows veneering and coating with melamine-impregnated paper. Edges can be edge coated so that the particleboard core is not visible anymore in the final product. The visual appearance of bamboo particleboard is identical with wood particleboard.

10.2.3.6 Bamboo Fiberboard

10.2.3.6.1 Medium Density Fiberboard

Medium density fiberboard (MDF) is another type of panel which is normally produced from wooden raw material sources. But, of course, also bamboo can be transformed into fiber material in a refiner process. The culm first has to be reduced to smaller particles which are fed into a refining process where under heat and some pressure, the material is separated into individual fibers. The fibers are then introduced into a blow pipe where the fiber material is dried and resinated. In a mat forming unit, the fibers are assembled in the form of a loose fiber mat which is pre-pressed before it is introduced into a continuous hot press, similar to those used for producing particleboard. At the end of the press, the boards are cut into sections, introduced into a cooling star, and after being cooled to approximately ambient temperature, the panels are calibrated in a sanding unit prior to delivery to customers.

MDF has a less pronounced density profile. Due to its homogeneous inner structure, MDF is best suited for processes where edges are round shaped or flat faces are carved. MDF can easily be coated and even be painted without prior coating. It, therefore, is mostly applied for making furniture or inner wall cladding.

10.2.3.6.2 High Density Fiberboard

Similar to MDF panels, High Density Fiberboard (HDF) panels can be produced. They are normally thinner and have a much higher density. Production process is almost identical as for MDF, but resin content may be higher. HDF panels are mostly used as core material for laminate flooring, comprising a melamine-impregnated paper face layer and a protective backside. HDF can also be produced from bamboo, but special attention must be given to fungal attack and dimensional stability, because HDF products, at least temporarily, will be exposed to humidity (e.g., during cleaning the laminate floor).

10.2.3.7 Nonwoven Strand Bamboo Products

A rather new type of bamboo product is nonwoven (compressed) strand bamboo. It is found on the market as plank, scantling (either four-side planed, profiled), tongue and groove element, parquet element, etc.

Nonwoven strand bamboo is produced by splitting the culm into segments which are fed through motor-driven rollers (crusher), which crash the bamboo into a type of mat. In some cases, the skin is removed by means of a preplaning process. In longitudinal direction, vascular bundles and surrounding parenchyma tissue are separated without destroying the vascular bundles. Mats are boiled to reduce sugar and starch content, dried, and subsequently resinated by dipping the mat into a resin bath (normally phenol-formaldehyde). The soaked mats are then transferred into pressing molds. In a hot press, the bamboo mats are compressed and densified to approx. 1.0 g/cm^3 or even higher while the resin is cured. After cooling and conditioning, the bamboo elements are planed, cut, or profiled to the desired dimensions. Nonwoven strand bamboo or compressed bamboo products (Fig. 10.39) exhibit extreme hardness, very high bending, and compression strength, comparable with or even better than the strongest wood species.

Due to its high density and high resin content (15–20 %), moisture uptake of nonwoven strand bamboo is slow. Nevertheless, long-term exposition to humid climate or liquid water will lead to moisture uptake and swelling. During swelling, the compressed bamboo will break resin bonds. As a consequence, the swelling of compressed lignocellulose material is not fully recoverable during shrinkage, which leads to unwanted deformations. For this reason, nonwoven strand bamboo products must be thoroughly protected against moisture uptake, either by coating with water repelling surface treatment or, even better, by not exposing it to long-term humid climate.

10.2.3.8 Bamboo Plastic Composites

Bamboo Plastic Composite (BPC) is a product derived from the wood sector. Bamboo material is ground into flour-like powder or very fine particles, which are mixed with plastic materials (normally PP or PVC) in a compounder. Here up to 60 % (weight basis) woody particles are thoroughly mixed with 40 % or more plastic material. The resulting compounded material is used in extrusion or injection processes. When bamboo is ground, it can easily be used instead of wood. Typical BPC products are extruded planks for decking (Fig. 10.40). Due to its water

Fig. 10.39 Various nonwoven compressed strand bamboo products (Waitkus)

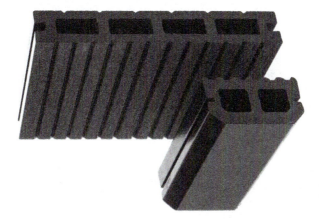

Fig. 10.40 Bamboo plastic composite (BPC) decking planks and substructure elements (Waitkus)

repellent properties, BPC planks are substitutes for pressure-treated wooden planks or wooden planks produced from wood species with high natural durability.

10.2.4 Paper and "Bamboo Textile"

10.2.4.1 Paper

Paper is a main product made from bamboo. Comprehensive literature deals with this topic. To provide certain information, the following is a shortened version from the GTZ book, "Bamboo-biology, silvics, properties, utilization" (Liese 1985). Dr. Othar Kordsachia kindly updated some data. Extended information is presented in the part on "Bamboo pulp and paper making" in the Handbook "Bamboo and Rattan in the World," 2007 edition by Jiang Zehui on pages 178–198.

In China, paper was handmade from bamboo already 2,000 years ago. During the fifth century AD, the inner parts of the bamboo were beaten into pulp and used in paper production. Today the annual output of bamboo, e.g., in India is about 3.23 million tons, and more than half of this production is consumed by the paper industry. Pulp mills using bamboo as raw material exist in most Asian countries. In India, 30–35 factories make paper from bamboo pulp. The present world production of pulp from bamboo is approximately 1.5 million air dry tons corresponding to an annual consumption of 7–8 million tons of raw bamboo.

The fiber length of bamboo varies from species to species and with internodal length. Mean values for species differ for fiber length from 1.5 to 4 mm, fiber diameter from 11 to 19 μm, lumen diameter from 2 to 4 μm, and wall thickness from 4 to 6 μm. Thus, the fibers of bamboo are well suited as raw material for pulp production. They are more slender than wood fibers, which contributes to the desired smoothness and flexibility. The high-cellulose content implies suitability for dissolving pulp and for rayon production. Since bamboo contains more impurities than wood, however, cooking is more costly and the pulp yield in the kraft process is adequate to hardwood cooking.

Since the pulp mills require a large amount of bamboo and extraction in a tropical climate is often limited to 6–8 months, a stack of bamboo for 6–9 months production has to be maintained. A mill producing 100 t per day requires about 43,000 t of bamboo in the yard. During storage, decay by fungi and beetle attack can cause serious damage. In the case of the common brown rot, the yield of pulp is considerably reduced and the kappa number is so high that the pulp becomes extremely difficult to be bleached. This is due to selective attack of the fungi on carbohydrates. Therefore, bamboo, subjected to brown rot fungi, is unsuitable for pulping. In the case of white rot, bamboo can be used for pulping, but lower yields of pulp and lower physical strength result with more need for bleaching chemicals. During storage for up to 12 months, about 20–25 % of the bamboo can be destroyed due to attack by wood destroying organisms. The storage loss could be reduced by preservative treatment, but sufficient drainage, aeration, and limitation of the size of the stacks also provide protection. Stored culms should have no soil contact. Precautions against fire hazard are necessary.

Because of its anatomical structure, bamboo is a suitable material for pulping. The cooking chemical can penetrate the large vessels and diffuse into the surrounding tissue. Formerly, the culms were first crushed and then chipped. New mills are using chippers only. Due to the hard and slippery skin of the culm, chipping of bamboo requires special techniques. The chips should be 18–20 mm long.

In early years, pulp mills used a two stage process for bamboo pulping. First, a weak alkaline solution removes the low polymer carbohydrates, followed by cooking with caustic soda and sodium sulfide. Presently, pulping is carried out mostly in single stage cooking. The kraft process is best suited for producing pulp chemically.

In a new mill with modern cooking technology, bamboo is pulped with 18–20 % effective alkali as NaOH at low temperature of 142–144 °C to produce pulps (50–55 % yield) well suitable for writing and printing papers. Based on kinetic data, regression equations have been developed for controlling the process of kraft pulping bamboo for producing pulps up to 70 % yield. To produce 1 t of unbleached pulp, 2 t of clean, chipped bamboo are needed, and the production of 1 t of bleached pulp requires about 2.5 t of air dry bamboos (4 t fresh). Although nodal portions give poorer kraft pulp, it is not considered feasible to separate nodes and internodes before digestion since the reduction in yield and quality due to nodal portions seems to be insignificant. Bamboo black liquor obtained by kraft cooking contains about 3–4 % silica, creating problems in the recovery cycle and requires partial damping of the lime mud.

Bamboo pulp is more resistant to bleaching than wood pulp because the chemicals cannot penetrate through pits into the cell lumen as in soft wood tracheids. This results in a poorer removal of lignin. Formerly, bleaching was conducted in a CEHH sequence applying relatively high chlorine charges. In this bleaching sequence, bamboo pulp could not be bleached beyond 78° brightness without seriously reducing its strength. Nowadays, modern ECF bleaching sequences such as OD (EOP) D P or ODO(DQ) (PO) are in industrial use. TCF bleaching according to OQ(EOP) Q(PO) was tested successfully at laboratory scale.

The quality of bamboo pulp is considered to be fairly good. It has a high tearing strength and compares well with softwood kraft pulp in this respect. Tensile and bursting strength are lower than those of softwood pulp but still on the same level as those of hardwood kraft pulps. For many purposes, bamboo pulp is mixed with pulp from other grasses, bagasse, wood, rag, or waste paper. By mixing it with ground wood pulp (60 and 40 % bamboo chemical pulp), it can be used in the production of newsprint. Bamboo pulp is suitable for a large variety of paper, as for writing, printing, wrapping, tissue, etc. The quality of paper made from bamboo compares favorably with conventional paper made from wood pulp. Bamboo is also used in rayon and cellophane production.

Fig. 10.41 "Die bambus
Socke" (Waikus)

10.2.4.2 Bamboo Textiles

Since some time "bamboo textiles" are marketed for underwear, t-shirts, socks, blankets, and many other textile products (Fig. 10.41). They are comfortable to wear, soft, and washed like any other cotton shirt. However, it has to be stated that this material has no direct origin in bamboo but comes from pure viscose. Such viscose can be obtained from any plant material by the rayon process.

Consequently, bamboo textiles are not natural fibers but come from a natural source. Such regenerated fibers are categorized as synthetic fibers (Shor 2006, 2007). Several Court decisions have ordered penalties for this wrong claim (Verbraucherzentrale Berlin 2008; FTC 2009).

10.2.5 Bamboo Charcoal

This contribution presents methods of preparation, properties, and the usage of bamboo charcoal with its important by-products bamboo, viz. vinegar, bamboo gas, and bamboo ash (Liese and Silbermann 2010).

Bamboo is one of the most important sources of energy for cooking and heating in many tropical and subtropical regions. The culms by themselves, however, do not form a good combustible material because they do not store well, burn fast, and tend to produce a dense smoke while burning if not seasoned properly before usage. Bamboo charcoal offers an alternative to bamboo culms for energy. For over 1,000 years, charcoal has been produced and utilized, primarily in China, and exported either in its basic form or as various manufactured products.

International organizations promote the production and use of bamboo charcoal and its by-products. For example, in about 800,000 ha of treeless grassland in Ethiopia, the lowland bamboo, *Oxytenanthera abyssinica*, is the only combustible

Fig. 10.42 Bamboo culms transported from far away for cooking, Ethiopia (Liese and Silbermann 2010)

Fig. 10.43 Bamboo burns fast, no glow (Liese and Silbermann 2010)

material available for cooking. The bamboo must be carried on the back in bundles from where it grows, many kilometers from the distant villages (Fig. 10.42). A 20 kg load will supply a family for 2 days. Only a limited number of culms can be stored because of their starch content and the lack of natural toxic substances and will rapidly be damaged by insects and fungi. Bamboo culms can be compared to solid wood, a short-lived combustible material: the pieces burn up rapidly, do not maintain the heat, new material must be added frequently (Fig. 10.43).

In Asia, bamboo is referred to as "black gold." Bamboo charcoal is available as culm segments, as chunks, compressed into briquettes, in a granular or powder form. In the basic form, it serves for cooking, heating, or smoke free grilling; however, for the most part, it is further processed into numerous products. By tradition, bamboo charcoal is primarily exported to Japan, South Korea, and Taiwan. Opening up new trade channels has brought bamboo charcoal to the USA and with increasing demand, to Europe.

International programs such as INBAR (International Network for Bamboo and Rattan) and ITTO (International Tropical Timber Organization) are currently active in promoting the production and use of bamboo charcoal and its by-products effectively. INBAR and the European Union are supporting the use of bamboo charcoal as an energy source instead of burning wood in Ethiopia and Ghana, working with local institutions and the University of Forestry in Nanjing, China, in a 4-year program (2009–2013) costing 1.3 million Euro. New industrial capacity for environmentally friendly products will arise through these international programs, especially in rural areas (Fu 2003).

10.2.5.1 Traditional Production

Bamboo charcoal is a black, lightweight, and porous material which comes from the pyrolysis of bamboo. The biomass is intensely heated in an oven with an oxygen free atmosphere, first under control, but then turned off. With this thermal treatment, the culms release their intrinsic water and volatile components. The end products are 30 % bamboo charcoal, 51 % bamboo vinegar, and 18 % bamboo gas with 1 % waste (Jiang 2007).

For moso bamboo, *Phyllostachys edulis*, the main species in China, culms of at least 4 years of age which exhibit extensive hardening of their fiber are suitable for industrial production. After felling, the culms are shortened to 80–120 cm, quartered, and air dried for weeks or months to the point of shrinkage from dehydration. The sticks are then exposed to 180–200 °C for 6–10 h in an industrial dehydrator. After dehydration, the water content of the material should amount to 15–20 %. Kilns are used instead of these specialized dehydrators in small, mostly rural, operations in which the bamboo is directly processed into charcoal after dehydration.

Different types of ovens are applied depending on the availability of bamboo culms, the local conditions, and financial circumstances. Traditionally, mud ovens are used, especially, by rural people. They consist of a heat-resistant shell with slanting pipes branching out, in which bamboo vinegar is condensed from the vapor and collected in a fastened container (Fig. 10.44). If financial means are available and production exceeds local needs, charcoal production takes place mostly in iron ovens, the manufacture of which also suffices as a small industry. Major industrial charcoal production utilizes ovens continuously fed with raw materials sourced from a wider area.

At the start of the process, the pieces of bamboo are stacked in the oven at right angles for good air circulation. After closing the mouth of the oven with stones and mud, it is stocked with wood or bamboo through which the air exhausts at the top opening. Then this opening is closed, leaving small side openings for regulating the air circulation. During further processing, these are plugged from the top down. As charcoal starts to be formed, individual openings are broken open to control the process with horizontal and vertical airflow, ensuring a uniform conversion to charcoal. The temperature in the burn chamber should be kept between 500 and 600 °C, since lower temperatures cause incomplete conversion to charcoal, or the

Fig. 10.44 Mud oven for charcoal with pipe for condensed vinegar (Liese and Silbermann 2010)

biomass starts to burn to ash as the end product instead of charcoal. After 2 days, all openings are closed, so that no fresh air can enter the burn chamber.

Cooling of the oven requires an additional day. After removal of the charcoal and ash, the oven is ready for reuse after cleaning. The entire process takes on average 5 days and minimizes softening of the charcoal. The yield of bamboo charcoal by this method is about 30 % of the dry initial weight of the culms used (Fig. 10.45).

Because no machinery is necessary for the manufacture of mud ovens, they can be built by experienced villagers. It is difficult to hold the temperature evenly high, since the pyrolysis temperature is higher than the ignition temperature of the bamboo. Faulty temperature control can lead to cracks in the oven wall, with the oven breaking apart or exploding. Iron ovens come with and without an integrated dehydrator. The higher temperatures required for carbonization are more easily reached and controlled than in the mud oven. This is necessary because during the charcoal processing, the temperature steadily climbs. It allows more throughputs per unit of time. The air circulation and carbonization temperature are easier to control, so greater carbon content is reached. This improves the quality of the bamboo charcoal and allows a better separation of bamboo vinegar and gas.

Fig. 10.45 Rows of charcoal ovens (Liese and Silbermann 2010)

10.2.5.2 Industrial Processing

By traditional methods used industrially, the oven has a conveyor belt or is made with a rotating body. A large quantity of bamboo charcoal, vinegar, and gas can be produced with a continuous use of an extrusion press. The culms are broken up into chips, which fractures the hard outer epidermis. This results in an inferior quality of the charcoal than the ovens described just previously, which use larger dimensioned pieces.

For bamboo charcoal making in Thailand, charcoal pieces are ground up, water and an adhesive are added, and compressed into six-sided tubes and dried. Compressed bamboo charcoal is very hard with fine pores, which promote sustained burning at high heat. The hexagonal shape is advantageous for shipping in cartons since the honeycomb structure is highly stable, sustaining little breakage and taking less packing space (Fig. 10.46). Because of the higher acquisition and maintenance costs, such ovens are used only for mass production of bamboo charcoal, whose raw material is harvested regionally and supplemented by industrial scraps.

10.2.5.3 Quality Control

There are no special international standards for the quality of bamboo charcoal. But, raw density, hardness, electrical resistance, and other properties make bamboo charcoal comparable to many wooden charcoal types. The "Hardgrove-Index" serves as a measure of hardness by grinding; the smaller the index, the harder the charcoal. Industrial analysis examines different characteristics, such as the content of moisture, ash, volatile components, and fixed carbon.

Fig. 10.46 Compressed
bamboo charcoal for easier
export, Thailand (Liese and
Silbermann 2010)

10.2.5.4   Properties and Utilization of Bamboo Charcoal

The preparation, properties, and possible uses of bamboo charcoal have long been
of intense interest in China. The Museum of Bamboo Charcoal in the Zhejiang
province gives an impressive depiction of the process for bamboo charcoal production, as well as documentation of the multiple products and their uses in
everyday and industrial domains (Fu and Chen 2009). The vitality of the market
for products made from bamboo charcoal is shown by the Internet, where Chinese
firms offer numerous products based on bamboo charcoal. The intensity of current
research is evidenced by the Chinese Academy of Forestry (Beijing) Journal
"World Bamboo and Rattan," volume 5 (2009), with 12 scientific articles
containing a broad diversity of topics concerning bamboo charcoal, which are the
results of a bamboo charcoal project.

Bamboo charcoal contains about 85 % carbons, 7 % gas, and from 2 to 4 % ash.
The pH is above 7.0 if the carbonization temperature is higher than 600 °C. It is
mostly used as combustible material because it contains a high heat value and does
not carry the risk of fungal or termite attack. Its carbon is in a finely granular form.
For easy storage and transportation, it is compressed.

The heat value as an important indication of its energy storage rises at a
combustion temperature from 500–800 °C to 7,400–8,000 Kcal (31–33 MJ), in
contrast to wood charcoal, the heat value of which varies from 7,000 to 7,800 Kcal
(29–33 MJ). With a burning time of 4 h, bamboo charcoal is especially suited for
use in a fireplace. Bamboo charcoal has a larger inner surface area than wood
charcoal and thus better absorptive properties. The density of bamboo charcoal at
300 °C temperature reaches 565 kg/m^3 and at 1,000 °C climbs to 720 kg/m^3
(Fu and Chen 2009).

The good absorptive quality of bamboo charcoal is utilized in various ways.
Charcoal absorbs harmful substances from room air such as formaldehyde,

ammonia, and benzene. Its high porosity binds moisture from humid air and releases it with decreasing humidity. Areas for such use include humid spaces like bathrooms and bedrooms, as well as for pillows and bedding. Bamboo charcoal is also mixed into walls, floors, and under houses. Granular or pulverized charcoal is incorporated into paper bags, pillows, or mattresses and is woven into outerwear. Depending upon the intensity of use, the effectiveness is about 3 weeks; after washing and drying, the material can again be put into use.

Bamboo charcoal binds with dangerous substances such as carbon monoxide, carbon dioxide, benzopyrene, and above all nicotine and tar. In China, cigarette filters contain bamboo charcoal and are said to absorb 95 % of the toxic substances.

In the kitchen, smoke filters often contain bamboo charcoal. Fruits and vegetables produce ethylenes in the refrigerator; these are bound by charcoal, lengthening the time the products stay fresh. The same applies to odors from fish and meat products. Bamboo charcoal is used to control body odor as well, in the form of shoe inserts and also as additive in soaps. Examples are given in Fig. 10.47.

10.2.5.5 Bamboo Vinegar

At the time of pyrolysis, bamboo vinegar and bamboo gas are by-products along with the charcoal. The escaping smoke condenses in the vents of the oven and drips into a receptacle. The dark brown condensate smells smoky and separates into a

Fig. 10.47 Bamboo charcoal is used for many commodities. The bottle contains vinegar as an important by-product (Waitkus)

light yellow layer (bamboo vinegar) and a dark bottom layer (bamboo tar). Depending on the temperature and type of bamboo, the condensed vinegar contains more than 200 organic substances, such as saturated and unsaturated acids, alcohols, aldehydes, and polyphenols.

The highest percentage of vinegar acid is produced at an extraction temperature of 300 °C, with an acid content of 8.7 % and a pH of 1.8. Because of this low pH, vinegar is applied to acidify alkaline soil. It is widely used in the area of medical products. Bamboo vinegar is used for skin infections and as an antimicrobial as in bamboo vinegar spray or soap and also in the cosmetic industries. A bamboo vinegar containing "Vital Patch" is said to eliminate unwanted substances from the skin.

Also the growth of plants is promoted by the use of bamboo vinegar, i.e., as fertilizer, especially for alkaline soils. Thus plants are protected from damaging microorganisms (Mu et al. 2004).

During pyrolysis of bamboo, about 7 % gas is produced, most importantly carbon dioxide, carbon monoxide, methane, ethylene, and hydrogen. This mixture of gases is used directly as fuel. At the end of the charcoal making process, about 2–4 % bamboo ash remains, depending upon the species of bamboo and the carbonization method used. The ash contains minerals such as silica, calcium, potassium, magnesium, sodium, iron, and manganese. They serve agriculture by protecting plants, acidifying the soil, and as fertilizer (Jiang 2007).

10.2.6 Special Uses

Besides the thousands of applications and uses for bamboo, there are constantly new ideas realized, some smaller for daily use, some impressive and of market value. The bamboo culm for carrying goods may be the first practical use; bamboo boats likewise are an early prerequisite for any movement and transport of masses. From the numerous means for transportation just two may be briefly touched: the bamboo bicycle and the "Fly boo."

Bamboo bicycles as simple constructions have been in use locally since long in bamboo countries. Already in the nineteenth century, they were constructed and patented in industrial countries. In recent years, bamboo bicycles become widely known in Germany and related countries, supported by the intention to use natural materials instead of diminishing recourses, like iron and aluminum. At a few places, like Lübek and Berlin, students established contacts with countries, like Ghana, where bamboo bicycles are used since long time.

From the predominant species, *Bambusa vulgaris*, mature culms with long internodes and thicker walls are cleaned, prepared to size, and assembled in the partner units. The prices of bamboo bicycles vary widely, up to $ 1.000 with an expected service time of about 20 years. There are quite a number of models on the market (Fig. 10.48). The combination of using natural resources with supporting social projects in the local communities is appealing. Details are provided in the internet as "bamboo bicycle" in general.

Fig. 10.48 Bamboo bikes (my Boo)

Fig. 10.49 Flyboo 3 at Costa Rica 1988 with its inventor Michel Abadie (Abadie)

To be mentioned is also "FLY BOO," a bamboo construction as an airplane. Its inventor, Michel Abadie, had the goal to create an airplane acting as ambassador to promote the use of bamboo as material for the future and also to drive a campaign. The project started in 1996 with an innovative design: four joints on a rhomboidal display. A scale model went to the wind tunnel in Toulouse with success and received the Henry Ford European Award for nature protection. Flyboo 1 exposed in Paris, Champs Elysees 1988 received four hundred thousand visitors. Flyboo 2 was presented in Costa Rica at the sixth World Bamboo Congress but was damaged during transportation. Flyboo 3 was exposed at the ninth WBC in Antwerp 2012 with a wingspan of 5.80 m and a length of 5.50 m (Fig. 10.49). However, the model needs refinements to fulfill its mission in the air.

Acknowledgement Our thanks are expressed to Prof. Dr. Dieter Eckstein and Dr. Othar Kordsachia for their review and valuable comments.

References

515store SA http://www.raumteiler-515.de/2er-Pack-raumteiler-Fenster-und-Bambus-dreiteilig-Paravent/p/4/10/3710/. Accessed 22 Feb 2014

Annonymus (2003) Hands-on-chinese style bamboo furniture. Manual on bamboo furniture making. INBAR. http://www.inbar.int/wp-content/uploads/2013/08/Bamboo-processing-Furni ture-Manual-PDF.pdf?7c424b. Accessed 27 Feb 2014

CAN (2013) Bamboo construction source book. Community Architects Network. http://community architectsnetwork.info/upload/opensources/public/file_14062013022345.pdf. Accessed 3 Mar 2014

Cusack V (1999) Bamboo world- the growing and use of clumping bamboos. Kangaroo, East Roseville, updated 2010

Federal Trade Commission (2009) FTC Charges companies with "Bamboo-zling" consumers with false product claims. http://www.ftc.gov/news-events/press-releases/2009/08/ftc-charges-com panies-bamboo-zling-consumers-false-product-claims. Accessed March 2014

Fu JH (2003) Bamboo charcoal products. Unpublished report, China

Fu JH, Chen W (2009) Profile of bamboo charcoal museum in Suichang County, Zhejiang 2009. World Bamb Rattan 7(5):1–45

Heinsdorff M (ed) (2010) The bamboo architecture – design with nature. Hirmer, München. ISBN 978-3-7774-2791-1

Janssen JA (2000) Designing and building with bamboo. International Network for Bamboo and Rattan (INBAR). Technical report no. 20. Beijing, China. http://www.fundeguadua.org/imagenes/DESARROLLOS%20TECNOLOGICOS/ARTICULOS%20Y%20PUBLICACIONES/INBAR_Technical_Report_No20.pdf. Accessed 3 Mar 2014

Jiang Z (2007) Bamboo pulp and paper making (chapter 11); Bamboo charcoal. Vinegar and active carbon (chapter 13). In: Bamboo and rattan in the world. China Forestry Publishing House, Beijing, pp 178–198, 214–226

Jiang S, Zhang Q, Jiang S (2002) On Structure, production, and market of bamboo-based panels in China. J For Res 13(2):151–156

Li X (2011) Untersuchungen zur Eignung von Bambus-Komposit-Kanteln für die Herstellung von Fenstern. Diplomarbeit, Universität Hamburg

Liese W (1985) Bamboos - biology, silvics, properties, utilization. Schriftenr. Dt Ges Techn Zusammenarbeit (GTZ), Nr. 180, TZ Verlagsges, Roßdorf

Liese W (1998a) The anatomy of bamboo culms. International Network for Bamboo and Rattan (INBAR), Technical report no. 18, Beijing, China

Liese W (1998b) Vom Nutzen der Bambusblätter. Bambus Brief 10; (1999) 2:20–23

Liese W (2003) *Oxytenanthera braunii*, der Wein-Bambus. EBS Schweiz 15:34–35

Liese W (2008) The blooming of *Melocanna baccifera* in Northeast India and its consequences. Bamb Bull 10(1):20–22 (Bamboo Society of Australia)

Liese W, Silbermann S (2010) Bamboo charcoal. Mag Am Bamb Soc 31(6):2–9

Malanit P (2009) The suitability of *Dendrocalamus asper* Backer for oriented strand lumber. Dissertation, Universität Hamburg

Minke G (2012) Building with bamboo. Verlag Birkhauser Architecture. ISBN 3034607482, 9783034607483

Mobilenewsblog.de http://www.mobilenewsblog.net/wp-content/uploads/2013/12/Bamboo-Wall-Paneling-From-Center-Slice.jpg. Accessed 28 Feb 2014

Mu J, Uehara T, Furuno T (2004) Effect of bamboo vinegar on regulation of germination and radide growth of seed plants II: composition of Moso bamboo vinegar at different collection temperature and its effects. J Wood Sci 50:470–476

Qisheng Z, Shenxue J, Yongyu T (2001) Industrial utilization on bamboo. Technical report no. 26. International Network for Bamboo and Rattan (INBAR)

Rainer Sturm pixelio.de. Accessed 25 Feb 2014

Rubio-Luna G (2007) Arte y Mañas de la Guadua – unaguiasobre el uso productive de un bambúgigante. Editor Info Art, Bogotá, Columbia

Schülke J (2009) Untersuchung der Eigenschaften von neuartigen Plattenwerkstoffen auf Basis von Bambus im Hinblick auf den Einsatz im Automotivebereich. Diplomarbeit, Universität Hamburg

Shor B (2006) Bamboo textiles: right or wrong? Mag Am Bamb Soc 27(5):4–6

Shor B (2007) More on textiles. Mag Am Bamb Soc 28(4):9–10

Tistl M, Velásquezgil JA (2001) Roofs instead of tents: a reconstruction project in the Colombian coffee zone after the earthquake of January 25th, 1999. In: Bamboo in Disaster Avoidance. Proceedings of the International Workshop on the Role of Bamboo in Disaster Avoidance. 6-8 August, 2001, Quayaquil, Ecuado

UNIDO and INBAR (2008) Cottage industry manuals: technique for plane woven bamboo products. http://www.unido.org/fileadmin/user_media/Services/Agro-Industries/FPPCs/Agro Support/smallmanual2-vb%20Cottage%20Industry%20Manuals%20-%20Techniques%20for %20Plane%20Woven%20Bamboo%20Products.pdf. Accessed 27 Feb 2014

Van der Lugt P, Van den Dobbelsteen AAJF, Janssen JJA (2005) An environmental, economic and practical assessment of bamboo as a building material for supporting structures. Construct Build Mater 20(9):648–656

Verbraucherzentrale Berlin (2008) Irreführende Materialbezeichnung bei Textilien. http://www. vz-berlin.de/Irrefüehrende-Materialbezeichnung-bei-Textilien. Accessed March 2014

Von Vengesack A, Kries M (eds) (2000) Grow your own house- Simon Vélez und die Bambusarchitektur. Vitra Design Museum und Fondation ZERI

Wegst UGK (2008) Bamboo and wood in musical instruments. Annu Rev Mater Res S38: 22.1–22.27

Widjaja EA (1980) The angklung and other west Japanese bamboo musical instruments. In: Lessard G, Chouinard A (eds) Bamboo research in Asia. IDRC, Ottawa, pp 201–204

Yu DW, Tan HW, Ruan Y (2011) A future bamboo-structure residential building prototype in China: life cycle assessment of energy use and carbon emission. Energy Build 43(10): 2638–2646

Zhang Q, Jiang S, Tang Y (2002) Industrial utilization on bamboo. INBAR technical report no. 26, Beijing

Index

A

Aboveground biomass, 85
Absorptive quality, 341
Aftercare, 156–158
  grazing and fire, 156
Age and collection time, 145
  best time, 145
  new aerial roots, 145
  rainy season, 145
  spring, 145
Age to harvest
  older than 6 years, insect damage, 195
  species dependent, 195
Aging, 242, 243, 245, 248, 249, 251
Agroforestry plantations, 158
  bamboo types, 158
  biogeochemical role, 160
  crops, 160
  intercropping, 159, 160
  mixed stands, 159
  models, 158
Aided Natural Regeneration, 132
  thick mat on the ground, 132
  weed suppression, 132
  wild seedlings, 132
Air canals, 241, 242
Anastomoses, 236, 240
Arundinaria spp.
  A. alpina, 39
  A. amabalis, 40
Arundinarieae. *See* Temperate woody bamboos
  (Arundinarieae)
Ash, 336, 339–341, 343
  content, 239, 245, 248
Attack, 309, 330, 332, 334, 341

B

Bamboo plastic composites (BPC), 332–333
Bamboos, 91, 299–344
  in arts and technology, 4
  beer, 300, 302, 305, 307
  biomass, 98, 101, 106–108
  branch and culm pests, 178, 183–184
  carbon sequestration, 101, 105–107
  characteristics, 32
  classification system, 5
  climate change, 105–108
  cytological and morphological data, 5
  diameter, 97–101, 103
  diseases, 185–190
  diversity
    Arundinarieae, 16
    ecological and economic importance,
     12
    and evolution, 4, 5, 8
    Olyreae, 9
    by tribe and subtribe, 12, 13
  drying rate, 288–295
  fiberboard, 331–332
  growth habits, 31
  habitats
    forest grasses, 6
    montane systems, 8
    natural tropical bamboo forests, 8
    Olyreae, 9
    secondary forest, 7
  in homestead, 169
    agricultural implements, 169
    construction works, 169
    integrated farming systems, 169
  leaf defoliators, 182–183, 189

© Springer International Publishing Switzerland 2015
W. Liese, M. Köhl (eds.), *Bamboo*, Tropical Forestry 10,
DOI 10.1007/978-3-319-14133-6

CPSIA information can be obtained
at www.ICGtesting.com
Printed in the USA
LVOW02*0748110716

495808LV00002B/7/P